Lecture Notes in Computer Scier

T0238934

Commenced Publication in 1973
Founding and Former Series Editors:
Gerhard Goos, Juris Hartmanis, and Jan van Leeuwen

Editorial Board

Yamine Ait Ameur
Klaus-Dieter Schewe (Eds.)

Abstract State Machines, Alloy, B, TLA, VDM, and Z

4th International Conference, ABZ 2014
Toulouse, France, June 2-6, 2014
Proceedings

 Springer

Volume Editors

Yamine Ait Ameur
INP-ENSEEIHT/IRIT
2 Rue Charles Camichel, BP 7122
31071 Toulouse Cedex 7, France
E-mail: yamine@enseeiht.fr

Klaus-Dieter Schewe
Software Competence Center Hagenberg
Softwarepark 21
4232 Hagenberg, Austria
E-mail: kd.schewe@scch.at

ISSN 0302-9743   e-ISSN 1611-3349
ISBN 978-3-662-43651-6   e-ISBN 978-3-662-43652-3
DOI 10.1007/978-3-662-43652-3
Springer Heidelberg New York Dordrecht London

Library of Congress Control Number: 2014939229

LNCS Sublibrary: SL 1 – Theoretical Computer Science and General Issues

Typesetting: Camera-ready by author, data conversion by Scientific Publishing Services, Chennai, India

Printed on acid-free paper

Springer is part of Springer Science+Business Media (www.springer.com)

Preface

The 4th international conference ABZ took place in Toulouse during June 2nd to June 6th. This conference records the latest development in refinement and proof state based formal methods. It follows the success of the London (2008), Orford (2010) and Pisa (2012) conferences.

This year's ABZ was marked by two major events. In addition to ASM, B, Z, Alloy and VDM, ABZ 2014 saw the introduction of TLA (Temporal Logic of Actions) as the 6th formal method covered by the scope of the conference. In order to emphasise the integration of TLA, Leslie Lamport was invited to be one of the keynote speakers. He agreed to give an invited talk entitled "TLA+ for Non-Dummies" the year he was distinguished by the Turing Award. Congratulations !

After the "Steam Boiler" case study raised 20 years ago, the second event highlighting the 4th ABZ conference was the introduction of a case study track. The aeronautic context offered by the Toulouse area pushed us to look for a case study issued from this domain. Frédéric Boniol and Virginie Wiels kindly and immediately accepted to propose a "landing gear system" to be modelled within proof and refinement state based methods in the scope of ABZ. A separate proceedings volume, also published by Springer Verlag, is dedicated to this case study.

ABZ 2014 received 81 submissions covering the whole formal methods in the scope of the conference: Alloy, ASM, B, TLA, VDM and Z. These papers ranged on a wide spectrum covering fundamental contributions, applications in industrial contexts, and tool developments and improvements. Each paper was reviewed by at least three reviewers and the Program Committee accepted 13 long papers and 19 short papers. Furthermore, 8 long and 3 short papers were accepted for the case study track published in another proceedings volume. This selection process led to an attractive scientific programme.

In addition to the invited talk of Leslie Lamport, ABZ 2014 invited two other speakers. Gerhard Schellhorn from the University of Augsburg, Germany gave a talk entitled "Development of a Verified Flash File System" centered towards the ASM formal method and Laurent Voisin from the Systerel company, France with a talk entitled "The Rodin Platform has turned ten" reporting the progress achieved within the Rodin platform supporting Event-B. We would like to thank the three invited speakers for their contributions to the success of ABZ 2014.

ABZ 2014 would not have succeeded without the deep investment and involvement of the Program Committee members and the external reviewers who contributed to review (more than 250 reviews) and select the best contributions. This event would not exist if authors and contributors did not submit their proposals. We address our thanks to every person, reviewer, author, Program

Committee member and Organization Committee member involved in the success of ABZ 2014.

The EasyChair system was set up for the management of ABZ 2014 supporting submission, review and volume preparation processes. It proved to be a powerful framework.

We wish to express our special thanks to Jean-Raymond Abrial, Frédéric Boniol, Egon Börger and Virginie Wiels for their valuable support.

Finally, ABZ 2014 received the support of several sponsors, among them Airbus, CNES, CNRS, CRITT Informatique, CS, ENSEEIHT Toulouse, FME, INP Toulouse, IRIT, Midi Pyrénées Region, ONERA, SCCH, University Paul Sabatier Toulouse. Many thanks for their support.

June 2014   Yamine Ait Ameur
                                                                           Klaus-Dieter Schewe

Organization

Program Committee

Jean-Raymond Abrial	Consultant, France
Yamine Ait Ameur	IRIT/INPT-ENSEEIHT, France
Richard Banach	University of Manchester, UK
Eerke Boiten	University of Kent, UK
Frederic Boniol	ONERA, France
Michael Butler	University of Southampton, UK
Egon Börger	University of Pisa, Italy
Ana Cavalcanti	University of York, UK
David Deharbe	Universidade Federal do Rio Grande do Norte, Brazil
John Derrick	University of Sheffield, UK
Juergen Dingel	Queen's University, Canada
Kerstin Eder	University of Bristol, UK
Roozbeh Farahbod	SAP Research, Germany
Mamoun Filali-Amine	IRIT-Toulouse, France
John Fitzgerald	Newcastle University, UK
Marc Frappier	University of Sherbrooke, Canada
Vincenzo Gervasi	University of Pisa, Italy
Dimitra Giannakopoulou	NASA Ames, USA
Uwe Glässer	Simon Fraser University, Canada
Stefania Gnesi	ISTI-CNR, Italy
Lindsay Groves	Victoria University of Wellington, New Zealand
Stefan Hallerstede	University of Düsseldorf, Germany
Klaus Havelund	California Institute of Technology, USA
Ian J. Hayes	University of Queensland, Australia
Rob Hierons	Brunel University, UK
Thai Son Hoang	Swiss Federal Institute of Technology Zurich, Switzerland
Sarfraz Khurshid	The University of Texas at Austin, USA
Regine Laleau	Paris Est Creteil University, France
Leslie Lamport	Microsoft Research, USA
Peter Gorm Larsen	Aarhus School of Engineering, Denmark
Thierry Lecomte	Clearsy, France
Michael Leuschel	University of Düsseldorf, Germany
Yuan-Fang Li	Monash University, Australia
Zhiming Liu	United Nations University - International Institute for Software Technology, Macao

Tiziana Margaria   University of Potsdam, Germany
Atif Mashkoor   Software Competence Center Hagenberg,
                          Austria
Dominique Mery   Université de Lorraine, LORIA, France
Stephan Merz   Inria Lorraine, France
Mohamed Mosbah   LaBRI - University of Bordeaux, France
Cesar Munõz   NASA Langley, USA
Uwe Nestmann   Technische Universität Berlin, Germany
Chris Newcombe   Amazon.com, USA
Jose Oliveira   Universidade do Minho, Portugal
Luigia Petre   Åbo Akademi University, Finland
Andreas Prinz   University of Agder, Norway
Alexander Raschke   Institute of Software Engineering and Compiler
                          Construction, Germany
Elvinia Riccobene   DTI - University of Milan, Italy
Ken Robinson   The University of New South Wales, Australia
Thomas Rodeheffer   Microsoft Research, USA
Alexander Romanovsky   Newcastle University, UK
Thomas Santen   European Microsoft Innovation Center in
                          Aachen, Germany
Patrizia Scandurra   DIIMM - University of Bergamo, Italy
Gerhard Schellhorn   University of Augsburg, Germany
Klaus-Dieter Schewe   Software Competence Center Hagenberg,
                          Austria
Steve Schneider   University of Surrey, UK
Colin Snook   University of Southampton, UK
Jing Sun   The University of Auckland, New Zealand
Mana Taghdiri   KIT, Germany
Margus Veanes   Microsoft Research, USA
Marcel Verhoef   Chess, The Netherlands
Friedrich Vogt   University of Technology Hamburg-Harburg,
                          Germany
Laurent Voisin   Systerel, France
Qing Wang   Information Science Research Centre,
                          New Zealand
Virginie Wiels   ONERA, France
Kirsten Winter   University of Queensland, Australia

Additional Reviewers

Arcaini, Paolo   Couto, Luís Diogo
Attiogbe, Christian   Cunha, Alcino
Barbosa, Haniel   Ernst, Gidon
Coleman, Joey   Esparza Isasa, José Antonio
Colvin, Robert   Fantechi, Alessandro

Gervais, Frederic
Herbreteau, Frédéric
Iliasov, Alexei
Kossak, Felix
Ladenberger, Lukas
Leupolz, Johannes
Macedo, Nuno
Mammar, Amel
Nalbandyan, Narek
Neron, Pierre

Pfähler, Jörg
Sandvik, Petter
Senni, Valerio
Singh, Neeraj
Tarasyuk, Anton
Tounsi, Mohamed
Treharne, Helen
Winter, Kirsten
Yaghoubi Shahir, Hamed

TLA+ for Non-Dummies

Leslie Lamport

Microsoft Research
Silicon Valley
USA

Abstract. I will discuss the motivation underlying TLA+ and some of the language's subtleties. Since Springer-Verlag requires a longer abstract, here is a simple sample TLA+ specification:

```
-------------------- MODULE Euclid --------------------
(*********************************************************)
(* This module specifies a version of Euclid's algorithm *)
(* for computing the  greatest common divisor of two     *)
(* positive integers.                                    *)
(*********************************************************)
EXTENDS Integers

CONSTANTS M, N

ASSUME /\ M \in Nat \ {0}
       /\ N \in Nat \ {0}

VARIABLES x, y

Init == (x = M)  /\  (y = N)

Next ==    (  x > y
           /\ x' = x - y
           /\ y' = y  )
        \/ (  y > x
           /\ y' = y - x
           /\ x' = x )

Spec == Init /\ [][Next]_<<x, y>>
=======================================================================
```

Table of Contents

The Rodin Platform Has Turned Ten

Laurent Voisin[1] and Jean-Raymond Abrial[2]

[1] Systerel, Aix-en-Provence, France
laurent.voisin@systerel.fr
http://www.systerel.fr
[2] Marseille, France
jrabrial@neuf.fr

Abstract. In this talk, we give an historical account of the development of the Rodin Platform during the last 10 years.

Keywords: Event-B, formal methods, tooling.

1 Introduction

Some time ago, when discussing about this paper, we "discovered" and "remembered" that the Rodin Platform Project started ten years ago. It seemed very strange to us as we had the impression that 2004 was just yesterday. However, if we think of the state of this platform now to be compared to that of it, say, eight years ago, it is clear that the difference is quite significant.

As a matter of fact, the Event-B notation and its theoretical background have not been modified during this period and will certainly not in the future. But, on the other hand, the integration of Event-B within an efficient tool is another story. Doing some Event-B modeling and proofs with pen and paper is easy, provided the model and its mathematical structure correspond to toy examples. But it is obviously impossible to do the same (pen and paper) on a serious (industrial) model: here the usage of a tool is absolutely indispensable.

This being said, it does not mean that any tool will do the job: we all know of some tools being so poor that the unfortunate user will soon forget it and return to a simplified (but unadapted) technique, thus quickly leading to the cancellation of the overall approach. In other words, right from the beginning we were very convinced that such a tool building is a very serious and difficult task that has to be performed by very skilled professional engineers.

The main functions of the Rodin platform [1] are twofold: (1) an Event-B modeling database, and (2) a proving machine.

The internal structure of the Event-B modeling database was facilitated by our early choice of Eclipse as a technical support for Rodin. The corresponding interactive editing of the models was quite difficult to master as we wanted the users to have a very friendly interface at his disposal.

For the proving machine, we borrowed some provers made by the French company Clearsy for the Atelier-B tool. Over the years, more provers were added to these initial facilities. More recently, some extremely powerful extension were

Y. Ait Ameur and K.-D. Schewe (Eds.): ABZ 2014, LNCS 8477, pp. 1–8, 2014.

made by incorporating some SMT solvers within Rodin. Another aspect of the proving machinery is the building of an interface allowing users to perform interactive proofs using a modern "clicking" approach, departing from more verbose techniques used in academic provers (Coq, HOL, Isabelle, PVS).

In the following almost independent sections, we develop more technically what has been briefly alluded so far. In section 2, we make clear the original funding structure of Rodin. In section 3, we cite the main people involved in the initial development of Rodin. Section 4 explains how some parts of the Rodin platform itself was formally developed and proved. Section 5 gives some information about the openness philosophy used in Rodin. Sections 6 and 7 show how Eclipse provided the basic ingredients needed to implement openness. In section 8, the main provers used within the Rodin platform are quickly described. Section 9 countains brief description of the main plug-ins that were developed by various partners. Section 10 shows how Rodin has spread in many places over the years. Finally, section 11 indicates how a large community of Rodin developers and users has gradually emerged. We conclude in section 12.

2   Funding

The initial development of the Rodin platform was co-funded by ETH Zurich and the eponymous European project (FP6-IST-511599, 2004-2007). Then the main development has been taken on by Systerel, still with a partial funding from the European Commission through two follow-up research projects: DEPLOY (FP7-IST-214158, 2008-2012) and ADVANCE (FP7-ICT-287563, 2011-2014).

The integration of SMT Solvers into the Rodin platform was partially funded by the French National Research Agency in the DECERT project (ANR-08-EMER-005, 2009-2012).

3   The Initial Team

Rodin started at ETH Zurich (Switzerland) from 2004 to 2008. Since then the development of the core platform is performed at Systerel (France). The original team was made of Laurent Voisin, Stefan Hallerstede, Farhad Mehta, and Thai Son Hoang. Later, François Terrier and Mathias Schmalz joined the initial ETH group. At Systerel, the main people involved in the development are Laurent Voisin, Nicolas Beauger and Thomas Muller.

It is interesting to note that the original team at ETH was partially funded by ETH itself (Laurent Voisin and Stefan Hallerstede were hired as ETH employees). The other people in the original ETH team were funded by the RODIN European Project.

The Rodin Platform software is open source. It is implemented on top of the Eclipse system. Laurent Voisin is the main architect: its duty was to deeply understand how Eclipse works and also to organize the team so that each member could work in parallel with others. Stefan Hallerstede implemented the static checker and the proof obligation generator. Farhad Mehta was in charge of developing some internal provers. Thai Son Hoang had the very difficult task of constructing a user friendly interface. Later, François Terrier and Mathias Schmalz also worked on the provers.

At Systerel, the team works partially on the Rodin Platform, implementing many demands made by the increasing number of users and external developers working on plug-ins for the platform.

A critical aspect of the development organisation was that the initial team in charge of designing and implementing the platform was not a random bunch of software developers. They were also direct end users of the tool, writing models, proving them correct and teaching students how to use Event-B and the tools at the same time. This proved essential for reaching an excellent level of usability. There was no dichotomy between users on the one hand and developers on the other hand.

4 Dog Fooding

As the Event-B notation was unstable during the platform design phase (the notation was finalised at the same time as the platform was developed), it was important to ensure that some theoretical parts were sound before coding them. This was the case notably for proof obligation generation. We have therefore developed formal models of the static checker and proof obligation generator to ensure their soundness before implementing them.

Having such formal models proved invaluable as it allowed to have on the table all the needed parts and to see how they arranged themselves. Thanks to this model, we could implement generic tools providing a framework for static checking and proof obligation generation, and then just fill the frameworks with small pieces of code corresponding to the events of the formal model.

5 The Openness Approach to Event-B and Rodin

One important technical philosophy of the Rodin Platform (and more generally of Event-B) is *openness*. To begin with, there is no such thing as an Event-B language dealing with a number of fixed keywords and closed syntactic structures. For entering and editing models into the tool, one uses directly the provided interface together with some wizards helping to simplify the job of the working user. This approach was very positive as extensions of the modelling and proving features could be done in a simple fashion, that is just by extending the interface. In fact, the only more formal notation is that of set theory: most of the time, the classical way used by mathematicians has been favored.

The choice of Eclipse also facilitates openness: further extensions of Event-B itself and corresponding extensions of the Rodin platform are performed by means of *plug-ins*. Among those plug-ins, an important one, called the *"Theory" plug-in* allows one to extend the mathematical notation by providing new generic operators together with some corresponding inference or rewriting rules able to enhance the provers.

6   The Choice of Eclipse: Core Architecture

When the Rodin project started, we add previous experience with Atelier B from ClearSy. Atelier B is a monolithic tool that has been fully developed from scratch. We knew then that developing a similar tool for Event-B from the ground up would be a tremendous task and would not fit the three-year project time-frame. We therefore looked for a firm basis on which to build Rodin, sparing us from redeveloping core services such as a project builder.

We therefore searched for a framework on which to build the Rodin tooling. We had the good fortune that a few months ago IBM had released Eclipse in the open by creating the Eclipse foundation. This looked very promising at the time. Eclipse was providing most of the needed basic blocks and was open enough to allow us to plug Rodin on top of it. Plus, the creation of the Eclipse foundation promised that the Eclipse platform would be vendor-neutral for the coming years.

7   The Choice of Eclipse: Influence on Openness

Similarly to the Event-B notation, which is very flexible, we wanted to have an open tool. By open, we mean here that the tool should allow anybody to customize it in several aspects, even in directions not initially foreseen. To achieve this goal, we first needed a plug-in mechanism for external developers to easily add new behavior to the platform. This is provided by the Eclipse platform.

But having this plug-in mechanism is not enough. We also had to take great care to design for openness [8]: as soon as a function seems amenable to customisation, publish an extension point so that plug-ins can extend or replace it with their own code.

Finally, we decided that the platform code should be open source. This brings several advantages: (1) Anybody can read the code and audit it if need be. (2) If one needs to customize the platform in a direction which was not foreseen (there is no corresponding extension point) then one can just copy the code and modify it. This is of course to be used only in last resort. But it has proved helpful sometimes, while waiting for the creation of a new extension point by the core platform maintainers. (3) Finally, open source ensures resiliency. If, for some reason, the core maintenance team disappears, anybody can take the development over.

8   The Internal and External Provers of Rodin

Like the rest of the Rodin platform, the prover has been designed for openness. The main code of the prover just maintains a proof tree in Sequent Calculus and does not contain any reasoning capability. It is extensible through reasoners and tactics. A reasoner is a piece of code that, given an input sequent, either fails or succeeds. In case of success, the reasoner produces a proof rule which is applied to the current proof tree node.

Reasoners could be applied interactively. However this would be very tedious. Reasoner application can thus be automated by using tactics that take a more global view of the proof tree and organise the running of reasoners. Tactics can also backtrack the proof tree, that is undo some reasoner application in case the prover entered a dead-end.

The core platform contains a small set of reasoners written in Java that either implement the basic proving rules (HYP, CUT, FALSE_L, etc.), or perform some simple clean-up on sequents such as normalisation or unit propagation (generalised modus-ponens). These reasoners allow to discharge the most simple proof obligations. They are complemented by reasoners that link the Rodin platform to external provers, such as those of Atelier B (ML and PP) and SMT solvers (Alt-Ergo, CVC3, VeriT, Z3, etc.) [5].

There is even a plug-in for translating Event-B sequents to an embedding in HOL which allows to perform proofs with the Isabelle proof assistant [14].

9 Some Important Plug-ins

There are more than forty plug-ins available for the Rodin platform. We have arbitrarily chosen to present some of them that we have found very useful in our modelling activities. We refer the reader to the Event-B Wiki[1] for a more complete list of available plug-ins.

AnimB. This plug-in [12] was developed by Christophe Métayer. It allows to perform animations of Event-B model. This is done by providing some values to the generic sets and constants defined in the various contexts of the model to animate. The plug-in computes the guards of the events and makes available those events with true guards. The user can then activate an enabled event and provide some values to the corresponding parameters, if any. One can see the state of the system as modified by event execution.

Decomposition. This plug-in [16] is used to decompose an Event-B model into two communicating models able to be later refined independently. Two distinct approaches are provided: one is organized around some shared variables while the other is organized around shared events. This plug-in was joint work of Renato Silva in the University of Southampton and Carine Pascal in Systerel.

Generic Instantiation. Two plug-ins are provided for generic instantiation. One has been developed in the University of Southampton [15] and the other [7] by ETH Zurich together with Hitachi. Both of them give the possibility to instantiate the generic elements (sets and constants) of the various contexts of an Event-B development. The corresponding instantiated axioms are transformed into theorems to be proved.

---

[1] http://wiki.event-b.org/index.php/Rodin_Plug-ins

Theory. This plug-in [3] allows to extend the basic mathematical operators of Event-B. These operators can be defined explicitly in terms of existing ones. But is is also possible to give some axiomatic definitions only. An interesting outcome of this last feature allows one to define the set of Real numbers axiomatically. Moreover, the user of this plug-in can add some corresponding theorems, and inference or rewriting rules able to extend the provers. It is also possible to define new (possibly recursive) datatypes. This very important plugin has been developed by Issam Maamria and Asieh Salehi in Southampton.

Graphical modelling. The UML-B [13] and iUML-B plug-ins provide a graphical modelling interface la UML on top of the Rodin platform. The graphical models are translated to plain Event-B machines and can be worked with like any Event-B model (proof, animation, etc.). These plug-ins make it particularily easy to model state-machines. Both plug-ins have been developed by Colin Snook in Southampton.

ProB. This plug-in [11] provides an animator and model checker. It comes as a complement to the Rodin prover for detecting modeling inconsistencies. Its animation facility allows to validate [10] Event-B models, i.e., to verify that the model behavior is indeed the intended one. ProB integrates BMotionStudio to easily generate domain specific graphical visualizations. ProB was developed by Michael Leuschel first in Southampton and then in Dusseldorf.

ProR. This Eclipse plug-in [9] supports requirements engineering. It integrates specific machinery that allow to link parts of Event-B models (guards, invariants, actions) to the requirements they model, thus allowing full traceability between a requirements document and an Event-B model. ProR has been developed by Michael Jastram in Dusseldorf.

B2LaTeX. This little plug-in [4] allows to translate an Event-B model to a LaTeX document. It has been developed by Kriangsak Damchoom in Southampton.

Camille. The Rodin platform comes with structured editors that expose directly the organisation of the Rodin database. As this was not appropriate to all users, the University of Dusseldorf has developed the Camille plug-in [2] that provides a plain-text editor on top of the Rodin database.

Code Generation. There are several code generation plug-ins that allow to translate some Event-B models to executable code. The most active one has been developed by Andy Edmunds [6] from the University of Southampton.

10   Spreading of the Rodin Platform

At the beginning, from 2004 to 2006, where the development of the Rodin platform was very much in progress, there was not so many users of this platform, except for some people working inside the Rodin European project itself. However, as time goes there are now more and more people using it. This increasing number is due to the organisation of many formal methods and Event-B courses taught around the world in Academia. Incorporating a tool like the Rodin Platform within a formal method course makes considerable changes in the students interest to this difficult topic: students can participate more actively in the practical aspect of the course and the mathematical notation used in Event-B becomes more familiar.

Courses using the platform are presently given in the following countries (sometimes in several sites): France, Switzerland, United Kingdom, Finland, Germany, Canada, Australia, Japan, China, India, Malaysia, Tunisia, Algeria, and probably some others.

Event-B and the Rodin platform are also used in Industry, especially in France by Systerel and in Japan by Hitachi.

11 The Rodin Community

The general availability of the Rodin platform on SourceForge has given rise to a vibrant community of researchers, teachers, users and developers. The main discussions take place in two mailing lists, one for users and the other for developers. All bug reports and feature requests are publicly available and can be entered by anyone. Almost all documentation is available in the Event-B wiki [2] which is also open to everyone.

Finally, the Rodin community meets physically every year at the Rodin User and Developer Workshop. This is the place to present recent advances and work-in-progress both as a platform user and as a plug-in developer.

12 Conclusion

In this short paper we have shown what has been developed over the years under the name of "Rodin Platform". What is quite remarkable about this project is the very large number of people involved. It is remarkable because all these people were all animated by a very strong *common spirit*, that of freely constructing a tool without any financial reward idea behind it, instead just the scientific and technical goal of producing the best instrument for future users, and we are absolutely certain that this spirit will continue to flourish in the years to come.

Let us take the opportunity of this paper to call for additional contributions from anyone interested in formal method practical outcome: we are all very open to enlarge our present very dynamic community of users and developers.

Acknowledgements. Laurent Voisin was partly funded by the FP7 ADVANCE Project (287563)[3] and the IMPEX Project (ANR-13-INSE-0001)[4].

---

[2] http://wiki.eventb.org

[3] http://www.advance-ict.eu

[4] http://impex.gforge.inria.fr

References

1. Abrial, J.-R., Butler, M., Hallerstede, S., Hoang, T.S., Mehta, F., Voisin, L.: Rodin: An open toolset for modelling and reasoning in Event-B. Int. J. Softw. Tools Technol. Transf. 12(6), 447–466 (2010)
2. Bendisposto, J., Fritz, F., Jastram, M., Leuschel, M., Weigelt, I.: Developing Camille, a text editor for Rodin. Software: Practice and Experience 41(2), 189–198 (2011)
3. Butler, M., Maamria, I.: Practical theory extension in Event-B. In: Liu, Z., Woodcock, J., Zhu, H. (eds.) Theories of Programming and Formal Methods. LNCS, vol. 8051, pp. 67–81. Springer, Heidelberg (2013)
4. Damchoom, K.: B2Latex, http://wiki.event-b.org/index.php/B2Latex
5. Déharbe, D., Fontaine, P., Guyot, Y., Voisin, L.: SMT solvers for Rodin. In: Derrick, J., Fitzgerald, J., Gnesi, S., Khurshid, S., Leuschel, M., Reeves, S., Riccobene, E. (eds.) ABZ 2012. LNCS, vol. 7316, pp. 194–207. Springer, Heidelberg (2012)
6. Edmunds, A., Rezazadeh, A., Butler, M.: Formal modelling for ada implementations: Tasking Event-B. In: Brorsson, M., Pinho, L.M. (eds.) Ada-Europe 2012. LNCS, vol. 7308, pp. 119–132. Springer, Heidelberg (2012)
7. Fürst, A., Hoang, T.S., Basin, D., Sato, N., Miyazaki, K.: Formal system modelling using abstract data types in Event-B. In: Ait Ameur, Y., Schewe, K.-D. (eds.) ABZ 2014. LNCS, vol. 8477, pp. 222–237. Springer, Heidelberg (2014)
8. Gamma, E., Beck, K.: Contributing to Eclipse: Principles, Patterns, and Plugins. Addison Wesley Longman Publishing Co., Inc., Redwood City (2003)
9. Hallerstede, S., Jastram, M., Ladenberger, L.: A Method and Tool for Tracing Requirements into Specifications. Science of Computer Programming (2013)
10. Hallerstede, S., Leuschel, M., Plagge, D.: Validation of formal models by refinement animation. Science of Computer Programming 78(3), 272–292 (2013)
11. Leuschel, M., Butler, M.: ProB: An automated analysis toolset for the B method. Software Tools for Technology Transfer (STTT) 10(2), 185–203 (2008)
12. Métayer, C.: AnimB: B model animator, http://www.animb.org
13. Said, M., Butler, M., Snook, C.: A method of refinement in UML-B. Software & Systems Modeling, 1–24 (2013)
14. Schmalz, M.: Formalizing the logic of Event-B. PhD thesis, ETH Zurich (2012)
15. Silva, R., Butler, M.: Supporting reuse of Event-B developments through generic instantiation. In: Breitman, K., Cavalcanti, A. (eds.) ICFEM 2009. LNCS, vol. 5885, pp. 466–484. Springer, Heidelberg (2009)
16. Silva, R., Pascal, C., Hoang, T.S., Butler, M.: Decomposition tool for Event-B. Software: Practice and Experience 41(2), 199–208 (2011)

Development of a Verified Flash File System

Gerhard Schellhorn, Gidon Ernst, Jörg Pfähler,
Dominik Haneberg, and Wolfgang Reif

Institute for Software & Systems Engineering
University of Augsburg, Germany
{schellhorn,ernst,joerg.pfaehler,
haneberg,reif}@informatik.uni-augsburg.de

Abstract. This paper gives an overview over the development of a formally verified file system for flash memory. We describe our approach that is based on Abstract State Machines and incremental modular refinement. Some of the important intermediate levels and the features they introduce are given. We report on the verification challenges addressed so far, and point to open problems and future work. We furthermore draw preliminary conclusions on the methodology and the required tool support.

1 Introduction

Flaws in the design and implementation of file systems already lead to serious problems in mission-critical systems. A prominent example is the Mars Exploration Rover Spirit [34] that got stuck in a reset cycle. In 2013, the Mars Rover Curiosity also had a bug in its file system implementation, that triggered an automatic switch to safe mode. The first incident prompted a proposal to formally verify a file system for flash memory [24,18] as a pilot project for Hoare's Grand Challenge [22].

We are developing a verified flash file system (FFS). This paper reports on our progress and discusses some of the aspects of the project. We describe parts of the design, the formal models, and proofs, pointing out challenges and solutions.

The main characteristic of flash memory that guides the design is that data cannot be overwritten in place, instead space can only be reused by erasing whole blocks. Therefore, data is always written to new locations, and sophisticated strategies are employed in order to make the file system efficient, including indexing and garbage collection. For the algorithmic problems, we base our design on UBIFS [23,20], which is a state-of-the-art FFS in the Linux kernel.

In order to tackle the complexity of the verification of an entire FFS, we refine an abstract specification of the POSIX file system interface [41] in several steps down to an implementation. Since our goal is an implementation that runs on actual hardware, besides functional correctness additional, nontrivial issues must be addressed, such as power cuts and errors of the hardware. A further requirement is that the models must admit generation of executable code. The derived code should correspond very closely to the models, thus simplifying an automated last refinement proof between the models and the actual code.

Y. Ait Ameur and K.-D. Schewe (Eds.): ABZ 2014, LNCS 8477, pp. 9–24, 2014.

The paper is structured as follows: Section 2 gives an overview over our approach and models. Section 3 provides more detail on the design of each of the models and their concepts. The section also highlights some of the problems that need to be tackled and describes our solutions. The formal models are also available at our web-presentation [11]. Section 4 explains how we generate Scala and C code from our model hierarchy and integrate it into Linux. In Sec. 5 we report on some of the lessons learned during the project. Section 6 concludes and points out some of the open challenges.

2   Overview

Figure 1 shows the high-level structure of the project. There are four primary conceptual layers. A top-level specification of the POSIX file system interface [41] defines functional correctness requirements. At the bottom is a driver interface model that encodes our assumptions about the hardware. Two layers in between constitute the main functional parts of the system: The file system (FS) implementation is responsible for mapping the high-level concepts found in POSIX (e.g., directories, files

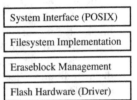

Fig. 1. High-level structure

and paths) down to an on-disk representation. A separate layer, the erase block management (EBM), provides advanced features on top of the hardware interface (e.g., logical blocks and wear-leveling).

Figures 2 and 3 show how the file system implementation is broken down to a refinement hierarchy. Boxes represent formal models. Refinements—denoted by dashed lines—ensure **functional correctness** of the final model that is the result of combining all models shaded in grey. From this combined model we generate executable Scala and C code.

The POSIX model for example is refined by a combination of a Virtual Filesystem Switch (VFS) and an abstract file system specification (AFS). The former realizes concepts common to all FS implementations such as path traversal. Such a VFS exists in Linux and other operating systems ("Installable File System" in Windows). The benefit of this approach is that concrete file systems such as UBIFS, Ext or FAT do not have to reimplement the generic functionality. Instead they satisfy a well-defined internal interface, denoted by the symbol —o— in Fig. 2. We capture the expected functional behaviour of a concrete FS by the AFS model. The modularization into VFS and AFS significantly reduces the complexity of the verification, since the refinement proof of POSIX to VFS is independent of the actual implementation, whereas the implementation needs to be verified against AFS only. The top-level layers are described in more detail in Sections 3.1 and 3.2. Although we provide our own verified VFS implementation, it will be possible to use the file system core with the Linux VFS instead.

Flash memory is not entirely reliable, for example, read operations may fail nondeterministically. Hence, **error handling** is taken into account from the

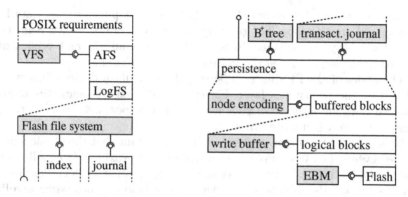

Fig. 2. Upper Layers of the File System   **Fig. 3.** Lower Layers of the File System

beginning. Concretely, the formal POSIX specification admits such errors as long as an unsuccessful operation has no observable effect. However, on intermediate layers failures may be visible.

A major concern that is orthogonal to functional correctness is **reset safety**: the file system guarantees recovery to a well-defined state after an unexpected power-cut. Reset-safety must be considered at different levels of abstraction. It is also critical that power-cuts are considered *during* an operation.

For example, the LogFS model introduces a distinction between persistent data stored on flash and volatile data in RAM. Only the latter is lost during power-cuts. New entries are first stored in a log and only later committed to flash. We prove abstractly that after a crash the previous RAM state can be reconstructed from the entries in the log and the flash state, as described in detail in Sec. 3.3. Atomicity of writes of several nodes is provided by another layer (transactional journal).

The core of the flash file system (represented by the correspondingly named box) implements strategies similar to UBIFS. This layer was also the first we have looked at [39]. The idea of UBIFS is to store a collection of data items, called "nodes", that represent files, directories, directory entries, and pages that store the content of files. This collection is unstructured and new data items are always written to a fresh location on flash, since overwriting in-place is not possible. Current versions of a node are referenced by an efficient **index**, implemented as a B^+ tree. The index exists both on flash and in memory. The purpose of the flash index is to speed up initialization of the file system, i.e., it is loaded on-demand. It is not necessary to scan the whole device during boot-up. Over time unreferenced nodes accrue and **garbage collection** is performed in the background to free up space by erasing the corresponding blocks.

Updates to the flash index are expensive due to the limitations of the hardware. Hence, these are collected up to a certain threshold and flushed periodically by a commit operation; one accepts that the flash index is usually outdated. Therefore, it is of great importance, that all information kept in RAM is actually redundant and can be recovered from flash. For this purpose, new data is kept in a special area, the *log*. The recovery process after a crash then reads the log to restore the previous state. In Sec. 3.3 we show an abstract version of this property. The log is part of the journal module in Fig. 2.

The nodes of the FFS have to be stored on the flash device. This is accomplished by the persistence layer. For space and time efficiency, the writes are buffered until an entire flash page can be written. Sec. 3.4 describes the problems that arise when writes are cached.

Besides correctness and reset safety, it is important that the physical medium is used evenly, i.e., erase cycles are distributed approximately equally between the blocks of the flash device, because erasing physically degrades the memory. To prolong the lifetime of the device, a technique called **wear-leveling** is implemented by the erase block management. Section 3.5 describes this layer.

3   Models

In this section we outline several formal models. There is a general schema. *Abstract* models, such as POSIX and AFS are as simple as possible. *Concrete* models, namely all grey components, are rather complex, in contrast.

Our tool is the interactive theorem prover KIV [35,12]. For a detailed description of the specification language see for example [13] (this volume) and [38]. It is based on *Abstract State Machines* [5] (ASMs) and ASM refinement [4]. We use algebraic specifications to axiomatize data types, and a weakest-precondition calculus to verify properties. This permits us to use a variety of data types, depending on the requirements. In the case study, lists, sets, multisets and maps (partial functions $\tau \nrightarrow \sigma$, application uses square brackets $f[a]$) are typically used in more abstract layers, whereas arrays are prevalent in the lower layers.

3.1   POSIX Specification and Model

Our formal POSIX model [15] defines the overall requirements of the system. It has been derived from the official specification [41] and is furthermore an evolution of a lot of existing related work to which we compare at the end of this section. Our model has been developed with two goals in mind:

1. It should be as abstract as possible, in particular, we chose an algebraic tree data structure to represent the directory hierarchy of the file system. The contents of files are stored as sequences of bytes.
2. It should be reasonably complete and avoid conceptual simplifications in the interface. Specifically, we support hard-links to files, handles to open files, and orphaned files (which are a consequence of open file handles as explained below); and the model admits nondeterministic failures in a well-specified way. These concepts are hard (or even impossible) to integrate in retrospect.

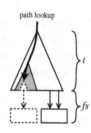

path lookup

Fig. 4. FS as a tree

t

fs

```
posix_create(path, md; err)
  choose err'
  with pre-create(path, md, t, fs, err')
  in err := err'

if err = ESUCCESS then
  choose fid with fid ∉ fs
  in t[path]   := fnode(fid)
     fs[fid]   := fdata(md, ⟨⟩)
```

Fig. 5. POSIX create operation

The model is easy to understand but also ensures that we will end up with a realistic and usable file system. Formally, the state of the POSIX model is given by the directory tree $t : Tree$ and a file store $fs : Fid \nrightarrow FData$ mapping file identifiers to the content.

We consider the following structural POSIX system-level operations: create, mkdir, rmdir, link, unlink, and rename. File content can be accessed by the operations open, close, read, write, and truncate. Finally, directory listings and (abstract) metadata can be accessed by readdir, readmeta (= stat), and writemeta (subsuming chmod/chown etc).

Structural system-level operations are defined at the level of *paths*. As an example, Fig. 4 shows the effect of create, where the grey part denotes the parent directory, and the parts with a dotted outline are newly created. The corresponding ASM rule is listed in Fig. 5.

The operation takes an absolute path *path* and some metadata *md* as input. After some error handling (which we explain below), a fresh file identifier *fid* is selected and stored in a file node (fnode) at path *path* in the tree. The file store *fs* is extended by a mapping from *fid* to the metadata and an empty content, where ⟨⟩ denotes an empty sequence of bytes. File identifiers serve as an indirection to enable hard-links, i.e., multiple references to the same file from different parent directories under possibly different names. An error code *err* is returned.

The converse operation is unlink(*path*; *err*), which removes the entry specified by *path* from the parent directory. In principle, it also deletes the file's content as soon as the last link is gone. File content can be read and written sequentially through *file handles fh*, which are obtained by open(*path*; *fh*, *err*) and released by close(*fh*; *err*). As a consequence, files are not only referenced from the tree, but also from open file handles, which store the identifier of the respective file. The POSIX standard permits to unlink files even if there is still an open file handle pointing to that file. The file's content is kept around until the last reference, either from the directory tree or a handle, is dropped.

Files that are no longer accessible from the tree are called *orphans*. The possibility to have orphans complicates the model and the implementation (see Sec. 3.3 on recovery after power-failure). However, virtually all Linux file systems provide this feature, since it is used by the operating system during package updates (application binaries are overwritten in place, while running applications still reference the previous binary image through a file handle) and by applications (e.g., Apache lock-files, MySQL temporary files).

Each POSIX operation returns an error code that indicates whether the operation was successful or not. Error codes fall into two categories. Preconditions are rigidly checked by the POSIX model in order to protect against unintended or malicious calls. For each possible violation it is specified which error codes may be indicated, but the order in which such checks occur is not predetermined. The other type of errors may occur anytime and corresponds to out-of-memory conditions or hardware failures. The model exhibits such errors nondeterministically, since on the given level of abstraction it is not possible to express a definite cause for such errors. The ASM rule in Fig. 5 simply chooses an error code err' that complies with the predicate `create-pre` and only continues in case of success, i.e., the precondition is satisfied and no hardware failure occurs.

We enforce the strong guarantee that an unsuccessful operation must not modify the state in any observable manner. It takes quite some effort to ensure this behavior in the implementation.

Our POSIX model takes a lot of inspiration from previous efforts to formalize file systems. An initial pen-and-paper specification that is quite complete is [28]. Mechanized models include [1,17,25,19,9,21]. These have varying degrees of abstraction and typically focus on specific aspects. Both [17] and [21] are path-based. The latter proves a refinement to a pointer representation, file content is treated as atomic data. Model [9] is based on a representation with parent pointers (compare Sec. 3.2). They prove multiple refinements and consider power-loss events abstractly. The efforts [1] and [25] focus on an analysis of reads and writes with pages, in the first, data is accessed a byte at a time, whereas [25] provides a POSIX-style interface for read/write while integrating some effect of power-loss. An in-depth technical comparison to our model can be found in [15].

3.2   VFS + AFS as a Refinement of POSIX

We have already motivated the purpose of the VFS and briefly described the integration with the AFS specification in Sec. 2 and [14]. We have proven our VFS correct wrt. the POSIX model relative to AFS [15]. As a consequence, any AFS-compliant file system can be plugged in to yield a correct system.

Conceptually, the VFS breaks down high-level operations to calls of operations of AFS. There are two major conceptual differences to the POSIX model:

1. The file system is represented as a pointer structure in AFS (as shown in Fig. 6) instead of an algebraic tree (compare Fig. 4), and
2. the content of files is represented as a sparse array of pages instead of a linear sequence of bytes.

These two aspects make the refinement proof challenging.

In VFS, directories and files are identified by "inode numbers" $ino : Ino$. The state of the AFS model is given by two (partial) functions $dirs : Ino \nrightarrow Dir$ and $files : Ino \nrightarrow File$. Directories Dir store a mapping from names of entries to inode

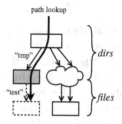

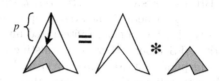

Fig. 7. Unfolding the tree abstraction, formally

Fig. 6. FS as pointer structure $\mathtt{tree}(t, ino) = \mathtt{tree}|_p(t, ino, ino') * \mathtt{tree}(t[p], ino')$

numbers. Files *File* store a partial mapping from page numbers to pages, which are byte-arrays. The refinement proof establishes a correspondence between the tree t and *dirs* and between *fs* and *files* and maintains it across operations.

The first issue can be solved elegantly using Separation Logic [36]. The main idea of Separation Logic is that disjointness of several parts of a pointer structure is captured in the syntax of formulas. The main connective is the separating conjunction $(\varphi * \psi)(h)$, which asserts that the two assertions φ and ψ on the structure of the heap h are satisfied in *disjoint* parts of h.

In our case, the heap is given by the directories of the file system *dirs*, and the abstraction to the algebraic tree is an assertion $\mathtt{tree}(t, ino)$. It specifies that *ino* is the root of the pointer structure represented by t. The abstraction directly excludes cycles and sharing ("aliasing") in *dirs*, and establishes that a modification at path p is *local*: directories located at other paths q are unaffected.

To perform an operation at path p the abstraction $\mathtt{tree}$ is unfolded as shown in Fig. 7. The directory tree *dirs* is split into two disjoint parts, the subtree $\mathtt{tree}(t[p], ino')$ located at path p and the remainder $\mathtt{tree}|_p(t, ino, ino')$, which results from cutting out the subtree at path p, with root *ino'*. A modification in the grey subtree is then guaranteed to leave the white subtree unaffected.

The second issue manifests itself in the implementation of the POSIX operations **read** and **write**, which access multiple pages sequentially. There are several issues that complicate the proofs but need to be considered: There may be gaps in the file, which implicitly represent zero-bytes, and the last page may exceed the file size. The first and the last page affected by a write are modified only partially, hence they have to be loaded first. Furthermore, writes may extend the file size or may even start beyond the end of a file.

Since reading and writing is done in a loop, the main challenge for the refinement proof is to come up with a good loop invariant. The trouble is that the high number of cases we have just outlined tends to square in the proof, if one considers the intermediate hypothetical size of the file. We have solved this problem by an abstraction that neglects the size of the file, namely, the content is mapped to an infinite stream of bytes by extension with zeroes at the end. The loop invariant for writing states that this stream can be decomposed into parts of the initial file at the beginning and at the end, with data from the input buffer in between.

3.3   From AFS to LogFS to Study Power Cuts

The UBIFS file system is *log-structured*, which means that there is an area on flash—the log—where all file system updates are written to in sequential order. Data in the log has not been integrated into the index that is stored on flash.

In the AFS model, there is no distinction between volatile state stored in RAM and persistent state stored on flash. Purpose of the LogFS model is to introduce this distinction on a very abstract level in order to study the effects of unexpected power cuts without getting lost in technical detail. The LogFS model thus bridges the gap between AFS and our FFS implementation.

The model simply assumes that there are two copies of the state, a current in-memory version *ramstate* and a possibly outdated flash version *flashstate*. The *log* captures the difference between the two states as a list of log entries.

Operations update the state in RAM and append entries to the log e.g., "delete file referenced by *ino*" or "write a page of data". The *flashstate* is not changed by operations. Instead, from time to time (when the log is big enough) an internal *commit* operation is executed by the file system, which overwrites the *flashstate* with *ramstate*, and resets the log to empty.

The effect of a power failure and subsequent recovery can now be studied: the crash resets the *ramstate*, recovery has to rebuild it starting from *flashstate* by recovering the entries of *log*. For crash safety we therefore have to preserve the recovery invariant

$$ramstate = \textsf{recover}(\mathit{flashstate}, \mathit{log}) \tag{1}$$

Note that although **recover** is shown here as a function, it is really a (rather complex) ASM rule in the model, implying that the invariant is not a predicate logic formula (but one of wp-calculus).

To propagate the invariant through operations it is necessary to prove that running the recovery algorithm on entries produced by one operation has the same effect on *ramstate* than the operation itself. This is non-trivial, since typical operations produce several entries and the intermediate *ramstate* after recovering a single entry is typically not a consistent state.

The possibility of orphans (Sec. 3.1) complicates recovery. Assume 1) a file is opened and 2) subsequently becoming an orphan, i.e., it is not accessible from the directory tree any more. The content of the file then can still be read and written through the file handle. Therefore, the file contents may only be erased from flash memory after the file is closed. However, if a crash occurs the process accessing the file is terminated and the file handle disappears. The file system should consequently ensure that the file content is actually deleted during recovery. There are two possibilities: the corresponding "delete file" is in the log before the crash, then no special means have to be taken. However, if a commit operation has happened before the crash but after 2), then the log has been cleared and consequently this entry will *not* be found.

To prevent that such a file occupies space forever, our file system records such files in an explicit set of flash-orphans *forphans* stored on the flash medium during commit.

Recovery thus starts with a modified flash state, namely *flashstate \ forphans*, and produces a state that contains no orphans wrt. the previous RAM state (denoted by *rorphans*). Invariant (1) becomes

$$ramstate \setminus rorphans = \texttt{recover}(flashstate \setminus forphans, log) \tag{2}$$

It is not correct to require that only the recorded orphans *forphans* have been deleted as a result of `recover`. Again referring to the example above, in the case that no commit has occurred since event 2), we have *ino* ∈ *rorphans* but *ino* ∉ *forphans*, but the file *ino* will be removed anyway, since the deletion entry is still in the log.

Technically, in order to prove (2), additional commutativity properties between deleting orphans and applying log entries must be shown. This is nontrivial, since several additional invariants are necessary: for example, that inode numbers *ino* ∈ *forphans* are not referenced from the directory tree of the RAM state, and that there is no "create" entry for *ino* in the log. Furthermore, since removal of directories and files is handled in a uniform way, directories can actually be orphans, too.

In our initial work [39] we have proved the recovery invariant on a lower level of abstraction, namely the file system core, but without orphans. Adding them to the model has lead to a huge increase in complexity, hence we switched to a more abstract analysis of the problem. It is then possible (although there are some technical difficulties) to prove that the recovery invariant of LogFS propagates to the concrete implementation by matching individual recovery steps.

The verification of recovery of a log in the given sense has been addressed in related work only by [31], which is in fact based on our model [39]. Their goal is to explore points-free relational modeling and proofs. They consider a simplified version of recovery that does not consider orphans, and restricts the log to the deletion of files only (but not creation).

Finally, we like to point out that the recovery invariant (2) needs to hold in every *intermediate* step of an operation—a power cut could happen anytime, not just in between operations. A fine-grained analysis is therefore necessary, e.g., based on a small-step semantics of ASMs, which we define in [33].

3.4   Node Persistence and Write Buffering

The persistence layer provides access to (abstract) erase blocks containing a list of nodes for the FFS core and its subcomponents. Clients can append nodes to a block and allocate or deallocate a block. This already maps well to the flash specific operations, i.e., appending is implemented by writing sequentially and deallocation leads to a deferred erase.

The node encoding layer stores these nodes as byte arrays on the flash device with the help of a write buffer. The implementation also ensures that nodes either appear to be written entirely or not at all. The write buffer provides a page-sized buffer in RAM that caches writes to flash until an entire page can be written. This improves the space-efficiency of the file system, since multiple nodes can potentially fit in one page. However, as a consequence of the caching of writes, it is possible that some nodes disappear during a crash. The challenge is to find a suitable refinement theory that includes crashes and provides strong guarantees in their presence. Clients should also be able to enforce that the write buffer is written. This is accomplished by adding *padding nodes*, i.e., nodes that span the remainder of the page. These nodes are invisible to a client of the persistence layer. The refinement between the persistence and the node encoding layer switches from indices in a list to offsets in a byte array as their respective input and output. The padding nodes complicate this refinement, since the abstraction relation from a list of nodes to the byte array is no longer functional.

For the sequential writes it is necessary that the layer stores how far an erase block is already written. However, it is difficult to update such a data structure efficiently, since only out-of-place updates are possible. Therefore, updates are performed in RAM and written to flash only during a commit operation similarly to the index. The recovery after a power failure has to reconstruct a consistent data structure and consider that some updates were not written to flash.

We formalize the byte encoding very abstractly, i.e., by just stating that decoding after encoded yields the original data. In future work we will consider whether we can use specificiation languages such as [2,27] to concretize the flash layout. Other models [25,9] neither consider the encoding nor buffer writes.

3.5 Erase Block Management

A raw flash device supports reading, sequential writing [10,16] and erasure of a block. Only after erasure it is possible to reuse a block. However, blocks degrade physically with each erase, i.e., typically after 10^4-10^6 erases a block becomes unreliable. There are two complementary techniques that deal with this problem and thereby increase the reliability of the device: wear-leveling and bad block management. The former means that necessary erasures are spread among the entire device evenly by moving stale data. Thus, bad blocks occur later in the lifetime of a device. Bad block management tries to detect which blocks are unusable and excludes them from future operations.

Both techniques are implemented transparently in the erase block management (EBM) by a mapping from *logical* to *physical* blocks. Fig. 8 shows this in-RAM mapping as arrows from logical to physical blocks. The client can only access logical blocks and the EBM keeps the mapping and allocates, accesses or deallocates physical blocks as needed. Wear-leveling copies the contents of a physical block to a new block and updates the mapping. Bad blocks are simply excluded from future allocations.

Another feature facilitated by the mapping is that clients may request an *asynchronous* erasure of a block. This increases the performance of the file system, because erasing a physical block is a slow operation. If the erasure is deferred, the client can already reuse the *logical* block. The EBM just allocates a different physical block.

The difficulty for the implementation and verification stems from interactions of the following factors (explained in more detail in [32]):

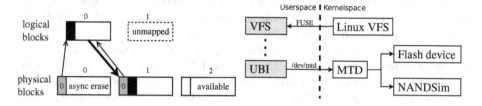

Fig. 8. Mapping in the EBM   **Fig. 9.** Integration with Linux

1. The mapping is stored inversely within each *physical* block (shown as the backwards arrows in Fig. 8) and must be written before the actual contents.[1]
2. The flash device may exhibit nondeterministic errors. Only writing of a single page is assumed to be atomic.
3. Asynchronous erasure defers an update of the on-disk mapping.
4. Wear-leveling must appear to be atomic, otherwise a power-cut in between would lead to a state, where half-copied data is visible to the client.

Due to 1) and 3), the recovery from a power-cut will restore some older version of the mapping. This is visible to the client and has to be handled by the client. This necessitates a refinement theory that allows us to specify the behavior that a model exhibits in response to a power-cut and shows that an implementation meets this specified reset behavior, i.e., is reset-safe.

Items 1) and 2) complicate 4), because they lead to an updated on-disk mapping referring to half-copied data. Additional measures (such as checksums) are necessary to ensure that a mapping produced during wear-leveling becomes only valid once the entire contents are written.

Other formal models of EBM [25,26,9] usually do not take the limitation to sequential writes within a block into account, do not store the mapping on disk, assume additional reliable bits per block to validate the mapping only after the pages have been programmed or store an inefficient page-based mapping (assumed to be updatable atomically). Our model of the flash hardware is ONFI-compliant [16] and based on the Linux flash driver interface MTD [29]. The flash hardware model [6] is below our model of a flash driver.

4   Code Generation

We currently generate Scala [30] code, which runs on the Java virtual machine. We chose Scala, because it is object-oriented, supports immutable data types and pattern matching and features a comprehensive library. As explained below, we can generate code that is very similar to the KIV models. Therefore, Scala is well-suited for early debugging and testing of the models. Furthermore, it is

---

[1] The reasons for this are rather technical, but follow from the limitations of the flash hardware to sequential writes within a block and performance considerations.

easier to integrate a visualization for the data structures and operations. The final code is derived from the models shaded in gray in Fig. 2 and 3, but Scala code generation is also supported for the other models.

The code generation for Scala currently supports free and non-free data types (potentially mutually recursive). Most of the KIV library data types (natural numbers, strings, arrays, lists, stores and sets) and algebraic operations on them are translated to corresponding classes and functions of the Scala library. Furthermore, an algebraic operation with a non-recursive or structurally-recursive definition is transformed into a function that uses pattern matching to discern the cases of a data type.

We mimic the refinement hierarchy shown in Fig. 2 and 3 by inheritance, i.e., for each of the abstract (white) layers an interface is generated with one method for each of the ASM rules. For the concrete (gray) layers we generate an implementing class that has the ASM state as a member. The usage of an interface (—◦— in the figure) is implemented by aggregation. Thus, the VFS layer for example has an additional reference to an AFS interface. In the final file system this reference points to an instance of the FFS.

This setup also allows us to test the VFS model based on a different implementation of the AFS interface. We optionally also generate an implementation of the AFS interface based the AFS ASM. This allows us to test a layer early on without having a refined model of all the used layers yet. These abstract layers however are rather nondeterministic, e.g., they heavily employ the following construct to choose an error code that may either indicate success or a low-level error (such as an I/O error).

```
choose err with err ∈ {ESUCCESS, EIO, ...} in { ... }
```

We currently simulate this by randomly choosing an element of the type and checking whether the element satisfies the given property.[2] This is sufficient when testing whether, e.g., VFS can handle the full range of behavior admitted by AFS.

However, in order to test whether a concrete model exhibits only behavior permitted by the abstract model this is insufficient. In that case it is necessary that the abstract model simulates the choices made by the concrete model. In future work, we will attempt to automatically derive an abstract model which is suitable for such tests. Each of the operations of this model may also take as input the output of the concrete layer and their goal is to construct a matching abstract execution. For the above example of error codes it is sufficient to just rewrite the abstract program by reusing the error code returned by the concrete model and checking that it satisfies the condition $err \in \{ESUCCESS, EIO, ...\}$. For more elaborate nondeterminism an SMT solver could be employed to make the proper choice.

We are currently also working on the generation of C code and expect approximately 10.000 lines of code. The C code generation will be limited to the models shaded in gray. In future work, we will investigate whether tools for the

---

[2] Note that for some forms of nondeterminism we can and do provide a better, deterministic implementation.

automated verification of C, such as VCC [7], could be employed to show the refinement between our last models and the generated C code automatically.

We are unaware of other file systems where code was generated automatically or the final code was verified against the models. [9] provides systematic translation rules that need to be applied manually to derive Java code.

The code generated from the models can be plugged into Linux via the *Filesystem in Userspace* (= FUSE) library [40], which allows mounting of a file system implemented in an application in user space. All calls to the mountpoint are then routed to the application by FUSE. This allows us to test and debug the file system more easily. The access to either a real flash device (we currently use a Cupid-Core Single-Board Computer with 256 MB Flash [8]) or to the simulator (NandSim) is realized via the Linux device /dev/mtd. Fig. 9 shows the entire setup.

5   Lessons Learned

In this section we report on the lessons we learned during the design and verification of the flash file system.

In our opinion, the file system challenge is interesting for the reason that a wide conceptual gap must be bridged. Abstractly, the file system is described as a tree indexed by paths, whereas the hardware interface is based on erase blocks and bytes. The strategies we have taken from the Linux VFS and UBIFS to map between these interfaces deal with many different concepts.

Capturing these concepts at the right degree of abstraction proved to be a major design challenge. Not all concepts that we have encountered have direct counterparts in the implementation. Here, we benefit from the refinement-based approach that supports well isolating difficult aspects of the verification in dedicated models. For example, the abstract log as specified in LogFS is actually encoded in the headers of nodes in UBIFS.

It proved beneficial that we have initially started with the core concepts of the UBIFS file system and derived an abstract model [39]. This model has served as an anchor-point to incrementally develop the rest of the model hierarchy.

Our experience is that models tend to change frequently. One reason for that is that requirements are clarified or assumptions are rectified. It also happens that some model is refactored, in order to e.g. maintain a new invariant. Such changes are typically non-local and propagate through the refinement hierarchy.

The prime reason for changes stems from the fact that there is a fragile balance between simple abstract specifications and the question which functional guarantees can be provided by the implementation. The technical cause lies in power cuts and hardware failures, which affect almost all layers and can leak through abstractions (nondeterministic errors in POSIX are one example). More specifically, one needs to determine to which extent failures can be hidden by the implementation, how this can be achieved, and which layer incorporates the concepts necessary to address a particular effect of a failure. For example, the flash file system core can not express partially written nodes, so a lower layer (namely, node encoding in Fig. 3) must be able to recognize and hide such partial nodes (e.g., by checksums). In general, dealing with power cuts and hardware failures has increased the complexity of the models and their verification by a signification amount.

In order to reduce the effort inherent in frequent changes, KIV provides an elaborate correctness management for proofs. The system tracks the dependencies of every proof, i.e., the axioms and theorems used in the proof. A proof becomes invalid only if a dependency has changed. Furthermore, KIV supports replaying of an old, invalid proof. User interactions are therefore mostly limited to parts of a proof affected by a change.

Similar to [3], we observed that strong abstraction capabilities of the used tools are essential. KIV supports arbitrary user-defined data types (given suitable axioms), which was for example exploited to abstract the pointer structure to an algebraic tree and to abstract the sparse pages of files to streams (see Sec. 3.2).

We found that for smaller invariants it would be useful to have stronger typing support in our tool, such as for example predicative subtypes [37]. This could be used for example to express the invariant that every erase block has a certain size in the type. Making such invariants implicit reduces the number of trivial preconditions for lemmas.

6   Conclusion and Open Challenges

We have given an overview over our approach to the development of a formally verified flash file system. We outlined our decomposition into a hierarchy of models, described some of the challenges posed in each of them and sketched the solutions we developed. We are currently finalizing the models and their correctness proofs for the various refinements, one rather challenging remains: verifying the implementation of indices by "wandering" B^+-trees, that are maintained in RAM, and are incrementally stored on flash.

Two major aspects remain future work:

We have not yet addressed concurrency. A realistic and efficient implementation, however, should perform garbage collection (in the flash file system model), erasure of physical blocks and wear-leveling (in the EBM model) concurrently in the background. In those instances, where the action is invisible to the client of the respective layer, the abstract behavior is trivial to specify. However, on top-level it is not clear which outcomes of two concurrent writes to overlapping regions are acceptable, and how to specify them (the POSIX standard does not give any constraint).

Another feature that is non-trivial to specify is the effect of caching when using the real VFS implementation. The difference between our non-caching VFS implementation and the one in Linux would be visible to the user if a power-cut occurs, as not all previous writes would have been persisted on flash. For two simpler caching related problems (the write buffer of Sec. 3.4 and the mapping of Sec. 3.5) we already successfully applied a refinement theory that incorporates power-cuts [33].

References

1. Arkoudas, K., Zee, K., Kuncak, V., Rinard, M.: Verifying a file system implementation. In: Davies, J., Schulte, W., Barnett, M. (eds.) ICFEM 2004. LNCS, vol. 3308, pp. 373–390. Springer, Heidelberg (2004)
2. Back, G.: DataScript - A Specification and Scripting Language for Binary Data. In: Batory, D., Blum, A., Taha, W. (eds.) GPCE 2002. LNCS, vol. 2487, pp. 66–77. Springer, Heidelberg (2002)
3. Baumann, C., Beckert, B., Blasum, H., Bormer, T.: Lessons Learned From Microkernel Verification – Specification is the New Bottleneck. In: SSV, pp. 18–32 (2012)
4. Börger, E.: The ASM Refinement Method. Formal Aspects of Computing 15(1-2), 237–257 (2003)
5. Börger, E., Stärk, R.F.: Abstract State Machines — A Method for High-Level System Design and Analysis. Springer (2003)
6. Butterfield, A., Woodcock, J.: Formalising Flash Memory: First Steps. In: IEEE Int. Conf. on Engineering of Complex Computer Systems, pp. 251–260 (2007)
7. Cohen, E., Dahlweid, M., Hillebrand, M., Leinenbach, D., Moskal, M., Santen, T., Schulte, W., Tobies, S.: VCC: A practical system for verifying concurrent C. In: Berghofer, S., Nipkow, T., Urban, C., Wenzel, M. (eds.) TPHOLs 2009. LNCS, vol. 5674, pp. 23–42. Springer, Heidelberg (2009)
8. http://www.garz-fricke.com/cupid-core_de.html
9. Damchoom, K.: An incremental refinement approach to a development of a flash-based file system in Event-B (October 2010)
10. Samsung Electronics. Page program addressing for MLC NAND application note (2009), http://www.samsung.com
11. Ernst, G., Pfähler, J., Schellhorn, G.: Web presentation of the Flash Filesystem (2014), https://swt.informatik.uni-augsburg.de/swt/projects/flash.html
12. Ernst, G., Pfähler, J., Schellhorn, G., Haneberg, D., Reif, W.: KIV - Overview and VerifyThis Competition. Software Tools for Technology Transfer (to appear, 2014)
13. Ernst, G., Pfähler, J., Schellhorn, G., Reif, W.: Modular Refinement for Submachines of ASMs. In: Ait Ameur, Y., Schewe, K.-D. (eds.) ABZ 2014. LNCS, vol. 8477, pp. 188–203. Springer, Heidelberg (2014)
14. Ernst, G., Schellhorn, G., Haneberg, D., Pfähler, J., Reif, W.: A Formal Model of a Virtual Filesystem Switch. In: Proc. of Software and Systems Modeling (SSV), pp. 33–45 (2012)
15. Ernst, G., Schellhorn, G., Haneberg, D., Pfähler, J., Reif, W.: Verification of a Virtual Filesystem Switch. In: Cohen, E., Rybalchenko, A. (eds.) VSTTE 2013. LNCS, vol. 8164, pp. 242–261. Springer, Heidelberg (2014)
16. Intel Corporation, et al.: Open NAND Flash Interface Specification (June 2013), http://www.onfi.org
17. Ferreira, M.A., Silva, S.S., Oliveira, J.N.: Verifying Intel flash file system core specification. In: Modelling and Analysis in VDM: Proc. of the Fourth VDM/Overture Workshop, School of Computing Science, Newcastle University, Technical Report CS-TR-1099, pp. 54–71 (2008)
18. Freitas, L., Woodcock, J., Butterfield, A.: POSIX and the Verification Grand Challenge: A Roadmap. In: ICECCS 2008: Proc. of the 13th IEEE Int. Conf. on Engineering of Complex Computer Systems (2008)
19. Freitas, L., Woodcock, J., Fu, Z.: Posix file store in Z/Eves: An experiment in the verified software repository. Sci. of Comp. Programming 74(4), 238–257 (2009)

20. Gleixner, T., Haverkamp, F., Bityutskiy, A.: UBI - Unsorted Block Images (2006), http://www.linux-mtd.infradead.org/doc/ubidesign/ubidesign.pdf

21. Hesselink, W.H., Lali, M.I.: Formalizing a hierarchical file system. Formal Aspects of Computing 24(1), 27–44 (2012)

22. Hoare, C.A.R.: The verifying compiler: A grand challenge for computing research. Journal of the ACM 50(1), 63–69 (2003)

23. Hunter, A.: A brief introduction to the design of UBIFS (2008), http://www.linux-mtd.infradead.org/doc/ubifs_whitepaper.pdf

24. Joshi, R., Holzmann, G.J.: A mini challenge: build a verifiable filesystem. Formal Aspects of Computing 19(2) (June 2007)

25. Kang, E., Jackson, D.: Formal Modeling and Analysis of a Flash Filesystem in Alloy. In: Börger, E., Butler, M., Bowen, J.P., Boca, P. (eds.) ABZ 2008. LNCS, vol. 5238, pp. 294–308. Springer, Heidelberg (2008)

26. Kang, E., Jackson, D.: Designing and Analyzing a Flash File System with Alloy. Int. J. Software and Informatics 3(2-3), 129–148 (2009)

27. McCann, P.J., Chandra, S.: Packet Types: Abstract Specification of Network Protocol Messages. SIGCOMM Comp. Comm. Rev. 30(4), 321–333 (2000)

28. Morgan, C., Sufrin, B.: Specification of the unix filing system. In: Specification Case Studies, pp. 91–140. Prentice Hall Ltd., Hertfordshire (1987)

29. Memory Technology Device (MTD) and Unsorted Block Images (UBI) Subsystem of Linux, http://www.linux-mtd.infradead.org/index.html

30. Odersky, M., Spoon, L., Venners, B.: Programming in Scala: A Comprehensive Step-by-step Guide, 1st edn. Artima Incorporation, USA (2008)

31. Oliveira, J.N., Ferreira, M.A.: Alloy Meets the Algebra of Programming: A Case Study. IEEE Transactions on Software Engineering 39(3), 305–326 (2013)

32. Pfähler, J., Ernst, G., Schellhorn, G., Haneberg, D., Reif, W.: Formal Specification of an Erase Block Management Layer for Flash Memory. In: Bertacco, V., Legay, A. (eds.) HVC 2013. LNCS, vol. 8244, pp. 214–229. Springer, Heidelberg (2013)

33. Pfähler, J., Ernst, G., Schellhorn, G., Haneberg, D., Reif, W.: Crash-Safe Refinement for a Verified Flash File System. Technical report, University of Augsburg (2014)

34. Reeves, G., Neilson, T.: The Mars Rover Spirit FLASH anomaly. In: Aerospace Conference, pp. 4186–4199. IEEE Computer Society (2005)

35. Reif, W., Schellhorn, G., Stenzel, K., Balser, M.: Structured specifications and interactive proofs with KIV. In: Bibel, W., Schmitt, P. (eds.) Automated Deduction—A Basis for Applications, vol. II, pp. 13–39. Kluwer, Dordrecht (1998)

36. Reynolds, J.C.: Separation logic: A logic for shared mutable data structures. In: Proc. of LICS, pp. 55–74. IEEE Computer Society (2002)

37. Rushby, J., Owre, S., Shankar, N.: Subtypes for Specifications: Predicate Subtyping in PVS. IEEE Transactions on Software Engineering 24(9), 709–720 (1998)

38. Schellhorn, G.: Completeness of Fair ASM Refinement. Science of Computer Programming 76(9) (2009)

39. Schierl, A., Schellhorn, G., Haneberg, D., Reif, W.: Abstract Specification of the UBIFS File System for Flash Memory. In: Cavalcanti, A., Dams, D.R. (eds.) FM 2009. LNCS, vol. 5850, pp. 190–206. Springer, Heidelberg (2009)

40. Szeredi, M.: File system in user space, http://fuse.sourceforge.net

41. The Open Group. The Open Group Base Specifications Issue 7, IEEE Std 1003.1, 2008 Edition (2008), http://www.unix.org/version3/online.html (login required)

Why Amazon Chose TLA+

Chris Newcombe

Amazon, Inc.

Abstract. Since 2011, engineers at Amazon have been using TLA+ to help solve difficult design problems in critical systems. This paper describes the reasons why we chose TLA+ instead of other methods, and areas in which we would welcome further progress.

1   Introduction

Why Amazon is using formal methods. Amazon builds many sophisticated distributed systems that store and process data on behalf of our customers. In order to safeguard that data we rely on the correctness of an ever-growing set of algorithms for replication, consistency, concurrency-control, fault tolerance, auto-scaling, and other coordination activities. Achieving correctness in these areas is a major engineering challenge as these algorithms interact in complex ways in order to achieve high-availability on cost-efficient infrastructure whilst also coping with relentless rapid business-growth[1]. We adopted formal methods to help solve this problem.

Usage so far. As of February 2014, we have used TLA+ on 10 large complex real-world systems. In every case TLA+ has added significant value, either preventing subtle serious bugs from reaching production, or giving us enough understanding and confidence to make aggressive performance optimizations without sacrificing correctness. Executive management are now proactively encouraging teams to write TLA+ specs for new features and other significant design changes. In annual planning, managers are now allocating engineering time to use TLA+.

We lack space here to explain the TLA+ language or show examples of our specifications. We refer readers to [18] for a tutorial, and [26] for an example of a TLA+ specification from industry that is similar in size and complexity to some of the larger specifications at Amazon.

What we wanted in a formal method. Our requirements may be roughly grouped as follows. These are not orthogonal dimensions, as business requirements and engineering tradeoffs are rarely crisp. However, these do represent the most important characteristics required for a method to be successful in our industry segment.

1. *Handle very large, complex or subtle problems.* Our main challenge is complexity in concurrent and distributed systems. As shown in Fig. 1, we often need to verify the interactions between algorithms, not just individual algorithms in isolation.

---

[1] As an example of such growth; in 2006 we launched S3, our Simple Storage Service. In the six years after launch, S3 grew to store 1 trillion objects [6]. Less than one year later it had grown to 2 trillion objects, and was regularly handling 1.1 million requests per second [7].

Y. Ait Ameur and K.-D. Schewe (Eds.): ABZ 2014, LNCS 8477, pp. 25–39, 2014.

System	Components	Line count	Benefit
S3	Fault-tolerant low-level network algorithm	804 PlusCal	Found 2 design bugs. Found further design bugs in proposed optimizations.
	Background redistribution of data	645 PlusCal	Found 1 design bug, and found a bug in the first proposed fix.
DynamoDB	Replication and group-membership systems (which tightly interact)	939 TLA^+	Found 3 design bugs, some requiring traces of 35 steps.
EBS	Volume management	102 PlusCal	Found 3 design bugs.
EC2	Change to fault-tolerant replication, including incremental deployment to existing system, with zero downtime	250 TLA^+ 460 TLA^+ 200 TLA^+	Found 1 design bug.
Internal distributed lock manager	Lock-free data structure	223 PlusCal	Improved confidence. Failed to find a liveness bug as we did not check liveness.
	Fault tolerant replication and reconfiguration algorithm	318 TLA^+	Found 1 design bug. Verified an aggressive optimization.

Fig. 1. Examples of applying TLA^+ to some of our more complex systems

Also, many published algorithms make simplifying assumptions about the operating environment (e.g. fail-stop processes, sequentially consistent memory model) that are not true for real systems, so we often need to modify algorithms to cope with the weaker properties of a more challenging environment. For these reasons we need expressive languages and powerful tools that are equipped to handle high complexity in these problem domains.

2. *Minimize cognitive burden.* Engineers already have their hands full with the complexity of the problem they are trying to solve. To help them rather than hinder, a new engineering method must be relatively easy to learn and easy to apply. We need a method that avoids esoteric concepts, and that has clean simple syntax and semantics. We also need tools that are easy to use. In addition, a method intended for specification and verification of designs must be easy to remember. Engineers might use a design-level method for a few weeks at the start of a project, and then not use it for many months while they implement, test and launch the system. During the long implementation phase, engineers will likely forget any subtle details of a design-level method, which would then cause frustration during the next design phase. (We suspect that verification researchers experience a very different usage pattern of tools, in which this problem might not be apparent.)

3. *High return on investment.* We would like a single method that is effective for the wide-range of problems that we face. We need a method that quickly gives useful results, with minimal training and reasonable effort. Ideally we want a method that also improves time-to-market in addition to ensuring correctness.

Comparison of methods. We preferred candidate methods that had already been shown to work on problems in industry, that seemed relatively easy to learn, and that we could apply to real problems while we were learning. We were less concerned about "verification philosophy" such as process algebra vs. state machines; we would have used any method that worked well for our problems.

We evaluated Alloy and TLA+ by trying them on real-world problems [29]. We did a smaller evaluation of Microsoft VCC. We read about Promela/Spin [13], Event-B, B, Z, Coq, and PVS, but did not try them, as we halted the investigation when we realized that TLA+ solved our problem.

2 Handle Very Large, Complex or Subtle Problems

2.1 Works on Real Problems in Industry

TLA+ has been used successfully on many projects in industry: complex cache-coherency protocols at Compaq [25] and Intel [8], the Paxos consensus algorithm [21], the Farsite distributed file system [9], the Pastry distributed key-value store [28], and others. Alloy has been used successfully to find design flaws in the Chord ring membership protocol [32] and to verify an improved version [33]. Microsoft VCC has been used successfully to verify the Microsoft Hypervisor [27].

Event-B [5] and B [1] have been used in industry, although most of the applications appear to be control systems, which is an interesting area but not our focus. We found some evidence [12] that Z has been used to specify systems in industry, but we did not find evidence that Z has been used to verify large systems. PVS has been used in industry but most of the applications appear to be low-level hardware systems [30]. We did not find any relevant examples of Coq being used in industry.

2.2 Express Rich Dynamic Structures

To effectively handle complex systems, a specification language must be able to capture rich concepts without tedious workarounds. For example, when modelling replication algorithms we need to be able to specify dynamic sequences of many types of nested records. We found that TLA+ can do this simply and directly. In contrast, all structures in Alloy are represented as relations over uninterpreted atoms, so modelling nested structures requires adding intermediate layers of identifiers. This limitation deters us from specifying some distributed algorithms in Alloy. In addition, Alloy does not have a built-in notion of time, so mutation must be modelled explicitly, by adding a timestamp column to each relation and implementing any necessary dynamic structures such as sequences. This has occasionally been a distraction.

Alloy's limited expressiveness is an intentional trade-off, chosen to allow analysis of very abstract implicit specifications[2]. In contrast, TLA+ was designed for clarity

---

[2] The book on Alloy [15, p. 302] says, "... the language was stripped down to the bare essentials of structural modelling, and developed hand-in-hand with its analysis. Any feature that would thwart analysis was excluded", and [15, p. 41], "The restriction to flat relations makes the logic more tractable for analysis."

of expression and reasoning, initially without regard to the consequences for analysis tools[3]. In practice, we have found that the model checker for TLA$^+$ is usually capable of model-checking instances of real-world concurrent and distributed algorithms at least as large as those checkable by the Alloy Analyzer. However, the Alloy Analyzer can perform important types of inductive analysis (e.g. [33, p. 9]) that the currently available tools for TLA$^+$ cannot handle. We have not yet tried inductive analysis of real-world algorithms, but are interested in doing so for reasons discussed later in this paper.

2.3 Easily Adjust the Level of Abstraction

To verify complex designs we need to be able to add or remove detail quickly, with reasonably small localized edits to the specification. TLA$^+$ allows this by supporting arbitrarily complicated data structures in conjunction with the ability to define powerful custom operators[4] that can abstract away the details of those structures. We found that Alloy is significantly less flexible in this area as it does not support recursive operators or functions [15, pp. 74,280], or higher-order functions. Alloy does compensate to some extent by having several built-in operators that TLA$^+$ lacks, such as transitive closure and relational join, but we found it easy to define these operators in TLA$^+$. VCC allows the user to write "ghost code" in a superset of the C programming language. This is an extremely powerful feature, but the result is usually significantly more verbose than when using TLA$^+$ or Alloy. We have not investigated the extensibility or abstraction features of other methods.

2.4 Verification Tools That Can Handle Complex Concurrent and Distributed Systems

The TLA$^+$ model-checker works by explicitly enumerating reachable states. It can handle large state-spaces at reasonable throughput. We have checked spaces of up to 31 billion states[5], and have seen sustained throughput of more than 3 million states per minute when checking complex specifications. The model checker can efficiently use multiple cores, can spool state to disk to check state-spaces larger than physical memory, and supports a distributed mode to take good advantage of additional processors, memory and disk. Distributed mode has been very important to us, but we have also obtained good results on individual 32-core machines with local SSD storage.

At the time of our evaluation, Alloy used off the shelf SAT solvers that were limited to using a single core and some fraction of the physical RAM of a single machine. For moderate finite scopes the Alloy Analyzer is extremely fast; much faster than the TLA$^+$

---

[3] The introductory paper on TLA$^+$ [19] says, "We are motivated not by an abstract ideal of elegance, but by the practical problem of reasoning about real algorithms ... we want to make reasoning as simple as possible by making the underlying formalism simple." The designer of TLA$^+$ has expressed surprise that the language could be model-checked at all [16]. The first model-checker for TLA$^+$ did not appear until approximately 5 years after the language was introduced.

[4] TLA$^+$ supports defining second-order operators [20, p. 318] and recursive functions and operators with a few restrictions.

[5] Taking approximately 5 weeks on a single EC2 instance with 16 virtual CPUs, 60 GB RAM and 2 TB of local SSD storage.

model checker. However, we found that the SAT solvers often crash or hang when asked to solve the size of finite model that is necessary for achieving reasonable confidence in a more complex concurrent or distributed system. When we were preparing this paper, Daniel Jackson told us [14] that Alloy was not intended or designed for model checking such algorithms. Jackson says, "Alloy is a good choice for direct analysis if there is some implicitness in the specification, e.g. an arbitrary topology, or steps that can make very unconstrained changes. If not, you may get Alloy to be competitive with model checkers, but only on short traces." However we feel that Jackson is being overly modest; we found Alloy to be very useful for understanding and debugging the significant subset of concurrent algorithms that have relatively few variables, shallow nesting of values, and for which most interesting traces are shorter than perhaps 12 to 14 steps. In such cases, Alloy can check sophisticated safety properties sufficiently well to find subtle bugs and give high confidence—as Pamela Zave found when using Alloy to analyze the Chord ring membership protocol [32,33].

VCC uses the Z3 SMT solver, which is also limited to a single machine and physical memory (we are not sure if it uses multiple cores). We have not yet tried the ProB model checker for B or Event B, but we believe that it is also limited to using a single core and physical RAM on a single machine.

Another limitation of Alloy is that the SAT-based analysis can only handle a small finite set of contiguous integer values. We suspect that this might prevent Alloy from effectively analyzing industrial systems that rely on properties of modulo arithmetic, e.g. Pastry. TLA+ was able to specify, model-check, and prove Pastry [28].

2.5   Good Tools to Help Users Diagnose Bugs

We use formal methods to help us find and fix very subtle errors in complex designs. When such an error is found, the cause can be difficult for the designer to diagnose. To help with diagnosis we need tools that allow us to comprehend lengthy sequences of complex state changes. This is one of the most significant differences between the formal methods that we tried. The primary output of the Alloy Analyzer tool is diagrams; it displays an execution trace as a graph of labelled nodes. The tool uses relatively sophisticated algorithms to arrange the graph for display, but we still found this output to be incomprehensible for systems with more than a few variables or time-steps. To work around this problem we often found ourselves exporting the execution trace to an XML file and then using a text editor and other tools to explore it. The Alloy Analyzer also has a feature that allows the user to evaluate arbitrary Alloy expressions in the context of the execution trace. We found the evaluator feature to be very useful (although the versions we used contain a bug that prevents copy/paste of the displayed output). The TLA+ Toolbox contains a similar feature called Trace Explorer that can evaluate several arbitrary TLA+ expressions in every state of an execution trace and then display the results alongside each state. We found that this feature often helped us to diagnose subtle design errors. At the time of writing, Trace Explorer still has some quirks: it is somewhat slow and clunky due to having to launch a separate short-lived instance of the model checker on every change, and the IDE sometimes does a poor job of managing screen space, which can cause a tedious amount of clicking, resizing, and scrolling of windows.

2.6   Express and Verify Complex Liveness Properties

While safety properties are more important than liveness properties, we have found that liveness properties are still important. For example, we once had a deadlock bug in a lock-free algorithm that we caught in stress testing after we had failed to find it during model-checking because we did not check liveness. TLA$^+$ has good support for liveness. The language can express fairness assumptions and rich liveness properties using operators from linear time temporal logic. The model checker can check a useful subset of liveness properties, albeit only for very small instances of the system, as the algorithm for checking liveness is restricted to using a single core and physical RAM. However, the current version of the model checker has a significant flaw: if a liveness property is violated then the model checker may report an error trace that is much more complicated than necessary, which makes it harder for the user to diagnose the problem. For deeper verification of liveness, the TLA$^+$ proof system will soon support machine-checked proofs using temporal logic. The Alloy book says that it supports checking liveness[6], but only on short traces[7]. When evaluating Alloy, Zave found [33], "There are no temporal operators in Alloy. Strictly speaking the progress property could be expressed in Alloy using quantification over timestamps, but there is no point in doing so because the Alloy Analyzer could not check it meaningfully ... For all practical purposes, progress properties cannot be asserted in Alloy." VCC can verify local progress properties, e.g. termination of loops and functions, but we don't know if VCC can express fairness or verify global liveness properties.

2.7   Enable Machine-Checked Proof

So far, the benefits that Amazon have gained from formal methods have arisen from writing precise specifications to eliminate ambiguity, and model-checking finite models of those specifications to try to find errors. However, we have already run into the practical limits of model-checking, caused by combinatorial state-explosion. In one case, we found a serious defect that was only revealed in an execution trace comprising 35 steps of a high-level abstraction of a complex system. Finding that defect took several weeks of continuous model-checking time on a cluster of 10 high-end machines, using carefully crafted constraints to bound the state-space. Even when using such constraints the model-checker still has to explore billions of states; we had to request several enhancements to the model checker to get even this far. We doubt that model-checking can go much further than this. Therefore, for our most critical algorithms we wish to also use proofs.

In industry, engineers are extremely skeptical of proofs. Engineers strongly doubt that proofs can scale to the complexity of real-world systems, so any viable proof method would need an effective mechanism to manage that complexity. Also, most

---

[6] The Alloy book [15, p. 302] says, "[Alloy] assertions can express temporal properties over traces", and [15, p. 179] "we'll take an approach that ... allows [the property 'some leader is eventually elected'] to be checked directly." See also [15, p. 186], "How does the expressiveness of Alloy's trace assertions compare to temporal logics?"

[7] [15, p. 187] "In an Alloy trace analysis, only traces of bounded length are considered, and the bound is generally small."

proofs are so intricate that there is more chance of an error in the proof than an error in the algorithm, so for engineers to have confidence that a proof is correct, we would need machine verification of the proof. However, most systems that we know of for machine-checked proof are designed for proving conventional theorems in mathematics, not correctness of large computing systems.

TLA⁺ has a proof system that addresses these problems. The TLA⁺ proof system (TLAPS) uses structured hierarchical proof, which we have found to be an effective method for managing very complex proofs. TLAPS works directly with the original TLA⁺ specification, which allows users to first eliminate errors using the model checker and then switch to proof if even more confidence is required. TLAPS uses a declarative proof language that emphasizes the basic mathematics of the proof and intentionally abstracts away from the particular languages and tactics of the various back-end proof checkers. This approach allows new proof checkers to be added over time, and existing checkers to evolve, without risk of requiring significant change to the structure or content of the proof. If a checker does happen to change in a way that prevents an existing step from being proved, the author can simply expand that local step by one or more levels.

There are several published examples of TLAPS proofs of significant algorithms [28,22,11]. We have tried TLAPS on small problems and found that it works well. However, we have not yet proved anything useful about a real system. The main barrier is finding inductive invariants of complex or subtle algorithms. We would welcome tools and training in this area. The TLA⁺ Hyperbook [18] contains some examples but they are relatively simple. We intend to investigate using Alloy to help debug more complex inductive invariants, as Alloy was expressly designed for inductive analysis.

While preparing this paper we learned that several proof systems exist for Alloy. We have not yet looked at any of these, but may do so in future.

VCC works entirely via proof but also provides many of the benefits of a model checker; if a proof fails then VCC reports a concrete counter-example (a value for each variable) that helps the engineer understand how to change the code or invariants to allow the proof to succeed. This is an immensely valuable feature that we would strongly welcome in any proof system. However, VCC is based on a low level programming language, so we do not know if it is a practical tool for proofs of high-level distributed systems.

3 Minimize Cognitive Burden

3.1 Avoid Esoteric Concepts

Methods such as Coq and PVS involve very complicated type-systems and proof languages. We found these concepts to be difficult to learn. TLA⁺ is untyped, and through using it we have become convinced that a type system is not necessary for a specification method, and might actually be a burden as it could constrain expressiveness. In a TLA⁺ specification, "type safety" is just another (comparatively shallow) property that the system must satisfy, which we verify along with other desired properties. Alloy has a very simple type system that can only prevent a few trivial kinds of error in a specification. In fact, the Alloy book says [15, p. 119], "Alloy can be regarded as an untyped language". VCC has a simple type system, but adds some unfamiliar concepts such as "ghost claims" for proving safe access to objects in memory. We did not use these features in our evaluation of VCC, so we don't know if they complicate or simplify verification.

3.2 Use Conventional Terminology

TLA+ largely consists of standard discrete math plus a subset of linear temporal logic. Alloy is also based on discrete math but departs from the standard terminology in ways that we initially found confusing. In Alloy, a set is represented as a relation of arity one, and a scalar is represented as a singleton set. This encoding is unfamiliar, and has occasionally been a distraction. When reasoning informally about a formula in Alloy, we find ourselves looking for context about the arity and cardinality of the various values involved in order to deduce the semantics.

3.3 Use Just a Few Simple Language Constructs

TLA+ uses a small set of constructs from set theory and predicate logic, plus a new but straightforward notation for defining functions [17]. TLA+ also includes several operators from linear temporal logic that have semantics that are unfamiliar to most engineers in industry. However, TLA+ was intentionally designed to minimize use of temporal operators, and in practice we've found that we rarely use them. Alloy has a single underlying logic based on the first order relational calculus of Tarski [14], so the foundations of Alloy are at least as simple as those of TLA+. However, the documentation for Alloy describes the logic in terms of three different styles: conventional first-order predicate calculus, "navigation expressions" (including operators for relational join, transpose, and transitive closure), and relational calculus (including operators for intersection and union of relations). All three styles of logic can be combined in the same formula. We found that learning and using this combined logic took a bit more effort than using the non-temporal subset of TLA+, perhaps due to the "paradox of choice" [31].

3.4 Avoid Ambiguous Syntax

TLA+ has a couple of confusing syntactic constructs [20, p. 289], but they rarely arise in practice. The one example that the author has encountered, perhaps twice in four years of use, is an expression of the form $\{x \in S : y \in T\}$. This expression is syntactically ambiguous in the grammar for TLA+ set constructors. The expression is given a useful meaning via an ad hoc rule, but the visual ambiguity still causes the reader a moment of hesitation.

Alloy has a significant amount of syntactic overloading: the "_[_]" operator is used for both applying predicates and functions to arguments [15, p. 123] and for a variation of relational join [15, p. 61]; the "_._" operator is used for both relational join [15, p. 57] and the "syntactic pun" of receiver syntax [15, pp. 124,126]; some formulas ('signature facts') look like normal assertions but contain implicit quantification and implicit relational joins [15, pp. 99,122]; some keywords are reused for different concepts, e.g. for quantifiers and multiplicities [15, p. 72]. Each of these language features is reasonable in isolation, but collectively we found that these details can be a burden for the user to remember. TLA+ has relatively few instances of syntax overloading. For instance, the CASE value-expression uses the symbol □ as a delimiter but elsewhere that symbol is used for the temporal logic operator that means "henceforth". However, we rarely use CASE value-expressions in our specifications, so this is a minor issue compared to the overloading of the primary operators in Alloy. In the TLA+ proof system, the CASE keyword is overloaded with a meaning entirely different to its use in specifications. However, the context of proof vs. specification is always clear so we have not found this overloading to be a problem.

In Alloy, variable names (fields) can be overloaded by being defined in multiple signatures [15, pp. 119,267], and function and predicate names can be overloaded by the types of their arguments. Disambiguation is done via the static type system. We try to avoid using these features as they can make specifications harder to understand. TLA+ forbids redefinition of symbols that are in scope, even if the resulting expression would be unambiguous [20, p. 32].

3.5 Simple Semantics

TLA+ has the clean semantics of ordinary mathematics. There are a few subtle corner cases: e.g. the ENABLED operator may interact in surprising ways with module instantiation, because module instantiation is based on mathematical substitution and substitution does not distribute over ENABLED. However, such cases have not been a problem for us (we suspect that the main impact might be on formal proof by refinement, which we have not yet attempted).

The Alloy language has some surprising semantics in the core language, e.g. for integers. The Alloy book states [15, p. 82,290] that the standard theorem of integer arithmetic,

$$S =< T \text{ and } T =< S \text{ implies } S = T$$

is not valid in Alloy. The book admits that this is "somewhat disturbing". In Alloy, when S and T each represent a set of integers, the integer comparison operator performs an implicit summation but the equality operator does not. (The equality operator considers two sets of integers to be equal if and only if they contain the same elements.) Alloy has a feature called 'signature facts' that we have found to be a source of confusion. The Alloy book [15, p. 122] says, "The implicit quantification in signature facts can have unexpected consequences", and then gives an example that is described as "perhaps the most egregiously baffling". Fortunately 'signature facts' are an optional feature that can easily be avoided, once the user has learned to do so.

VCC has semantics that are an extension of the C programming language, which is not exactly simple, but is at least familiar to engineers. At the time of our evaluation, VCC had several constructs with unclear semantics (e.g. the difference between 'pure ghost' functions and 'logic' functions), as the language was still evolving, and up-to-date documentation and examples were scarce.

3.6 Avoid Distorting the Language in Order to Be More Accessible

Most engineers find formal methods unfamiliar and odd at first. It helps tremendously if a language has some facilities to help bridge the gap from how engineers think (in terms of programming language concepts and idioms) to declarative mathematical specifications. To that end, Alloy adopts several syntactic conventions from object-oriented programming languages. However, the adoption is superficial, and we have found that the net effect is to obscure what is really going on, which actually makes it slightly harder to think in Alloy. For example, the on-line tutorial for Alloy describes three distinct ways to think about an Alloy specification [4]:

- "The highest level of abstraction is the OO paradigm. This level is helpful for getting started and for understanding a model at a glance; the syntax is probably familiar to you and usually does what you'd expect it to do.
- "The middle level of abstraction is set theory. You will probably find yourself using this level of understanding the most (although the OO level will still be useful).
- "The lowest level of abstraction is atoms and relations, and corresponds to the true semantics of the language. However, even advanced users rarely think about things this way."

But the page goes on to say, "... the OO approach occasionally [leads to errors]." While we doubt there is a single 'right' way to think, we have found that it helps to think about Alloy specifications in terms of atoms and relations (including the leftmost signature column), as this offers the most direct and accurate mental model of all of the various language constructs.

TLA$^+$ takes a different approach to being accessible to engineers. TLA$^+$ is accompanied by a second, optional language called PlusCal that is similar to a C-style programming language, but much more expressive as it uses TLA$^+$ for expressions and values. PlusCal is automatically translated to TLA$^+$ by a tool, and the translation is direct and easy to understand. PlusCal users do still need to be familiar with TLA$^+$ in order to write rich expressions, and because the model-checker works at the TLA$^+$ level. PlusCal is intended to be a direct replacement for conventional pseudo-code, and we have found it to be effective in that role. Several engineers at Amazon have found that they are more productive in PlusCal than TLA$^+$, particularly when first exposed to the method.

3.7 Flexible and Undogmatic

So far we have investigated three techniques for verifying safety properties using state-based formal methods:

- *Direct:* Write safety properties as invariants, using history variables if necessary. Then check that every behavior of the algorithm satisfies the properties.
- *Refinement:* Write properties in the form of very abstract system designs, with their own variables and actions. These abstract systems are very far from implementable but are designed to be easy to understand and "obviously correct". Then check that the real algorithm implements the abstract design, i.e. that every legal behavior of the real algorithm is a legal behavior of the abstract design.

– *Inductive Invariance:* Find an invariant that implies the desired safety properties. Then check that the invariant is always preserved by every possible atomic step of the algorithm, and that the invariant is also true in all possible initial states.

The first technique (direct) is by far the easiest, and is the only method we have used so far on real systems. We are interested in the second technique (refinement) because we anticipate that such "ladders of abstraction" may be a good way to think about and verify systems that are very complex. We have performed one small experiment with refinement and we intend to pursue this further. We have tried the third technique (inductive invariance), but have found it difficult to apply to real-world systems because we don't yet have good tools to help discover complex inductive invariants.

TLA$^+$ supports the direct method very well. TLA$^+$ also explicitly supports verification of refinement between two levels of abstraction, via a mechanism based on substitution of variables in the high-level spec with expressions using variables from the lower-level spec. We find this mechanism to be both elegant and illuminating, and intend to use it more in the future. (For good examples, see the three levels of abstraction in [24], and the four levels of abstraction in [22].) The TLA$^+$ proof system is designed to use inductive invariants, but the TLA$^+$ model-checker has very limited support for debugging inductive invariants, which is currently a significant handicap.

Alloy supports the direct method fairly well. The Alloy book contains an example of checking refinement [15, p. 227] but the method used in that example appears to require each type of event in the lower-level specification to be analyzed separately, in addition to finding an overall state abstraction function. We have not yet tried to use Alloy to check refinement. Daniel Jackson says [14] that Alloy was designed for inductive analysis, and Pamela Zave used Alloy to perform inductive analysis on the Chord ring membership protocol [32,33]. However, so far we have found inductive analysis of concurrent and distributed algorithms to be very difficult; it currently seems to require too much effort for most algorithms in industry.

Event-B seems to focus on refinement, in the form of derivation. The book [2] and papers [3] on Event-B place strong emphasis on "correctness by construction" via many steps of top-down derivation from specification to algorithm, with each step being verified by incremental proof. While we are interested in exploring verification by refinement (upwards, from algorithm to specification in very few steps), we are skeptical of derivation as a design method. We have never applied this method ourselves, or seen a convincing example of it being applied elsewhere to a system in our problem domain. In general, our correctness properties are both high-level and non-trivial, e.g. sequential consistency, linearizability, serializability. We doubt that engineers can design a competitive industrial system by gradually deriving down from such high-level correctness properties in search of efficient concurrent or fault-tolerant distributed algorithm. Even if it were feasible to derive competitive algorithms from specifications, formal proof takes so much time and effort that we doubt that incremental formal proof can be a viable part of a design method in any industry in which time-to-market is a concern. Our established method of design is to first specify the desired correctness properties and operating environment, and then invent an algorithm using a combination of experience, creativity, intuition, inspiration from engineering literature, and informal reasoning. Once we have a design, we then verify it. For almost all verification tasks we have found that model checking dramatically beats proof, as model checking gives high confidence with reasonable effort. In addition to model checking we occasionally use informal proof, and are keen to try informal hierarchical proof [23]. We do have one or two algorithms that are so critical that they justify verification by formal proof (for which we are investigating the TLA$^+$ proof system). But we doubt that we would use incremental formal proof as a design technique even for those algorithms.

VCC takes a slightly different approach to verification. VCC enables engineers to verify complex global "two-state invariants" (predicates on atomic state transitions) by performing local verification of individual C functions. To achieve this, VCC imposes certain admissibility conditions on the kinds of invariants that may be declared [10]. We don't know if these admissibility conditions may become an inconvenience in practice as our evaluation of VCC was limited to small sequential algorithms and did not verify a global invariant. VCC is based on proof rather than model-checking, but a significant innovation is that the proof tool is guided solely by assertions and "ghost code" added alongside the executable program code. The assertions and ghost code are written in a rich superset of C, so guiding the prover feels like normal programming. VCC seems to support just this one style of verification. We found this style to be engaging and fun when applied to the set of simple problems that we attempted in our evaluation. However, we don't know how well this style works for more complex problems.

4   High Return on Investment

4.1   Handle All Types of Problems

We don't have time to learn multiple methods, so we want a single method that works for many types of problem: lock-free and wait-free concurrent algorithms, conventional concurrent algorithms (using locks, condition variables, semaphores), fault-tolerant distributed systems, and data-modelling. We have used TLA$^+$ successfully for all of these. Alloy is good for data modelling but while preparing this paper we learned that Alloy was not designed for checking concurrent or distributed algorithms [14]. VCC excels at verifying the design and code of low-level concurrent algorithms, but we don't know how to use VCC to verify high-level distributed systems.

4.2   Quick to Learn and Obtain Useful Results on Real Problems

We found that Alloy can give very rapid results for problems of modest complexity. TLA$^+$ is likewise easy to adopt; at Amazon, many engineers at all levels of experience have been able to learn TLA$^+$ from scratch and get useful results in 2 to 3 weeks, in some cases just in their personal time on weekends and evenings, and without help or training. We found that VCC is significantly harder to learn because of relatively sparse documentation, small standard library, and very few examples.

4.3   Improve Time to Market

After using TLA$^+$ to verify a complex fault-tolerant replication algorithm for Dynamo DB, an engineer in Amazon Web Services remarked that, had he known about TLA$^+$

before starting work on Dynamo DB, he would have used it from the start, and this would have avoided a significant amount of time spent manually checking his informal proofs. For this reason we believe that using TLA$^+$ may improve time-to-market for some systems in addition to improving quality. Similarly, Alloy may improve time-to-market in appropriate cases. However, Alloy was not designed to check complex distributed algorithms such as those in Dynamo DB, so in those cases we believe that using Alloy would take much longer than using TLA$^+$, or might be infeasible. VCC is based on proof, and even with VCC's innovations proof still requires more effort than model-checking. We did find that, when developing small sequential algorithms, proof by VCC can take less effort than thorough manual testing of the algorithm. However, we don't know if those results translate to complex concurrent systems.

5   Conclusion

We were impressed with all of the methods that we considered; they demonstrably work very well in various domains.

We found that Alloy is a terrific tool for concurrent or distributed algorithms of modest complexity, despite not being designed for such problems. The analyzer tool has an engaging user interface, the book and tutorials are helpful, and Alloy offers the fastest path to useful results for engineers who are new to formal methods. However, Alloy's limited expressiveness, slightly complicated language, and limited analysis tool, result in a method that is not well suited to large complex systems in our problem domain.

We found Microsoft VCC to be a compelling tool for verifying low-level C programs. VCC does have some features for abstraction but we could not see how to use these features to verify high-level designs for distributed systems. It may be possible, but it seemed difficult and we could not find any relevant examples.

We found that Coq, PVS, and other tools based solely on interactive proof assistants, are too complicated to be practical for our combination of problem domain and time constraints.

Event-B seems a promising method, but we were deterred from actually trying it due to the documentation's strong emphasis on deriving algorithms from specifications by top-down refinement and incremental proof, which we believe to be impractical in our problem domain. However, we later learned that Event-B has a model checker, so may support a more practical development process.

TLA$^+$ is a good fit for our needs, and continues to serve us well. TLA$^+$ is simple to learn, simple to apply, and very flexible. The IDE is quite useable and the model-checker works well. The TLA$^+$ Proof System is a welcome addition that we are still evaluating.

We would welcome further progress in the following areas:

- Model checking: Ability to check significantly larger system-instances.
- Verify the code: Improved tools for checking that executable code meets its high-level specification. We already use conventional tools for static code analysis, but are disappointed that most such tools are shallow or generate a high rate of false-positive errors.

- Proof: Improved support for standard types and operators, e.g. sequences and TLA⁺'s CHOOSE operator (Hilbert's epsilon). More libraries of lemmas for common idioms, plus examples of how to use them in proofs. Education and tools to help find complex inductive invariants.
- Methods for modelling and analyzing performance: For instance, predicting the distribution of response latency for a given system throughput. We realize that this is an entirely different problem from logical correctness, but in industry performance is almost as important as correctness. We speculate that model-checking might help us to analyze statistical performance properties if we were able to specify the cost distribution of each atomic step.

We would like to thank Leslie Lamport, Daniel Jackson, Stephan Merz, and the anonymous reviewers from the ABZ2014 program committee, for comments that significantly improved this paper.

References

1. Abrial, J.-R.: Formal methods in industry: achievements, problems, future. In: 28th Intl. Conf. Software Engineering (ICSE), Shanghai, China, pp. 761–768. ACM (2006)
2. Abrial, J.-R.: Modeling in Event-B. Cambridge University Press (2010)
3. Abrial, J.-R., et al.: Rodin: an open toolset for modelling and reasoning in Event-B. STTT 12(6), 447–466 (2010)
4. Alloy online tutorial: How to think about an alloy model: 3 levels, http://alloy.mit.edu/alloy/tutorials/online/sidenote-levels-of-understanding.html
5. Event-B wiki: Industrial projects, http://wiki.event-b.org/index.php/Industrial_Projects
6. Barr, J.: Amazon S3 – the first trillion objects. Amazon Web Services Blog (June 2012), http://aws.typepad.com/aws/2012/06/amazon-s3-the-first-trillion-objects.html
7. Barr, J.: Amazon S3 – two trillion objects, 1.1 million requests per second. Amazon Web Services Blog (March 2013), http://aws.typepad.com/aws/2013/04/amazon-s3-two-trillion-objects-11-million-requests-second.html
8. Batson, B., Lamport, L.: High-level specifications: Lessons from industry. In: de Boer, F.S., Bonsangue, M.M., Graf, S., de Roever, W.-P. (eds.) FMCO 2002. LNCS, vol. 2852, pp. 242–261. Springer, Heidelberg (2003)
9. Bolosky, W.J., Douceur, J.R., Howell, J.: The Farsite project: a retrospective. Operating Systems Reviews 41(2), 17–26 (2007)
10. Cohen, E., Moskal, M., Schulte, W., Tobies, S.: Local verification of global invariants in concurrent programs. In: Touili, T., Cook, B., Jackson, P. (eds.) CAV 2010. LNCS, vol. 6174, pp. 480–494. Springer, Heidelberg (2010)
11. Douceur, J., et al.: Memoir: Formal specs and correctness proof (2011), http://research.microsoft.com/pubs/144962/memoir-proof.pdf
12. Hall, A.: Seven myths of formal methods. IEEE Software 7(5), 11–19 (1990)
13. Holzmann, G.: Design and Validation of Computer Protocols. Prentice Hall, New Jersey (1991)
14. Jackson, D.: Personal communication (2014)
15. Jackson, D.: Software Abstractions, revised edition. MIT Press (2012), http://www.softwareabstractions.org/

16. Lamport, L.: Comment on the history of the TLC model checker,
    http://research.microsoft.com/en-us/um/people/lamport/
    pubs/pubs.html#yuanyu-model-checking
17. Lamport, L.: Summary of TLA⁺,
    http://research.microsoft.com/en-us/um/people/lamport/tla/summary.pdf
18. Lamport, L.: The TLA⁺ Hyperbook,
    http://research.microsoft.com/en-us/um/people/lamport/tla/hyperbook.html
19. Lamport, L.: The Temporal Logic of Actions. ACM Trans. Prog. Lang. Syst. 16(3), 872–923 (1994)
20. Lamport, L.: Specifying Systems. Addison-Wesley (2002),
    http://research.microsoft.com/
    en-us/um/people/lamport/tla/book-02-08-08.pdf
21. Lamport, L.: Fast Paxos. Distributed Computing 19(2), 79–103 (2006)
22. Lamport, L.: Byzantizing Paxos by refinement. In: Peleg, D. (ed.) DISC 2011. LNCS, vol. 6950, pp. 211–224. Springer, Heidelberg (2011)
23. Lamport, L.: How to write a 21st century proof. Fixed Point Theory and Applications (2012)
24. Lamport, L., Merz, S.: Specifying and verifying fault-tolerant systems. In: Langmaack, H., de Roever, W.-P., Vytopil, J. (eds.) FTRTFT 1994 and ProCoS 1994. LNCS, vol. 863, pp. 41–76. Springer, Heidelberg (1994)
25. Lamport, L., Sharma, M., Tuttle, M., Yu, Y.: The wildfire challenge problem (2001),
    http://research.microsoft.com/
    en-us/um/people/lamport/pubs/wildfire-challenge.pdf
26. Lamport, L., Tuttle, M., Yu, Y.: The wildfire verification challenge problem [example of a specification from industry], http://research.microsoft.com/
    en-us/um/people/lamport/tla/wildfire-challenge.html
27. Leinenbach, D., Santen, T.: Verifying the Microsoft Hyper-V Hypervisor with VCC. In: Cavalcanti, A., Dams, D.R. (eds.) FM 2009. LNCS, vol. 5850, pp. 806–809. Springer, Heidelberg (2009)
28. Lu, T., Merz, S., Weidenbach, C.: Towards verification of the Pastry protocol using TLA⁺. In: Bruni, R., Dingel, J. (eds.) FMOODS/FORTE 2011. LNCS, vol. 6722, pp. 244–258. Springer, Heidelberg (2011)
29. Newcombe, C.: Debugging designs. Presented at the 14th Intl. Wsh. High-Performance Transaction Systems (2011), http://hpts.ws/papers/2011/
    sessions_2011/Debugging.pdf and associated specifications:
    http://hpts.ws/papers/2011/sessions_2011/amazonbundle.tar.gz
30. Owre, S., et al.: Combining specification, proof checking, and model checking. In: Alur, R., Henzinger, T.A. (eds.) CAV 1996. LNCS, vol. 1102, pp. 411–414. Springer, Heidelberg (1996)
31. Schwartz, B.: The paradox of choice,
    http://www.ted.com/talks/barry_schwartz_on_the_paradox_of_choice.html
32. Zave, P.: Using lightweight modeling to understand Chord. Comp. Comm. Reviews 42(2), 49–57 (2012)
33. Zave, P.: A practical comparison of Alloy and Spin. Formal Aspects of Computing (to appear, 2014), http://www2.research.att.com/~pamela/compare.pdf

Translating B to TLA$^+$ for Validation with TLC

Dominik Hansen and Michael Leuschel

Institut für Informatik, Universität Düsseldorf*
Universitätsstr. 1, 40225 Düsseldorf
{hansen,leuschel}@cs.uni-duesseldorf.de

Abstract. The state-based formal methods B and TLA$^+$ share the common base of predicate logic, arithmetic and set theory. However, there are still considerable differences, such as the way to specify state transitions, the different approaches to typing, and the available tool support. In this paper, we present a translation from B to TLA$^+$ to validate B specifications using the model checker TLC. We provide translation rules for almost all constructs of B, in particular for those which are not built-in in TLA$^+$. The translation also includes many adaptations and optimizations to allow efficient checking by TLC. Moreover, we present a way to validate liveness properties for B specifications under fairness conditions. Our implemented translator, TLC4B, automatically translates a B specification to TLA$^+$, invokes the model checker TLC, and translates the results back to B. We use PROB to double check the counter examples produced by TLC and replay them in the PROB animator. We also present a series of case studies and benchmark tests comparing TLC4B and PROB.

Keywords: B-Method, TLA$^+$, Tool Support, Model Checking, Animation.

1 Introduction and Motivation

B [1] and TLA$^+$ [8] are both state-based formal methods rooted in predicate logic, combined with arithmetic, set theory and support for mathematical functions. However, as already pointed out in [5], there are considerable differences:

- B is strongly typed, while TLA$^+$ is untyped. For the translation, it is obviously easier to translate from a typed to an untyped language than vice versa.
- The concepts of modularization are quite different.
- Functions in TLA$^+$ are total, while B supports relations, partial functions, injections, bijections, etc.
- B is limited to invariance properties, while TLA$^+$ also allows the specification of liveness properties.

---

* Part of this research has been sponsored by the EU funded FP7 project 287563 (ADVANCE) and the DFG funded project Gepavas.

Y. Ait Ameur and K.-D. Schewe (Eds.): ABZ 2014, LNCS 8477, pp. 40–55, 2014.

– The structure of a B machine or development is prescribed by the B-method, while in a TLA⁺ specification any formula can be considered as a system specification.

As far as tool support is concerned, TLA⁺ is supported by the explicit state model checker TLC [13] and more recently by the TLAPS prover [3]. TLC has been used to validate a variety of distributed algorithms (e.g. [4]) and protocols. B has extensive proof support, e.g., in the form of the commercial product AtelierB [2] and the animator, constraint solver and model checker PROB [9,10]. Both AtelierB and PROB are being used by companies, mainly in the railway sector for safety critical control software. In an earlier work [5] we have presented a translation from TLA⁺ to B, which enabled applying the PROB tool to TLA⁺ specifications. In this paper we present a translation from B to TLA⁺, this time with the main goal of applying the model checker TLC to B specifications. Indeed, TLC is a very efficient model checker for TLA⁺ with an efficient disk-based algorithm and support for fairness. PROB has an LTL model checker, but it does not support fairness (yet) and is entirely RAM-based. The model checking core of PROB is less tuned than TLA⁺. On the other hand, PROB incorporates a constraint solver and offers several features which are absent from TLC, in particular an interactive animator with various visualization options. One feature of our approach is to replay the counter-examples produced by TLC within PROB, to get access to those features but also to validate the correctness of our translation. In this paper, we also present a thorough empirical evaluation between TLC and PROB. The results show that for lower-level, more explicit formal models, TLC fares better, while for certain more high-level formal models PROB is superior to TLC because of its constraint solving capabilities. The addition of a lower-level model checker thus opens up many new application possibilities.

2 Translation

The complete translation process from B to TLA⁺ and back to B is illustrated in Fig. 1.

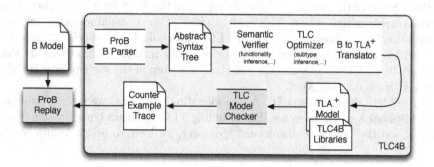

Fig. 1. The TLC4B Translation and Validation Process

Before explaining the individual phases, we will illustrate the translation with an example and explain the various phases based on that example. Translation rules of all data types and operators can be found in the extended version of our paper [6]. More specific implementation details will be covered in Section 4.

2.1 Example

Below we use a specification (adapted from [10]) of a process scheduler (Fig. 2). The specified system allows at most one process to be active. Each process can qualify for being selected by the scheduler by entering a FIFO queue. The specification contains two variables: a partial function *state* mapping each process to its state (a process must be created before it has a state) and a FIFO queue modeled as an injective sequence of processes. In the initial state no process is created and the queue is empty. The specification contains various operations to create (*new*), delete (*del*), or add a process to queue (*addToQueue*). Additionally, there are two operations to load a process into the processor (*enter*) and to remove a process from the processor (*leave*). The specification contains two safety conditions beside the typing predicates of the variables:

- At most one process should be active.
- Each process in the FIFO queue should have the state *ready*.

The translated TLA$^+$- specification is shown in Fig. 3. At the beginning of the module some standard modules are loaded via the *EXTENDS* statement. These modules contain several operators used in this specification. The enumerated set *STATE* is translated as a constant definition. The definition itself is renamed (into *STATE_1*) by the translator because *STATE* is keyword in TLA$^+$. The invariant of the B specification is divided into several definitions in the TLA$^+$ module. This enables TLC to provide better feedback about which part of the invariant is violated. We translate each B operation as a TLA$^+$ action. Substitutions are translated as before-after predicates where a primed variable represents the variable in the next state. Unchanged variables must be explicitly specified. Note that a parameterized operation is translated as existential quantification. The quantification itself is located in the next-state relation *Next*, which is a disjunction of all actions. Some of the operations' guards appear in the *Next* definition rather than in the corresponding action. This is an instance of our translator optimizing the translation for the interpretation with TLC. The whole TLA$^+$ specification is described by the *Spec* definition. A valid behavior for the system has to satisfy the *Init* predicate for the initial state and then each step of the system must satisfy the next-state relation *Next*.

To validate the translated TLA$^+$ specification with TLC we have to provide an additional configuration file (Fig 4) telling TLC the main (specification) definition and the invariant definitions. Moreover, we have to assign values to all

```
MODEL Scheduler
SETS
  PROCESSES;
  STATE = {idle, ready, active}
VARIABLES
  state,
  queue
INVARIANT
  state ∈ PROCESSES ⇸ STATE
  & queue ∈ iseq(PROCESSES)
  & card(state⁻¹[{active}]) ≤ 1
  & !x.(x ∈ ran(queue) ⇒ state(x) = ready)
INITIALISATION
  state := {} ∥ queue := [ ]
OPERATIONS
  new(p) = SELECT p ∉ dom(state)
        THEN state := state ∪ {(p ↦ idle)} END;

  del(p) = SELECT p ∈ dom(state) ∧ state(p) = idle
        THEN state := {p} ⩤ state END;

  addToQueue(p) = SELECT p ∈ dom(state) ∧ state(p) = idle
        THEN state(p) := ready ∥ queue := queue ← p END;

  enter = SELECT queue ≠ [ ] ∧ state⁻¹[{active}] = {}
        THEN state(first(queue)) := active ∥ queue := tail(queue) END;

  leave(p) = SELECT p ∈ dom(state) ∧ state(p) = active
        THEN state(p) := idle END
END
```

Fig. 2. MODEL Scheduler

constants of the module.[1] In this case we assign a set[2] of model values to the constant *PROCESSES* and single model values to the other constants. In terms of functionality, model values correspond to elements of a enumerated set in B. Model values are equal to themselves and unequal to all other model values.

2.2 Translating Data Values and Functionality Inference

Due to the common base of B and TLA⁺, most data types exist in both languages, such as sets, functions and numbers. As a consequence, the translation of these data types is almost straightforward.

---

[1] We translate a constant as variable if the axioms allow several solutions for this constant. All possible solution values will be enumerated in the initialization and the variable will be kept unchanged by all actions.

[2] The size of the set is a default number or can be specified by the user.

────────────────────── MODULE *Scheduler* ──────────────────────

EXTENDS *Naturals, FiniteSets, Sequences, Relations, Functions,*
  FunctionsAsRelations, SequencesExtended, SequencesAsRelations

CONSTANTS *PROCESSES, idle, ready, active*

VARIABLES *state, queue*

$STATE\_1 \triangleq \{idle, ready, active\}$

$Invariant1 \triangleq state \in RelParFuncEleOf(PROCESSES, STATE\_1)$
$Invariant2 \triangleq queue \in ISeqEleOf(PROCESSES)$
$Invariant3 \triangleq Cardinality(RelImage(RelInverse(state), \{active\})) \leq 1$
$Invariant4 \triangleq \forall x \in Range(queue) : RelCall(state, x) = ready$

$Init \triangleq \ \wedge state \ = \{\}$
$\qquad\qquad \wedge queue = \langle\rangle$

$new(p) \triangleq \ \wedge state' = state \cup \{\langle p, idle\rangle\}$
$\qquad\qquad\quad \wedge \text{UNCHANGED } \langle queue\rangle$

$del(p) \ \triangleq \ \wedge RelCall(state, p) = idle$
$\qquad\qquad\quad \wedge state' = RelDomSub(\{p\}, state)$
$\qquad\qquad\quad \wedge \text{UNCHANGED } \langle queue\rangle$

$addToQueue(p) \triangleq \ \wedge RelCall(state, p) = idle$
$\qquad\qquad\qquad\qquad \wedge state' \ = RelOverride(state, \{\langle p, ready\rangle\})$
$\qquad\qquad\qquad\qquad \wedge queue' = Append(queue, p)$

$enter \triangleq \ \wedge queue \neq \langle\rangle$
$\qquad\qquad \wedge RelImage(RelInverse(state), \{active\}) = \{\}$
$\qquad\qquad \wedge state' \ = RelOverride(state, \{\langle Head(queue), active\rangle\})$
$\qquad\qquad \wedge queue' = Tail(queue)$

$leave(p) \triangleq \ \wedge RelCall(state, p) = active$
$\qquad\qquad\quad \wedge state' = RelOverride(state, \{\langle p, idle\rangle\})$
$\qquad\qquad\quad \wedge \text{UNCHANGED } \langle queue\rangle$

$Next \triangleq \ \vee \exists p \in PROCESSES \setminus RelDomain(state) : new(p)$
$\qquad\qquad \vee \exists p \in RelDomain(state) : del(p)$
$\qquad\qquad \vee \exists p \in RelDomain(state) : addToQueue(p)$
$\qquad\qquad \vee enter$
$\qquad\qquad \vee \exists p \in RelDomain(state) : leave(p)$

$vars \triangleq \langle state, queue\rangle$

$Spec \triangleq Init \wedge \Box[Next]_{vars}$

Fig. 3. Module Scheduler

```
SPECIFICATION Spec
INVARIANT Invariant1, Invariant2, Invariant3, Invariant4
CONSTANTS
PROCESSES = {PROCESSES1, PROCESSES2, PROCESSES3}
idle = idle
ready = ready
active = active
```

Fig. 4. Configuration file for the module Scheduler

TLA+ has no built-in concept of Relations[3], but TLA+ provides all necessary data types to define relations based on the model of the B-Method. We represent a relation in TLA+ as a set of tuples (e.g. $\{\langle 1, TRUE\rangle, \langle 1, FALSE\rangle \ \langle 2, TRUE\rangle\}$). The drawback of this approach is that in contrast to B, TLA+'s own functions and sequences are not based on the relations defined is this way. As an example, we cannot specify a TLA+ built-in function as a set of pairs; in B it is usual to do this as well as to apply set operators (e.g. the union operator as in $f \cup \{2 \mapsto 3\}$) to functions or sequences. To support such a functionality in TLA+, functions and sequences should be translated as relations if they are used in a "relational way". It would be possible to always translate functions and sequences as relations. But in contrast to relations, functions and sequences are built-in data types in TLA+ and their evaluation is optimized by TLC (e.g. lazy evaluation). Hence we extended the B type-system to distinguish between functions and relations. Thus we are able to translate all kinds of relations and to deliver an optimized translation.

We use a type inference algorithm adapted to the extended B type-system to get the required type information for the translation. Unifying a function type with a relation type will result in a relation type (e.g. $\mathbb{P}(\mathbb{Z} \times \mathbb{Z})$ for both sides of the equation $\lambda x.(x \in 1..3|x + 1) = \{(1,1)\}$). However, there are several relational operators preserving a function type if they are applied to operands with a function type (e.g. *ran*, *front* or *tail*). For these operators we have to deliver two translation rules (functional vs relational). Moreover the algorithm verifies the type correctness of the B specification (i.e. only values of the same type can be compared with each other).

2.3 Translating Operators

In TLA+ some common operators such as arithmetic operators are not built-in operators. They are defined in separate modules called standard modules that can be extended by a specification.[4] We reuse the concept of standard modules to include the relevant B operators. Due to the lack of relations in TLA+ we have to provide a module containing all relational operators (Fig. 5).

---

[3] Relations are not mentioned in the language description of [8]. In [7] Lamport introduces relations in TLA+ only to define the transitive closure.

[4] TLC supports operators of the common standard modules Integers and Sequences in a efficient way by overwriting them with Java methods.

——————— MODULE *Relations* ————————

EXTENDS *FiniteSets, Naturals, TLC*
$Relation(X, Y) \triangleq$ SUBSET $(X \times Y)$
$RelDomain(R) \triangleq \{x[1] : x \in R\}$
$RelRange(R) \triangleq \{x[2] : x \in R\}$
$RelInverse(R) \triangleq \{\langle x[2], x[1] \rangle : x \in R\}$
$RelDomRes(S, R) \triangleq \{x \in R : x[1] \in S\}$ Domain restriction
$RelDomSub(S, R) \triangleq \{x \in R : x[1] \notin S\}$ Domain subtraction
$RelRanRes(R, S) \triangleq \{x \in R : x[2] \in S\}$ Range restriction
$RelRanSub(R, S) \triangleq \{x \in R : x[2] \notin S\}$ Range subtraction
$RelImage(R, S) \triangleq \{y[2] : y \in \{x \in R : x[1] \in S\}\}$
$RelOverride(R1, R2) \triangleq \{x \in R : x[1] \notin RelDomain(R2)\} \cup R2$
$RelComposition(R1, R2) \triangleq \{\langle u[1][1], u[2][2] \rangle : u \in$
    $\{x \in RelRanRes(R1, RelDomain(R2)) \times RelDomRes(RelRange(R1), R2) :$
        $x[1][2] = x[2][1]\}\}$

$\vdots$

Fig. 5. An extract of the Module Relations

Moreover B provides a rich set of operators on functions such as all combinations of partial/total and injective/surjective/bijective. In TLA$^+$ we only have total functions. We group all operators on functions together in an additional module (Fig. 6). Sometimes there are several ways to define an operator. We choose the definition which can be best handled by TLC.[5]

——————— MODULE *Functions* ————————

EXTENDS *FiniteSets*
$Range(f) \triangleq \{f[x] : x \in$ DOMAIN $f\}$
$Image(f, S) \triangleq \{f[x] : x \in S\}$
$TotalInjFunc(S, T) \triangleq \{f \in [S \to T] :$
    $Cardinality($DOMAIN $f) = Cardinality(Range(f))\}$
$ParFunc(S, T) \triangleq$ UNION $\{[x \to T] : x \in$ SUBSET $S\}$
$ParInjFunc(S, T) \triangleq \{f \in ParFunc(S, T) :$
    $Cardinality($DOMAIN $f) = Cardinality(Range(f))\}$

$\vdots$

Fig. 6. An extract of the Module Functions

Some operators exists in both languages but their definitions differ slightly. For example, the B-Method requires that the first operand for the modulo operator must be a natural number. In TLA$^+$ it can be also a negative number.

---

[5] Note that some of the definitions are based on the *Cardinality* operator that is restricted to finite sets.

Operator	B-Method	TLA$^+$
$a\ modulo\ b$	$a \in \mathbb{N} \wedge b \in \mathbb{N}_1$	$a \in \mathbb{Z} \wedge b \in \mathbb{N}_1$

To verify B's well-definedness condition for modulo we use TLC's ability to check assertions. The special operator $Assert(P, out)$ throws a runtime exception with the error message out if the predicate P is false. Otherwise, $Assert$ will be evaluated to TRUE. The B modulo operator can thus be expressed in TLA$^+$ as follows:

$$Modulo(a,\ b) \triangleq \text{IF } a \geq 0 \text{ THEN } a \% b \text{ ELSE } Assert(\text{FALSE, "WD ERROR"})$$

We also have to consider well-definedness conditions if we apply a function call to a relation:

$$RelCall(r,\ x) \triangleq \text{IF } Cardinality(r) = Cardinality(RelDom(r)) \wedge x \in RelDom(r)$$
$$\text{THEN } (\text{CHOOSE } y \in r : y[1] = x)[2]$$
$$\text{ELSE } Assert(\text{FALSE, "WD ERROR"})$$

In summary, we provide the following standard modules for our translation:

- Relations
- Functions
- SequencesExtended (Some operators on sequences which are not included the standard module Sequences)
- FunctionsAsRelations (Defines all function operators on sets of pairs ensuring their well-definedness conditions)
- SequencesAsRelations (Defines all operators on sequences which are represented as sets of pairs.)
- BBuiltins (Miscellaneous operators e.g. modulo, min and max)

2.4 Optimizations

Subtype Inference. Firstly, we will describe how TLC evaluates expressions: In general TLC evaluates an expression from left to right. Evaluating an expression containing a bound variable such as an existential quantification ($\exists x \in S : P$), TLC enumerates all values of the associated set and then substitutes them for the bound variable in the corresponding predicate. Due to missing constraint solving techniques, TLC is not able to evaluate another variant of the existential quantification without an associated set ($\exists x : P$). This version is also a valid TLA$^+$ expression and directly corresponds to the way of writing an existential quantification in B ($\exists x.(P)$). However, we confine our translations to the subset of TLA$^+$ which is supported by TLC. Thus the translation is responsible for making all required adaptations to deliver an executable TLA$^+$ specification. For the existential quantification (or all other expressions containing bound variables), we use the inferred type τ of the bound variable as the associated set ($\exists x \in \tau_x : P$.) However, in most cases, it is not performant to let TLC enumerate over a type of a variable, in particular TLC aborts if it has to enumerate a infinite set. Alternatively, it is often possible to restrict the type of the bound variable based on a static analysis of the corresponding (typing) predicate. We use a pattern matching algorithm to find the following kind of expressions[6] where x is a bound variable, e is an expression, and S is ideally a subset of the type: $x = e$, $x \in S$, $x \subseteq S$, $x \subset S$ or $x \notin S$. In case of the last pattern we build the set difference of the type of the variable x and the set S:

B-Method	TLA$^+$
$\exists x.(x \notin S \wedge P)$	$\exists x \in (\tau_x \setminus S) : P$

If more than one of the patterns can be found for one variable, we build the intersection to keep the associated set as small as possible:

B-Method	TLA$^+$
$\exists x.(x = e \wedge x \in S_1 \wedge x \subseteq S_2 \wedge P)$	$\exists x \in (\{e\} \cap S_1 \cap SUBSET\ S_2) : P$

This reduces the number of times TLC has to evaluate the predicate P.

Lazy Evaluation. Sometimes TLC can use heuristics to evaluate an expression. For example TLC can evaluate $\langle 1, 2, 1 \rangle \in Seq(\{1, 2\})$ to true without evaluating the infinite set of sequences. We will show how we can use these heuristics to generate an optimized translation. As mentioned before functions have to be translated as relations if they are used in a relational way in the B specification. How then should we translate the set of all total functions $(S \to T)$? The easiest way is to convert each function to a relation in TLA$^+$:

$$MakeRel(f) \triangleq \{\langle x, f[x]\rangle : x \in DOMAIN\ f\}$$

The resulting operator for the set of all total functions is:

$$RelTotalFunctions(S, T) \triangleq \{MakeRel(f) : f \in [S \to T]\}$$

However this definition has a disadvantage, if we just want to check if a single function is in this set the whole set will be evaluated by TLC. Using the following definition TLC avoids the evaluation of the whole set:

$$RelTotalFunctionsEleOf(S, T) \triangleq \{f \in SUBSET(S \times T) :$$
$$\wedge\ Cardinality(RelDomain(f)) = Cardinality(f)$$
$$\wedge\ RelDomain(f) = S\}$$

In this case, TLC only checks if a function is a subset of the cartesian product (the whole Cartesian product will not be evaluated) and the conditions are checked only once. Moreover this definition fares well even if S or T are sets of functions (e.g. $S \to V \to W$ in B). The advantage of the first definition is that it is faster when the whole set must be evaluated. As a consequence, we use both definitions for our translation and choose the first if TLC has to enumerate the set (e.g. $\exists x \in RelTotalFunctions(S, T) : P$) and the second testing if a function belongs to the set (e.g. $f \in RelTotalFunctionsEleOf(S, T)$ as an invariant).

---

[6] The B language description in [2] requires that each (bound) variable must be typed by one of these patterns before use.

3   Checking Temporal Formulas

One of the main advantages of TLA$^+$ is that temporal properties can be specified directly in the language itself. Moreover the model checker TLC can be used to verify such formulas. But before we show how to write temporal formulas for a B specification we first have to describe a main distinction between the two formal methods. In contrast to B, the standard template of a TLA$^+$ specification ($Init \wedge \Box[Next]_{vars}$) allows stuttering steps at any time.[7] This means that a regular step of a TLA$^+$ specification is either a step satisfying one of the actions or a stuttering step leaving all variables unchanged. When checking a specification for errors such as invariant violations it is not necessary to consider stuttering steps, because such an error will be detected in a state and stuttering steps only allow self transitions and do not add additional states. For deadlock checking stuttering steps are also not regarded by TLC, but verifying a temporal formula with TLC often ends in a counter-example caused by stuttering steps. For example, assuming we have a very simple specification of a counter in TLA$^+$ with a single variable c:

$$Spec \;\; \triangleq \;\; c = 1 \wedge \Box[c' = c + 1]_c$$

We would expect that the counter will eventually reach 10 ($\Diamond(c = 10)$). However TLC will report a counter-example, saying that at a certain state (before reaching 10), an infinite number of stuttering steps occurs and 10 will never reached. From the B side we do not want to deal with these stuttering steps. TLA$^+$ allows to add fairness conditions to the specification to avoid infinite stuttering steps. Adding weak fairness for the next-state relation ($WF_{vars}(Next)$) would prohibit an infinite number of stuttering steps if a step of the next-state relation is possible (i.e. $Next$ is always enabled):

$$WF_{vars}(A) \;\; \triangleq \;\; \vee \; \Box\Diamond(\langle A \rangle_{vars})$$
$$\vee \; \Box\Diamond(\neg \; \text{ENABLED} \; \langle A \rangle_{vars})$$

However this fairness condition is too strong: It asserts that either the action A will be executed infinitely often changing the state of the system (A must not be a stuttering step)

$$\langle A \rangle_{vars} \;\; \triangleq \;\; A \wedge vars' \neq vars$$

or A will be disabled infinitely often. Assuming weak fairness for the next state relation will also eliminate user defined stuttering steps. User defined stuttering steps result from B operations which do not change the state of the system (e.g. skip or call operations). These stuttering steps may cause valid counter-examples and should not be eliminated. Hence, the translation should retain user defined stuttering steps in the translated TLA$^+$ specfication and should disable stuttering steps which are implicitly included. In [12], Richards describes a way to distinguish between these two kinds of stuttering steps in TLA$^+$. We use his definition of "Very Weak Fairness" applied to the next state relation ($VWF_{vars}(Next)$) to disable implicit stuttering steps and allow user defined stuttering steps in the TLA$^+$ specification:

---

[7] $[Next]_{vars} \;\; \triangleq \;\; Next \vee UNCHANGED \; vars.$

$$VWF_{vars}(A) \triangleq \lor\ \Box\Diamond(\langle A\rangle_{vars})$$
$$\lor\ \Box\Diamond(\neg\ \text{ENABLED}\ \langle A\rangle_{vars})$$
$$\lor\ \Box\Diamond(\text{ENABLED}\ (A \land \text{UNCHANGED}\ \ vars))$$

The definition of $VWF_{vars}(A)$ is identical to $WF_{vars}(A)$ except for an additional third case allowing infinite stuttering steps if A is a stuttering action ($A \land$ UNCHANGED $vars$). We define the resulting template of the translated TLA$^+$ specification as follows:

$$Init \land \Box[Next]_{vars} \land VWF_{vars}(Next)$$

We allow the B user to use following temporal operators to define liveness conditions for a B specification:[8]

- $\Box f$ (Globally)
- $\Diamond f$ (Finally)
- $ENABLED(op)$ (Check if the operation op is enabled)
- $\exists x.(P \land f)$ (Existential quantification)
- $\forall x.(P \Rightarrow f)$ (Universal quantification)
- $WF(op)$ (Weak Fairness will be translated to VWF)
- $SF(op)$ (Strong Fairness will be translated to "Almost Strong Fairness"[9])

4   Implementation and Experiments

Our translator, called TLC4B, is implemented in Java and it took about six months to develop the initial version. Figure 1 in Section 2 shows the translation and validation process of TLC4B. After parsing the specification TLC4B performs some static analyses (e.g. type checking or checking the scope of the variables) verifying the semantic correctness of the B specification. Moreover, as explained in Section 2, TLC4B extracts required information from the B specification (e.g. subtype inference) to generate an optimized translation. Subsequently, TLC4B creates a TLA$^+$ module with an associated configuration file and invokes the model checker TLC. The results produced by TLC are translated back to B. For example, a goal predicate is translated as a negated invariant. If this invariant is violated, a "Goal found" message is reported. We expect TLC to find the following kinds of errors in the B specification:

- Deadlocks
- Invariant violations
- Assertion errors
- Violations of constants properties (i.e., axioms over the B constants are false)
- Well-definedness violations
- Violations of temporal formulas

---

[8] We demonstrate the translation of a liveness condition with a concrete example in the extended version of this paper [6].

[9] Analogically Richards defines "Almost Strong Fairness" (ASF) as a weaker version of strong fairness (SF) reflecting the different kinds of stuttering steps.

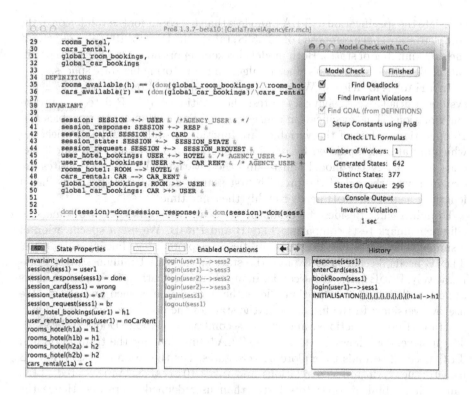

Fig. 7. PROB animator

For certain kinds of errors such as a deadlock or an invariant violation, TLC reports a trace leading to the state where the error occurs. A trace is a sequence of states where each state is a mapping from variables to values. TLC4B translates the trace back to B (as a list of B state predicates). TLC4B has been integrated into PROB as of version 1.3.7-beta10: The user needs no knowledge of TLA$^+$ because the translation is completely hidden. As shown in figure 7 counter-examples found by TLC are automatically replayed in the PROB animator (displayed in the history pane) to give the user an optimal feedback.

We have successfully validated several existing models from the literature (Table 1). The following examples show some fields of application of TLC4B. The experiments were all run on a Macbook Air with Intel Core i5 1,8 GHz processor, running TLC Version 2.05 and Prob version 1.3.7-beta9. The source code of the examples can be found at http://www.stups.uni-duesseldorf.de/w/Pub:TLC4B_Benchmarks.

Can-Bus. One example is a 314 line B specification of the Controller Area Network (CAN) Bus, containing 18 variables and 21 operations. The specification is rather low level, i.e., the operations consist of simple assignments of concrete values to variables (no constraint solving is required). TLC4B needs 1.5 seconds[10] to

---

[10] Most of this time is required to start the JVM and to parse the B specification.

translate the specification to TLA$^+$ and less than 6 seconds for the validation of the complete state space composed of 132,598 states. PROB needs 192 seconds to visit the same number of states. Both model checkers report no errors. For this specification TLC benefits from its efficient algorithm for storing big state spaces.

Invariant violations. We use a defective specification of a travel agency system (CarlaTravelAgencyErr) to test the abilities of TLC4B detecting invariant violations. The specification consists of 295 line of B code, 11 variables and 10 operations. Most of the variables are functions (total, partial and injective) which are also manipulated by relational operators. TLC4B needs about 3 seconds to translate the model and to find the invariant violation. 377 states are explored with the aid of the breadth first search and the resulting trace has a length of 5 states. PROB needs roughly the same time.

Benchmarks. Besides the evaluation of real case studies, we use some specific benchmark tests comparing TLC4B and PROB. We use a specification of a simple counter testing TLC4B's abilities to explore a big (linear) state space. TLC4B needs 3 seconds to explore the state space with 1 million states. Comparatively, PROB takes 204 seconds. In another specification, the states of doors are controlled. The specification allows the doors to be opened and closed. We use two versions: In the first version the state of the doors are represented as a function (Doors_Functions) and in the second as a relation (Doors_Relations). The first version allows TLC4B to use TLA$^+$ functions for the translation and TLC needs 2 seconds to explore 32,768 states. For the second version TLC4B uses the newly introduced relations and takes 10 seconds. As expected, TLC can evaluate built-in operators faster than user defined operators. Hence the distinction TLC4B has between functions and relations can make a significant difference in running times. PROB needs about 100 seconds to explore the state space of both specifications. However, PROB needs less than a second using symmetry reduction.[11]

In summary, PROB is substantially better than TLC4B when constraint solving is required (NQueens, SumAndProduct, GraphIsomorphism[12]) or when naive enumeration of operation arguments is inefficient (GardnerSwitchingPuzzle). For some specifications (not listed in the table) TLC was not able to validate the translated TLA$^+$ specification because TLC had to enumerate an infinite set. On the other hand, TLC4B is substantially better than PROB for lower-level specifications with a large state space.

5   Correctness of the Translation

There are several possible cases where our validation of B models using TLC could be unsound: there could be a bug in TLC, there could be a bug in our TLA$^+$ library for the B operators, there could be a bug in our implementation

---

[11] TLC's symmetry reduction does not scale for large symmetric sets.

[12] See `http://www.data-validation.fr/data-validation-reverse-engineering/` for larger industrial application of this type of task.

Table 1. Empirical Results: Running times of Model Checking (times in seconds)

Model	Lines	Result	States	Transitions	PROB	TLC4B	$\frac{ProB}{Tlc4B}$
Counter	13	No Error	1000000	1000001	186.5	3.7	50.653
Doors_Functions	22	No Error	32768	983041	103.2	3.3	31.194
Can-Bus	314	No Error	132598	340265	191.8	7.2	26.624
KnightsTour[1]	28	Goal	508450	678084	817.5	34.1	23.998
USB_4Endpoints	197	NoError	16905	550418	72.5	5.7	12.632
Countdown	67	Inv. Viol.	18734	84617	31.4	2.8	11.073
Doors_Relations	22	No Error	32768	983041	103.3	11.6	8.926
Simpson_Four_Slot	78	No Error	46657	11275	33.7	4.3	7.874
EnumSetLockups	34	No Error	4375	52495	6.5	2.1	3.105
TicTacToe[1]	16	No Error	6046	19108	7.5	3.1	2.435
Cruise_finite1	604	No Error	1360	25696	6.2	3.2	1.954
CarlaTravelAgencyErr	295	Inv. Viol.	377	3163	3.3	3.1	1.069
FinalTravelAgency	331	No Error	1078	4530	4.7	4.4	1.068
CSM	64	No Error	77	210	1.4	1.6	0.859
SiemensMiniPilot_Abrial[1]	51	Goal	22	122	1.5	1.7	0.849
JavaBC-Interpreter	197	Goal	52	355	1.7	2.4	0.708
Scheduler	51	No Error	68	205	1.4	2.1	0.682
RussianPostalPuzzle	72	Goal	414	1159	1.7	2.8	0.588
Teletext_bench	431	No Error	13	122	1.8	3.7	0.496
WhoKilledAgatha	42	No Error	6	13	1.5	5.2	0.295
GardnerSwitchingPuzzle	59	Goal	206	502	2.5	11.7	0.213
NQueens_8	18	No Error	92	828	1.4	23.2	0.062
JobsPuzzle	66	Deadlock	2	2	1.6	29.3	0.053
SumAndProduct[1]	51	No Error	1	1	9.7	420.8	0.023
GraphIsomorphism	21	Deadlock	512	203	1.8	991.5	0.002

[1] Without Deadlock Check

of the translation from B to TLA+, there could be a fundamental flaw in our translation.

We have devised several approaches to mitigate those hazards. Firstly, when TLC finds a counter example it is replayed using PROB. In other words, every step of the counter example is double checked by PROB and the invariant or goal predicate is also re-checked by PROB. This makes it very unlikely that we produce incorrect counter examples. Indeed, PROB, TLC, and our translator have been developed completely independently of each other and rely on different technology. Such independently developed double chains are often used in industry for safety critical tools.

The more tricky case is when TLC finds no counter example and claims to have checked the full state space. Here we cannot replay any counter example and we have the added difficulty that, contrary to PROB, TLC stores just fingerprints of states and that there is a small probability that not all states have been checked. We have no simple solution in this case, apart from re-checking the model using either PROB or formal proof. In addition, we have conducted extensive tests to validate our approach. For example, we use a range of models encoding mathematical laws to stress test our translation. These have proven to be very useful for detecting bugs in our translation and libraries (mainly bugs involving operator precedences). In addition, we have uncovered a bug in TLC relating to the cartesian product.[13] Moreover, we use a wide variety of benchmarks, checking that PROB and TLC produce the same result and generate the same number of states.

6   More Related Work, Discussion and Conclusion

Mosbahi et al. [11] were the first to provide a translation from B to TLA$^+$. Their goal was to verify liveness conditions on B specifications using TLC. Some of their translation rules are similar to the rules presented in this paper. For example, they also translate B operations into TLA$^+$ actions and provide straightforward rules for operators which exist in both languages. However, there are also significant differences:

- Our main contribution is that we deliver translation rules for almost all B operators and in particular for those which are not build-in operators in TLA$^+$. E.g., we specified the concept of relations including all operators on relations.
- Moreover, we also consider subtle differences between B and TLA$^+$ such as different well-definedness conditions and provide an appropriate translation.
- Regarding temporal formulas, we provide a way that a B user does not have to care about stuttering steps in TLA$^+$.
- We restrict our translation to the subset of TLA$^+$ which is supported by the model checker TLC. Furthermore, we made many adaptations and optimizations allowing TLC to validate B specification efficiently.
- The implemented translator is fully automatic and does not require the user to know TLA$^+$.

In the future, we would like to improve our automatic translator:

- Supporting modularization and the refinement techniques of B.
- Improving the performance of TLC by implementing Java modules for the new standard modules.
- Integrating TLC4B into Rodin and supporting Event-B specifications.

In conclusion, by making TLC available to B models, we have closed a gap in the tool support and now have a range of complementary tools to validate B models: Atelier-B (or Rodin) providing automatic and interactive proof support, PROB being able to animate and model check high-level B specifications and providing constraint-based validation, and now TLC providing very efficient model checking of lower-level B specifications. The latter opens up many new possibilities,

---

[13] TLC erroneously evaluates the expression $\{1\} \times \{\} = \{\} \times \{1\}$ to *FALSE*.

such as exhaustive checking of hardware models or sophisticated protocols. A strong point of our approach is the replaying of counter examples using PROB. Together with the work in [5] we have now constructed a two-way bridge between TLA$^+$ and B, and also hope that this will bring both communities closer together.

Acknowledgements. We are grateful to Ivaylo Dobrikov for various discussions and support in developing the Tcl/Tk interface of TLC4B. Finally, we are thankful to the anonymous referees for their useful feedback.

References

1. Abrial, J.-R.: The B-Book. Cambridge University Press (1996)
2. ClearSy. B language reference manual, http://www.tools.clearsy.com/resources/Manrefb_en.pdf (accessed: November 10, 2013)
3. Cousineau, D., Doligez, D., Lamport, L., Merz, S., Ricketts, D., Vanzetto, H.: TLA$^+$ proofs. In: Giannakopoulou, D., Méry, D. (eds.) FM 2012. LNCS, vol. 7436, pp. 147–154. Springer, Heidelberg (2012)
4. Gafni, E., Lamport, L.: Disk Paxos. Distributed Computing 16(1), 1–20 (2003)
5. Hansen, D., Leuschel, M.: Translating TLA$^+$ to B for validation with PROB. In: Derrick, J., Gnesi, S., Latella, D., Treharne, H. (eds.) IFM 2012. LNCS, vol. 7321, pp. 24–38. Springer, Heidelberg (2012)
6. Hansen, D., Leuschel, M.: Translating B to TLA+ for validation with TLC (2013), http://www.stups.uni-duesseldorf.de/w/Special:Publication/HansenLeuschel_TLC4B_techreport
7. Lamport, L.: The TLA+ hyperbook, http://research.microsoft.com/en-us/um/people/lamport/tla/hyperbook.html (accessed: October 30, 2013)
8. Lamport, L.: Specifying Systems, The TLA+ Language and Tools for Hardware and Software Engineers. Addison-Wesley (2002)
9. Leuschel, M., Butler, M.: ProB: A model checker for B. In: Araki, K., Gnesi, S., Mandrioli, D. (eds.) FME 2003. LNCS, vol. 2805, pp. 855–874. Springer, Heidelberg (2003)
10. Leuschel, M., Butler, M.J.: ProB: an automated analysis toolset for the B method. STTT 10(2), 185–203 (2008)
11. Mosbahi, O., Jemni, L., Jaray, J.: A formal approach for the development of automated systems. In: Filipe, J., Shishkov, B., Helfert, M. (eds.) ICSOFT (SE), pp. 304–310. INSTICC Press (2007)
12. Reynolds, M.: Changing nothing is sometimes doing something. Technical Report TR-98-02, Department of Computer Science, King's College London (February 1998)
13. Yu, Y., Manolios, P., Lamport, L.: Model checking TLA+ specifications. In: Pierre, L., Kropf, T. (eds.) CHARME 1999. LNCS, vol. 1703, pp. 54–66. Springer, Heidelberg (1999)

αRby—An Embedding of Alloy in Ruby

Aleksandar Milicevic, Ido Efrati, and Daniel Jackson

Computer Science and Artificial Intelligence Laboratory
Massachusetts Institute of Technology
{aleks,idoe,dnj}@csail.mit.edu

Abstract. We present αRby—an embedding of the Alloy language in Ruby—
and demonstrate the benefits of having a declarative modeling language (backed
by an automated solver) embedded in a traditional object-oriented imperative programming language. This approach aims to bring these two distinct paradigms
(imperative and declarative) together in a novel way. We argue that having the
other paradigm available within the same language is beneficial to both the modeling community of Alloy users and the object-oriented community of Ruby programmers. In this paper, we primarily focus on the benefits for the Alloy community, namely, how αRby provides elegant solutions to several well-known,
outstanding problems: (1) mixed execution, (2) specifying partial instances, (3)
staged model finding.

1 Introduction

A common approach in formal modeling and analysis is to use (1) a declarative language (based on formal logic) for writing *specifications*, and (2) an automated constraint solver for finding valid *models* of such specifications[1]. Such models are most
often either examples (of states or execution traces), or counterexamples (of correctness claims).

In many practical applications, however, the desired analysis involves more than a
single model finding step. At the very least, a tool must convert the generated model into
a form suitable for showing to the user; in the case of Alloy [8], this includes projecting
higher-arity relations so that the model can be visualized as a set of snapshots. In some
cases, the analysis may involve repeating the model finding step, e.g., to find a minimal
model by requesting a solution with fewer tuples [13].

To date, these additional analyses have been hard-coded in the analysis tool. The key
advantage of this approach is that it gives complete freedom to the tool developer. The
disadvantage is that almost no freedom is given to the modeler, who must make do with
whatever additional processing the tool developer chose to provide.

This paper explores a different approach, in which, rather than embellishing the analysis in an ad hoc fashion in the implementation of the tool, the modeling language itself
is extended so that the additional processing can be expressed directly by the end user.
An imperative language seems most suitable for this purpose, and the challenge therefore is to find a coherent merging of two languages, one declarative and one imperative.
We show how this has been achieved in the merging of Alloy and Ruby.

---

[1] Throughout this paper, we use the term 'model' in its mathematical sense, and never to mean
the artifact being analyzed, for which we use the term 'specification' instead.

Y. Ait Ameur and K.-D. Schewe (Eds.): ABZ 2014, LNCS 8477, pp. 56–71, 2014.
© Springer-Verlag Berlin Heidelberg 2014

This challenge poses two questions, one theoretical and one practical. Theoretically, a semantics is needed for the combination: what combinations are permitted, and what is their meaning? Practically, a straightforward way to implement the scheme is needed. In particular, can a tool be built without requiring a new parser and engine that must handle both languages simultaneously?

This project focuses on the combination of Alloy and Ruby. In some respects, these choices are significant. Alloy's essential structuring mechanism, the signature [7], allows a relational logic to be viewed in an object-oriented way (in just the same way that instance variables in an object-oriented language can be viewed as functions or relations from members of the class to members of the instance variable type). So Alloy is well suited to an interpretation scheme that maps it to object-oriented constructs. Ruby is a good choice because, in addition to being object-oriented and providing (like most recent scripting languages) a powerful reflection interface, it offers a syntax that is flexible enough to support an almost complete embedding of Alloy with very few syntactic modifications.

At the same time, the key ideas in this paper could be applied to other languages; there is no reason, in principle, that similar functionality might not be obtained by combining the declarative language B [1] with the programming language Racket, for example.

The contributions of this paper include: (1) an argument for a new kind of combination of a declarative and an imperative language, justified by a collection of examples of functionality implemented in a variety of tools, all of which are subsumed by this combination, becoming expressible by the end-user; (2) an embodiment of this combination in αRby, a deep embedding of Alloy in Ruby, along with a semantics, a discussion of design challenges, and an open-source implementation [3] for readers to experiment with; and (3) an illustration of the use of the new language in a collection of small but non-trivial examples (out of 23 examples available on GitHub [4], over 1400 lines of specification in total).

2   Motivations

Analysis of a declarative specification typically involves more than just model finding. In this section, we outline the often needed additional steps.

Preprocessing   The specification or the analysis command may be updated based on user input. For example, in an analysis of Sudoku, the size of the board must be specified. In Alloy, this size would be given as part of the 'scope', which assigns an integer bound to each basic type. For Sudoku, we would like to ensure that the length of a side is a perfect square; this cannot be specified directly in Alloy.

Postprocessing   Once a model has been obtained by model finding, some processing may be needed before it is presented to the user. A common application of model finding in automatic configuration is to cast the desired configuration constraints as the specification, then perform the configuration steps based on the returned solution.

Partial instances   A partial instance is a partial solution that is known (given) upfront. In solving a Sudoku problem, for example, the model finder must be given not only the rules of Sudoku but also the partially filled board. It is easy to encode a partial solution as a formula that is then just conjoined to the specification. But although this approach is conceptually simple, it is not efficient in practice, since the model finder must work to reconstruct from this formula information (namely the partial solution) that is already known, thus needlessly damaging performance.

Kodkod (the back-end solver for Alloy) explicitly supports partial instances: it allows the user to specify relation *bounds* in terms of tuples that <u>must</u> (*lower bound*) and <u>may</u> (*upper bound*) be included in the final value. Kodkod then uses the bounds to shrink the search space, often leading to significant speedups [18]. At the Alloy level, however, this feature is not directly available[2].

Staged model finding   Some analyses involve repeated calls to the model finder. In the simplest case, the bounds on the analysis are iteratively increased (when no counterexample has been found, to see if one exists within a larger scope), or decreased (when a counterexample has already been found, to see if a smaller one exists). Sometimes this is used as a workaround when a desired property cannot be expressed directly because it would require higher order logic.

Mixed execution   Model finding can be used as a step in a traditional program execution. In this case, declarative specifications are executed 'by magic', as if, in a conventional setting, the interpreter could execute a program assertion by making it true despite the lack of any explicit code to establish it [10,9]. Alternatively, flipping the precedence of the two paradigms, the interpreter can be viewed as a declarative model finder that uses imperative code to setup a declarative specification to be solved. In this paper, we are primarily concerned with the latter direction, which has not been studied in the literature as much.

The Alloy Analyzer—the official and the most commonly used IDE for Alloy—does not currently provide any scripting mechanisms around its core model finding engine. Instead, its Java API must be used to automate even the most trivial scripting tasks. Using the Java API, however, is inconvenient; the verbosity and inflexibility of the Java language leads to poor transparency between the API and the underlying Alloy specification, making even the simplest scripts tedious and cumbersome to write. As a result, the official API is rarely used in practice, and mostly by expert users and researchers building automated tools on top of Alloy. This is a shame, since a simple transparent scripting shell would be, in many respects, beneficial to the typical Alloy user—the user who prefers to stay in an environment conceptually similar to that of Alloy and not have to learn a second, foreign API.

This is exactly what αRby provides—an embedding of the Alloy language in Ruby. Thanks to Ruby's flexibility and a very liberal parser, the αRby language manages to offer a syntax remarkably similar to that of Alloy, while still being syntactically correct Ruby. To reduce the gap between the two paradigms further, instead of using a separate AST, αRby maps the core Alloy concepts are onto the corresponding core concepts in Ruby (e.g., sigs are classes, fields are instance variables, atoms are objects, functions and predicates are methods, most operators are the same in both languages). αRby

---

[2] The Alloy Analyzer recognizes certain idioms as partial instances; some extensions (discussed in Section 6) support explicit partial instance specification.

automatically interoperates with the Alloy back end, so all the solving and visualization features of the Alloy Analyzer can be easily invoked from within an αRby program. Finally, the full power of the Ruby language is at the user's disposal for other tasks unrelated to Alloy.

3   αRby for Alloy Users

A critical requirement for embedding a modeling language in a programming language is that the embedding should preserve enough of the syntax of the language for users to feel comfortable in the new setting. We first introduce a simple example to illustrate how αRby achieves this for Alloy. Next, we address the new features brought by αRby, highlighted in Section 2, which are the primary motivation for the embedding.

Consider using Alloy to specify directed graphs and the *Hamiltonian Path* algorithm. *Signatures* are used to represent unary sets: Node, Edge, and Graph. *Fields* are used to represent relations between the signatures: val mapping each Node to an integer value; src and dst mapping each Edge to the two nodes (source and destination) that it connects; and nodes and edges mapping each Graph to its sets of nodes and edges.

A standard Alloy model for this is shown in Fig. 1(b), lines 2–4; the same declarations are equivalently written in αRby as shown in Fig. 1(a), lines 2–4.

To specify a Hamiltonian path (that is, a path visiting every node in the graph exactly once), a predicate is defined; lines 6–12 in Figs. 1(b) and 1(a) show the Alloy and αRby syntax, with equivalent semantics. This predicate asserts that the result (path) is a sequence of nodes, with the property that it contains all the nodes in the graph, and that, for all but the last index i in that sequence, there is an edge in the graph connecting the nodes at positions i and i+1. A *run* command is defined for this predicate (line 18), which, when executed, returns a satisfying instance.

Just as a predicate can be run for examples, an assertion can be checked for counterexamples. Here we assert that starting from the first node in a Hamiltonian path and transitively following the edges in the graph reaches all other nodes in the graph (lines 13–17). We expect this check (line 19) to return no counterexample.

From the model specification in Fig. 1(a), αRby dynamically generates the class hierarchy in Fig. 1(c). The generated classes can be used to freely create and manipulate graph instances, independent of the Alloy model.

In Alloy, a command is executed by selecting it in the user interface. In αRby, execution is achieved by calling the exe_cmd method. Fig. 1(d) shows a sample program that calls these methods, which includes finding an arbitrary satisfying instance for the hampath predicate and checking that the reach assertion indeed cannot be refuted.

This short example is meant to include as many different language features as possible and illustrate how similar αRby is to Alloy, despite being embedded in Ruby. We discuss syntax in Section 5.1; a summary of main differences is given in Table 1.

4   Beyond Standard Analysis

Sudoku has become a popular benchmark for demonstrating constraint solvers. The solver is given a partially filled $n \times n$ grid (where n must be a square number, so that

(a) Graph specification in αRby

```
1  alloy :GraphModel do
2    sig Node [val: (lone Int)]
3    sig Edge [src, dst: (one Node)] {src != dst}
4    sig Graph[nodes:(set Node), edges:(set Edge)]
5
6    pred hampath[g: Graph, path: (seq Node)] {
7      path[Int] == g.nodes and
8      path.size == g.nodes.size and
9      all(i: 0...path.size-1) |
10       some(e: g.edges) {
11         e.src == path[i] && e.dst == path[i+1] }
12   }
13   assertion reach {
14     all(g: Graph, path: (seq Node)) |
15       if hampath(g, path)
16         g.nodes.in? path[0].*((~src).dst)
17       end }
18   run :hampath, 5, Graph=>exactly(1), Node=>3
19   check :reach, 5, Graph=>exactly(1), Node=>3
20 end
```

(b) Equivalent Alloy specification

```
1  module GraphModel
2  sig Node {val: lone Int}
3  sig Edge {src, dst: one Node}{src != dst}
4  sig Graph{nodes: set Node, edges: set Edge}
5
6  pred hampath[g: Graph, path: seq Node] {
7    path[Int] = g.nodes
8    #path = #g.nodes
9    all i: Int | i >= 0 && i < minus[#path,1] => {
10     some e: g.edges |
11       e.src = path[i] && e.dst = path[plus[i,1]] }
12 }
13 assert reach {
14   all g: Graph, path: seq Node |
15     hampath[g, path] =>
16       g.nodes in path[0].*(~src.dst)
17 }
18 run hampath for 5 but exactly 1 Graph, 3 Node
19 check reach for 5 but exactly 1 Graph, 3 Node
20
```

⟺ (between lines 9–10)

⇃⇃

(c) Automatically generated Ruby classes

```
module GraphModel
  class Node;  attr_accessor :val end
  class Edge;  attr_accessor :src, :dst end
  class Graph; attr_accessor :nodes, :edges end

  def self.hampath(g, path) #same as above end
  def self.reach()          #same as above end
  def self.run_hampath() exe_cmd :hampath end
  def self.check_reach() exe_cmd :reach end
end
```

(d) Running hampath, checking reach

```
1  # find an instance satisfying the :hampath pred
2  sol = GraphModel.run_hampath
3  assert sol.satisfiable?
4  g, path = sol["$hampath_g"], sol["$hampath_path"]
5  puts g.nodes # => e.g., {<Node$0>, <Node$1>}
6  puts g.edges # => e.g., {<Node$1, Node$0>}
7  puts path    # => {<0, Node$1>, <1, Node$0>}
8  # check that the "reach" assertion holds
9  sol = GraphModel.check_reach
10 assert !sol.satisfiable?
```

Fig. 1. *Hamiltonian Path* example

the grid is perfectly divided into n times $\sqrt{n} \times \sqrt{n}$ sub-grids), and is required to fill the empty cells with integers from $\{1, \ldots, n\}$ so that all cells within a given row, column, and sub-grid have distinct values.

Implementing a Sudoku solver directly in Alloy poses a few problems. A practical one is that such an implementation cannot easily be used as a stand-alone application, e.g., to read a puzzle from some standard format and display the solution in a user-friendly grid. A more fundamental problem is the inability to express the information about the pre-filled cell values as a partial instance; instead, the given cell values have to be enforced with logical constraints, resulting in significant performance degradation [18]. The αRby solution in Fig. 2 addresses both of these issues: on the left is the formal αRby specification, and on the right is the Ruby code constructing bounds and invoking the solver for a concrete puzzle.

Mixed Execution   The imperative statements (lines 2, 7, 8) used to dynamically produce a Sudoku specification for a given size would not be directly expressible in Alloy. A concrete Ruby variable N is declared to hold the size, and can be set by the user before the specification is symbolically evaluated. Another imperative statement calculates the square root of N (line 6); that value is later embedded in the symbolic expression specifying uniqueness within sub-grids (line 13). For illustration purposes, a lambda function is defined (line 7) and used to compute sub-grid ranges (line 14).

(a) Sudoku specification in αRby	(b) Solving the specification for a partial instance

```
1  alloy :SudokuModel do
2  SudokuModel::N = 9
3
4  sig Sudoku[grid: Int ** Int ** (lone Int)]
5
6  pred solved[s: Sudoku] {
7    m   = Integer(Math.sqrt(N))
8    rng = lambda{|i| m*i...m*(i+1)}
9
10   all(r: 0...N) {
11     s.grid[r][Int] == (1..N) and
12     s.grid[Int][r] == (1..N)
13   } and
14   all(c, r: 0...m) {
15     s.grid[rng[c]][rng[r]] == (1..N)
16   }
17  }
18 end
```

```
1  class SudokuModel::Sudoku
2    def pi
3      bnds = Arby::Ast::Bounds.new
4      inds = (0...N)**(0...N) - self.grid.project(0..1)
5      bnds[Sudoku]          = self
6      bnds.lo[Sudoku.grid] = self ** self.grid
7      bnds.hi[Sudoku.grid] = self ** inds ** (1..N)
8      bnds.bound_int(0..N)
9    end
10   def solve() SudokuModel.solve :solved, self.pi end
11   def display() puts grid end
12   def self.parse(s) Sudoku.new grid:
13     s.split(/;\s*/).map{|x| x.split(/,/).map(&:to_i)}
14   end
15 end
16 SudokuModel.N = 4
17 s = Sudoku.parse "0,0,1; 0,3,4; 3,1,1; 2,2,3"
18 s.solve(); s.display(); # => {<0,0,1>, <0,1,3>, ...}
```

Fig. 2. A declarative *Sudoku* solver using αRby with partial instances

Partial Instances   Fig. 2(b) shows how the bounds are computed for a given Sudoku puzzle, using a Ruby function pi (for "partial instance"). Remember that bounds are just tuples (sequences of atoms) that a relation must or may include; since signature definitions in αRby are turned into regular Ruby classes, instances of those classes will be used as atoms. The Sudoku signature is bounded by a singleton set containing only the self Sudoku object (line 5). Tuples that must be included in the grid relation are the values currently present in the puzzle (line 6); additional tuples that may be included are values from 1 to N for the empty cells (line 7; empty cell indexes computed in line 4). We also bound the set of integers to be used by the solver; Alloy, in contrast, only allows a cruder bound, and would include all integers within a given bitwidth. Finally, a Sudoku instance can be parsed from a string, and the solver invoked to find a solution satisfying the solved predicate (lines 17–18). When a satisfying solution is found, if a partial instance was given, fields of all atoms included in that partial instance are automatically populated to reflect the solution (confirmed by the output of line 18). This particular feature makes for seamless integration of *executable specifications* into otherwise imperative programs, since there is no need for any manual back and forth conversion of data between the program and the solver.

Staged Model Finding   Consider implementing a Sudoku puzzle generator. The goal is now to find a partial assignment of values to cells such that the generated puzzle has a unique solution. Furthermore, the generator must be able to produce various difficulty levels of the same puzzle by iteratively decrementing the number of filled cells (while maintaining the uniqueness property). This is a higher-order problem that cannot be solved in one step in Alloy. With αRby, however, it takes only the following 8 lines to achieve this with a simple search algorithm on top of the already implemented solver:

```
1  def dec(sudoku, order=Array(0...sudoku.grid.size).shuffle)
2    return nil if order.empty? # all possibilities exhausted
3    s_dec = Sudoku.new grid: sudoku.grid.delete_at(order.first) # delete a tuple at random position
4    sol   = s_dec.clone.solve() # clone so that "s_dec" doesn't get updated if a solution is found
5    (sol.satisfiable? && !sol.next.satisfiable?) ? s_dec : dec(sudoku, order[1..-1])
6  end
7  def min(sudoku) (s1 = dec(sudoku)) ? min(s1) : sudoku end
8  s = Sudoku.new; s.solve(); s = min(s); puts "local minimum found: #{s.grid.size}"
```

The strategy here is to generate a solved puzzle (line 8), and keep removing one tuple from its grid at a time until a local minimum is reached (line 7); the question is which one can be removed without violating the uniqueness property. The algorithm first generates a random permutation of all existing grid tuples (line 1) to determine the order of trials. It then creates a new Sudoku instance with the chosen tuple removed (line 3) and runs the solver to find a solution for it. It finally calls next on the obtained solution (line 5) to check if a different solution exists; if it does not, a decremented Sudoku is found, otherwise moves on to trying the rest of the tuples. On a commodity machine, on average it takes about 8 seconds to minimize a Sudoku of size 4 (generating 13 puzzles in total, number of filled cells ranging from 16 down to 4), and about 6 minutes to minimize a puzzle of size 9 (55 intermediate puzzles), number of filled cells ranging from 81 down to 27).

5   The αRby Language

αRby is implemented as a domain-specific language in Ruby, and is (in standard parlance) "deeply embedded". *Embedded* means that all syntactically correct αRby programs are syntactically correct Ruby programs; *deeply* means that αRby programs exist as an AST that can be analyzed, interpreted, and so on. Ruby's flexibility makes it possible to create embedded languages that look quite different from standard Ruby. αRby exploits this, imitating the syntax of Alloy as closely as possible. Certain differences are unavoidable, mostly because of Alloy's infix operators that cannot be defined in Ruby.

The key ideas behind our approach are: (1) mapping the core Alloy concepts directly to those of object-oriented programming (OOP), (2) implementing keywords as methods, and (3) allowing mixed (concrete and symbolic) execution in αRby programs.

Mapping Alloy to OOP is aligned with the general intuition, encouraged by Alloy's syntax, that signatures can be understood as classes, atoms as objects, fields as instance variables, and all function-like concepts (functions, predicates, facts, assertions, commands) as methods [2].

Implementing keywords as methods works because Ruby allows different formats for specifying method arguments. αRby defines many such methods (e.g., **sig**, **fun**, **fact**, etc.) that (1) mimic the Alloy syntax and (2) dynamically create the underlying Ruby class structure using the standard Ruby metaprogramming facilities. For an example of the syntax mimicry, compare Figs. 1(a) and 1(b); for an example of metaprogramming, see Fig. 1(c).

Note that the meta information that appears to be lost in Fig. 1(c) (for example, the types of fields) is actually preserved in separate *meta* objects and made available via the *meta* methods added to each of the generated modules and classes (e.g., `Graph.meta.field("nodes").type`).

Mixed execution, implemented on top of the standard Ruby interpreter, translates αRby programs into symbolic Alloy models. Using the standard interpreter means adopting the Ruby semantics of name resolution and operator precedence (which is inconvenient when it conflicts with Alloy's); a compensation, however, is the benefit of being able to mix symbolic and concrete code. We override all the Ruby operators in our symbolic expression classes to match the semantics of Alloy, and using a couple of other tricks (Section 5.3), are able to keep both syntactic (Section 5.1) and semantic (Section 5.2) differences to a minimum.

5.1 Syntax

A grammar of αRby is given in Fig. 3 and examples of principal differences in Table 1. In a few cases (e.g., function return type, field declaration, etc.) Alloy syntax has to be slightly adjusted to respect the syntax of Ruby (e.g., by requiring different kind of brackets). More noticeable differences stem from the Alloy operators that are illegal or cannot be overridden in Ruby; as a replacement, either a method call (e.g., `size` for cardinality) or a different operator (e.g., `**` for cross product) is used.

The difference easiest to overlook is the equality sign: `==` versus `=`. Alloy has no assignment operator, so the single equals sign always denotes boolean equality; αRby, in contrast, supports both concrete and symbolic code, so we must differentiate between assignments and equality checks, just as Ruby does.

The tokens for the join and the two closure operators (`.`, `^` and `*`) exist in Ruby, but have fundamentally different meanings than in Alloy (object dereferencing and an infix binary operator in Ruby, as opposed to an infix binary and a prefix unary operator in Alloy). Despite this, αRby preserves Alloy syntax for many idiomatic expressions. Joins in Alloy are often applied between an expression and a field whose left-hand type matches the type of the expression (ie, in the form `e.f`, where `f` is a field from the type of `e`). This corresponds closely to object dereferencing, and is supported by αRby (e.g., `g.nodes` in Fig. 1(a)). In other kinds of joins, the right-hand side must be enclosed in parentheses. Closures are often preceded by a join in Alloy specifications. Those constructs yield *join closure* expressions of the form `x.*f`. In Ruby, this translates to calling the `*` method on object x passing `f` as an argument, so we simply override the `*` method to achieve the same semantics (e.g., line 15, Fig. 1(a)).

This grammar is, for several reasons, an under-approximation of programs accepted by αRby: (1) Ruby allows certain syntactic variations (e.g., omitting parenthesis in method calls, etc.), (2) αRby implements special cases to enable the exact Alloy syntax for certain idioms (which do not always generalize), and (3) αRby provides additional methods for writing expression that have more of a Ruby-style feel.

5.2 Semantics

This section formalizes the translation of αRby programs into Alloy. We provide semantic functions (summarized in Fig. 5) that translate the syntactic constructs of Fig. 3

```
spec       ::= "alloy" cname "do" [open*] paragraph* "end"
open       ::= "open" cnameID
paragraph ::= factDecl  | funDecl | cmdDecl | sigDecl
sigQual    ::= "abstract" | "lone"  |  "one"   | "some"  | "ordered"
sigDecl    ::= sigQual* "sig"  cname,+ ["extends" cnameID] ["[" rubyHash "]"] [block]
factDecl   ::= "fact"        [fname] block
funDecl    ::= "fun"      fname "[" rubyHash "]"  "[" expr "]" block
           | "pred"      fname ["[" rubyHash "]"]        block
cmdDecl    ::= ("run"|"check") fname "," scope
           | ("run"|"check") "(" scope ")" block
expr       ::= ID | rubyInt | rubyBool | "(" expr ")"
           | unOp expr       | unMeth "(" expr ")"
           | expr binOp expr   | expr "[" expr "]"     | expr "if" expr
           | expr "." "(" expr ")"             // relational join
           | expr "." (binMeth | ID) "(" expr,* ")"  // function/predicate call
           | "if" expr "then" expr ["else" expr] "end"
           | quant "(" rubyHash ")" block
quant      ::= "all" | "no" | "some" | "lone" | "one" | "sum" | "let" | "select"
binOp      ::= "||" | "or" | "&&" | "and" | "**" | "&" | "+" | "-" | "*" | "/" | "%"
           | "<<" | ">>" | "==" | "<=>" | "!=" | "<" | ">" | "<=" | ">="
binMeth    ::= "closure" | "rclosure" | "size" | "in?" | "shr" | "<" | ">" | "*" | "^"
unOp       ::= "!" | "~" | "not"
unMeth     ::= "no" | "some" | "lone" | "one" | "set" | "seq"
block      ::= "{" stmt* "}" | "do" stmt* "end"
stmt       ::= expr | rubyStmt
scope      ::= rubyInt "," rubyHash  // global scope, individual sig scopes
ID         ::= cnameID | fnameID
cname      ::= cnameID | '"'cnameID'"' | "'"cnameID"'" | ":"cnameID
fname      ::= fnameID | '"'fnameID'"' | "'"fnameID"'" | ":"fnameID
cnameID    ::= constant identifier in Ruby (starts with upper case)
fnameID    ::= function identifier in Ruby (starts with lower case)
```

Fig. 3. Core αRby syntax in BNF. Productions starting with: ruby are defined by Ruby.

Table 1. Examples of differences in syntax between αRby and Alloy

description	Alloy	αRby
equality	`x = y`	`x == y`

sigs and fields	`sig S {`
	`  f: lone S -> Int`
	`}`
		`sig S [`
		`  f: lone(S) ** Int`
		`]`

fun return type declaration	`fun f[s: S]: set S {}`	`fun f[s: S][set S] {}`
set comprehension	`{s: S	p1[s]}`	`S.select{|s| p1(s)}`

quantifiers	`all s: S {`
	`  p1[s]`
	`  p2[s]`
	`}`
		`all(s: S) {`
		`  p1(s) and`
		`  p2(s)`
		`}`

illegal Ruby operators	`x in y, x !in y`
	`x !> y`
	`x -> y`
	`x . y`
	`#x`
	`x => y`
	`x => y else z`
	`S <: f, f >: Int`
		`x.in?(y), x.not_in?(y)`
		`not x > y`
		`x ** y`
		`x.(y)`
		`x.size`
		`y if x`
		`if x then y else z`
		`S.< f, f.> Int`

operator arity mismatch	`^x, *x`	`x.closure, x.rclosure`
fun/pred calls	`f1[x]`	`f1(x)`

```
Expr  = VarExpr(name: String, domain: Expr | Type)
      | IntExpr(value: Int) | BoolExpr(value: Bool)
      | UnExpr(sub: Expr) | BinExpr(lhs: Expr, rhs: Expr)
      | CallExpr(target: Expr, fun: FunDecl, args: Expr*)
      | QuantExpr(kind: String, vars: VarExpr*, body: Expr)
Decl  = Spec(name: String, opens: Spec*, sigs: SigDecl*, funs: FunDecl*)
      | SigDecl(name: String, parent: SigDecl, fields: VarExpr*, inv: FunDecl)
      | FunDecl(name: String, params: VarExpr*, ret: Expr, body: Expr)
Type  = Univ | None | Int | SigDecl | ProductType(lhs: Type, rhs: Type)
Store = {name: String; binding: Expr | Decl}
```

Fig. 4. Overview of the semantic domains. (`Expr` and `Decl` correspond directly to the Alloy AST)

$\mathcal{A}$:	specification → Store → Spec		$\mathcal{E}$:	expr → Store → Expr
ξ:	sigDecl → Store → SigDecl		β:	block → Store → Expr
ϕ:	funDecl → Store → FunDecl		δ:	decl* → Store → (VarExpr*, Store)

Fig. 5. Overview of the semantic functions which translate grammar rules to semantic domains

to Alloy AST elements defined in Fig. 4. A store, binding names to expressions or declarations, is maintained throughout, representing the current evaluation context.

Expressions   The evaluation of the αRby expression production rules (expr) into Alloy expressions (`Expr`) is straightforward for the most part (Fig. 6). Most of the unary and binary operators have the same semantics as in Alloy; exceptions are ** and **if**, which translate to -> and => (lines 5–11). For the operators that do not exist in Ruby, an equivalent substitute method is used (lines 12–20). A slight variation of this approach is taken for the ^ and * operators (lines 21–22), to implement the "join closure" idiom (explained in Section 5.1).

The most interesting part is the translation of previously undefined method calls (lines 23–27). We first apply the τ function to obtain the type of the left-hand side expression, and then the $\oplus$ function to extend the current store with that type (line 23). In a nutshell, this will create a new store with added bindings for all fields and functions defined for the range signature of that type (the $\oplus$ function is formally defined in Fig. 8 and discussed in more detail shortly). Afterward, we look up meth as an identifier in the new store (line 24) and, if an expression is found (line 25), the expression is interpreted as a join; if a function declaration is found (line 26), it is interpreted a function call; otherwise, it is an error.

For quantifiers (lines 29-30), quantification domains are evaluated in the context of the current store (using the δ helper function, defined in Fig. 8) and the body is evaluated (using the β function, defined in Fig. 7) in the context of the new store with added bindings for all the quantified variables (returned previously by δ).

Blocks   The semantics of αRby blocks differs from Alloy's. An Alloy block (e.g., a quantifier body) containing a sequence of expressions is interpreted as a conjunction of all the constituent constraints (a feature based on Z [17]). In αRby, in contrast, such as sequence evaluates to the meaning of the last expression in the sequence. This was

$\mathcal{E} :$ expr $\rightarrow$ Store $\rightarrow$ Expr	
1. $\mathcal{E}[\![\texttt{ID}]\!]\sigma$	$\equiv$ $\sigma[\texttt{ID}]$
2. $\mathcal{E}[\![\texttt{rubyInt}]\!]\sigma$	$\equiv$ $\texttt{IntExpr(rubyInt)}$
3. $\mathcal{E}[\![\texttt{rubyBool}]\!]\sigma$	$\equiv$ $\texttt{BoolExpr(rubyBool)}$
4. $\mathcal{E}[\![(e)]\!]\sigma$	$\equiv$ $\mathcal{E}[\![e]\!]\sigma$
5. $\mathcal{E}[\![\texttt{unOp }e]\!]\sigma$	$\equiv$ $\texttt{UnExpr(unOp, }\quad\mathcal{E}[\![e]\!]\sigma\texttt{)}$
6. $\mathcal{E}[\![\texttt{unMeth}(e)]\!]\sigma$	$\equiv$ $\texttt{UnExpr(unMeth, }\mathcal{E}[\![e]\!]\sigma\texttt{)}$
7. $\mathcal{E}[\![e_1 \texttt{ ** } e_2]\!]\sigma$	$\equiv$ $\texttt{BinExpr("->", }\quad\mathcal{E}[\![e_1]\!]\sigma\texttt{, }\mathcal{E}[\![e_2]\!]\sigma\texttt{)}$
8. $\mathcal{E}[\![e_1 \texttt{ binOp } e_2]\!]\sigma$	$\equiv$ $\texttt{BinExpr(binOp, }\mathcal{E}[\![e_1]\!]\sigma\texttt{, }\mathcal{E}[\![e_2]\!]\sigma\texttt{)}$
9. $\mathcal{E}[\![e_1[e_2]]\!]\sigma$	$\equiv$ $\texttt{BinExpr("[]", }\quad\mathcal{E}[\![e_1]\!]\sigma\texttt{, }\mathcal{E}[\![e_2]\!]\sigma\texttt{)}$
10. $\mathcal{E}[\![e_1 \texttt{ if } e_2]\!]\sigma$	$\equiv$ $\texttt{BinExpr("=>", }\quad\mathcal{E}[\![e_2]\!]\sigma\texttt{, }\mathcal{E}[\![e_1]\!]\sigma\texttt{)}$
11. $\mathcal{E}[\![e_1.(e_2)]\!]\sigma$	$\equiv$ $\texttt{BinExpr(".", }\quad\mathcal{E}[\![e_1]\!]\sigma\texttt{, }\mathcal{E}[\![e_2]\!]\sigma\texttt{)}$
12. $\mathcal{E}[\![e.\texttt{binMeth}]\!]\sigma$	$\equiv$ $\textbf{match }\texttt{binMeth}\textbf{ with}$
13.	$\mid$ closure $\rightarrow$ $\texttt{UnExpr("^", }\quad\mathcal{E}[\![e]\!]\sigma\texttt{)}$
14.	$\mid$ rclosure $\rightarrow$ $\texttt{UnExpr("*", }\quad\mathcal{E}[\![e]\!]\sigma\texttt{)}$
15.	$\mid$ size $\rightarrow$ $\texttt{UnExpr("#", }\quad\mathcal{E}[\![e]\!]\sigma\texttt{)}$
16. $\mathcal{E}[\![e.\texttt{binMeth}(a_1)]\!]\sigma$	$\equiv$ $\textbf{match }\texttt{binMeth}\textbf{ with}$
17.	$\mid$ in? $\rightarrow$ $\texttt{BinExpr("in", }\quad\mathcal{E}[\![e]\!]\sigma\texttt{, }\mathcal{E}[\![a_1]\!]\sigma\texttt{)}$
18.	$\mid$ shr $\rightarrow$ $\texttt{BinExpr(">>>", }\mathcal{E}[\![e]\!]\sigma\texttt{, }\mathcal{E}[\![a_1]\!]\sigma\texttt{)}$
19.	$\mid$ < $\rightarrow$ $\texttt{BinExpr("<:", }\quad\mathcal{E}[\![e]\!]\sigma\texttt{, }\mathcal{E}[\![a_1]\!]\sigma\texttt{)}$
20.	$\mid$ > $\rightarrow$ $\texttt{BinExpr(":>", }\quad\mathcal{E}[\![e]\!]\sigma\texttt{, }\mathcal{E}[\![a_1]\!]\sigma\texttt{)}$
21.	$\mid$ ^ $\rightarrow$ $\mathcal{E}[\![e.(a_1.\texttt{closure})]\!]\sigma$
22.	$\mid$ * $\rightarrow$ $\mathcal{E}[\![e.(a_1.\texttt{rclosure})]\!]\sigma$
23. $\mathcal{E}[\![e.\texttt{ID}(a_1, \ldots)]\!]\sigma$	$\equiv$ $\textbf{let }\sigma_{sub} = \sigma \oplus \tau(e)\textbf{ in}$
24.	$\textbf{match }\sigma_{sub}[\texttt{ID}]\textbf{ as }x\textbf{ with}$
25.	$\mid$ Expr $\rightarrow$ $\texttt{BinExpr(".", }\mathcal{E}[\![e]\!]\sigma\texttt{, }x\texttt{)}$
26.	$\mid$ FunDecl $\rightarrow$ $\texttt{CallExpr(}\mathcal{E}[\![e]\!]\sigma\texttt{, }x\texttt{, }\mathcal{E}[\![a_1]\!]\sigma\texttt{, }\ldots\texttt{)}$
27.	$\mid$ $\rightarrow$ $fail$
28. $\mathcal{E}[\![\textbf{if }e_1\textbf{ then }e_2\textbf{ else }e_3\textbf{ end}]\!]\sigma$	$\equiv$ $\mathcal{E}[\![(e_2\textbf{ if }e_1)\textbf{ and }(e_3\textbf{ if }!e_1)]\!]\sigma$
29. $\mathcal{E}[\![\texttt{quant}(d_*)\texttt{ block}]\!]\sigma$	$\equiv$ $\textbf{let }v_*,\ \sigma_b = \delta(d_*)\sigma\textbf{ in}$
30.	$\texttt{QuantExpr(quant, }v_*\texttt{, }\beta[\![\texttt{block}]\!]\sigma_b\texttt{)}$

Fig. 6. Evaluation of αRby expressions (expr production rules) into Alloy expressions (Expr)

a design decision, necessary to support mixed execution (as in Fig. 2(a), lines 7–16). Since Ruby is not a pure functional language, previous statements can affect the result of the last statement by mutating the store, which effectively gives us the opportunity to easily mix concrete and symbolic execution.

This behavior is formally captured in the β function (Fig. 7). Statements $(s_1, \ldots, s_n)$ are evaluated in order (line 32). If a statement corresponds to one of the expression rules from the αRby grammar (line 34), it is evaluated using the previously defined $\mathcal{E}$ function; otherwise (line 35), it is interpreted by Ruby (abstracted as a call to the $\mathcal{R}$ functions). Statements interpreted by Ruby may change the store, which is then passed on to the subsequent statements.

Function Declarations The evaluation function (ϕ, Fig. 7, lines 37–38) is similar to quantifier evaluation, except that the return type is different. The semantics of other function-like constructs (predicates, facts, etc.) is analogous, and is omitted for brevity.

β : block → Store ⇸ Expr

31. $\beta[\![\mathbf{do}\ s_1;\ldots;s_n\ \mathbf{end}]\!]\sigma\ \equiv\ \beta[\![\{\ s_1;\ldots;s_n\ \}]\!]\sigma\ \equiv\ \sigma_{curr} = \sigma, res = \mathsf{nil}$
32. $\qquad\qquad\qquad\qquad\qquad\qquad\qquad\qquad\qquad\mathbf{for}\ s_i\colon \{s_1,\ \ldots,\ s_n\}\ \mathbf{do}$
33. $\qquad\qquad\qquad\qquad\qquad\qquad\qquad\qquad\qquad\quad \mathbf{match}\ s_i\ \mathbf{with}$
34. $\qquad\qquad\qquad\qquad\qquad\qquad\qquad\qquad\qquad\quad\ \mid\ expr\ \rightarrow\ \ res \leftarrow \mathcal{E}[\![s_i]\!]\sigma_{curr}$
35. $\qquad\qquad\qquad\qquad\qquad\qquad\qquad\qquad\qquad\quad\ \mid\ \qquad\ \rightarrow\ \ res, \sigma_{curr} \leftarrow \mathcal{R}(s_i)\sigma_{curr}$
36. $\qquad\qquad\qquad\qquad\qquad\qquad\qquad\qquad\qquad \mathbf{return}\ res$

ϕ : funDecl → Store ⇸ FunDecl

37. $\phi[\![\mathbf{fun}\ \mathsf{fname}[d_*][e_{ret}]\ \mathsf{block}]\!]\sigma\ \equiv\ \mathbf{let}\ v_*,\ \sigma_b\ =\ \delta(d_*)\sigma\ \mathbf{in}$
38. $\qquad\qquad\qquad\qquad\qquad\qquad\qquad \mathsf{FunDecl}(\mathsf{fname},\ v_*,\ \mathcal{E}[\![e_{ret}]\!]\sigma,\ \beta[\![\mathsf{block}]\!]\sigma_b)$

ξ : sigDecl → Store ⇸ SigDecl

39. $\xi[\![\mathbf{sig}\ \mathsf{cname}\ \mathbf{extends}\ \mathsf{sup}\ [d_*]\ \mathsf{block}]\!]\sigma\ \equiv$
40. $\quad \mathbf{let}\ s_p\qquad = \tau(\sigma[\mathsf{sup}])\ \mathbf{in}$
41. $\quad \mathbf{let}\ fld_*,\ \_ = \delta(d_*)\sigma\ \mathbf{in}$
42. $\quad \mathbf{let}\ this\quad\ = \mathsf{VarExpr}(\text{"this"},\ \mathsf{cname})\ \mathbf{in}$
43. $\quad \mathbf{let}\ tfld_*\quad = map(\lambda f_i \cdot \mathsf{BinExpr}(\text{"."},\ this,\ f_i), fld_*)\ \mathbf{in}$
44. $\quad \mathbf{let}\ \sigma_s\qquad = \sigma \oplus \mathsf{SigDecl}(\mathsf{cname},\ s_p,\ tfld_*,\ \mathsf{BoolExpr}(\mathsf{true}))\ \mathbf{in}$
45. $\quad\ \mathsf{SigDecl}(\mathsf{cname},\ s_p,\ fld_*,\ \beta[\![\mathsf{block}]\!]\sigma_s[\text{"this"} \mapsto this])$

$\mathcal{A}$: spec → Store ⇸ Spec

46. $\mathcal{A}[\![\mathbf{alloy}\ \mathsf{cname}\ \mathbf{do}\ \mathsf{open}*\ \mathsf{paragraph}*\ \mathbf{end}]\!]\sigma\ \equiv$
47. $\quad \mathbf{let}\ opn_*\quad = map(\lambda \mathsf{cnameID} \cdot \sigma[\mathsf{cnameID}], \mathsf{open}*)\ \mathbf{in}$
48. $\quad \mathbf{let}\ sig_*\qquad = map(\xi, filter(\mathsf{sigDecl}, \mathsf{paragraph}*))\ \mathbf{in}$
49. $\quad \mathbf{let}\ fun_*\qquad = map(\phi, filter(\mathsf{funDecl}, \mathsf{paragraph}*))\ \mathbf{in}$
50. $\quad \mathbf{let}\ a\qquad\quad = \mathsf{Spec}(\mathsf{cname}, opn_*,\ sig_*,\ fun_*)\ \mathbf{in}$
51. $\quad\ \ \mathbf{if}\ resolved(a)\ \mathbf{then}\ a$
52. $\quad\ \mathbf{elsif}\ \sigma \oplus a \neq \sigma\ \mathbf{then}\ \mathcal{A}[\![\mathbf{alloy}\ \mathsf{cname}\ \mathbf{do}\ \mathsf{open}*\ \mathsf{paragraph}*\ \mathbf{end}]\!]\sigma \oplus a\ \mathbf{else}\ fail$

Fig. 7. Evaluation of blocks and all declarations

Signature Declarations   The evaluation function (function ξ, Fig. 7, lines 39–45) is conceptually straightforward: as before, functions δ and β can be reused to evaluate the field name-domain declarations and the appended facts block, respectively. The caveat is that appended facts in Alloy must be evaluated in the context of Alloy's *implicit this*, meaning that the fields from the parent signature should be implicitly joined on the left with an implicit this keyword. To achieve this, we create a variable corresponding to this and a new list of fields with altered domains (lines 42–43). A temporary SigDecl containing those fields is then used to extend the current store (line 44). A binding for this is also added and the final store is used to evaluate the body (line 45). The temporary signature is created just for the convenience of reusing the semantics of the $\oplus$ operator (explained shortly).

Top-Level Specifications   Evaluation of an αRby specification (function $\mathcal{A}$, Fig. 7, lines 46–52) uses the previously defined semantic functions to evaluate the nested signatures and functions. Since declaration order does not matter in Alloy, multiple passes may be needed until everything is resolved or a fixed point is reached (lines 51–52).

Name-Domain Declaration Lists   Name-domain lists are used in several places (for fields, method parameters, and quantification variables); common functionality is

δ : decl* → Store → (VarExpr*, Store)
53. $\delta(v_1\colon e_1,\ \ldots,\ v_n\colon e_n)\sigma\ \equiv\ \mathbf{let}\ vars\ =\ \bigcup_{1<=i<=n}\mathsf{VarExpr}(v_i,\ \mathcal{E}[\![e_i]\!]\sigma)\ \mathbf{in}$
54. $[vars,\ \sigma\oplus vars]$

$\oplus$: Store → Any → Store
55. $\sigma\oplus x\ \equiv\ \mathbf{match}\ x\ \mathbf{with}$
56. | VarExpr$(n,\_)$ | FunDecl$(n,\_)$ → $\sigma[n\mapsto x])$
57. | VarExpr* | FunDecl* → $fold(\oplus,\sigma,x)$
58. | ProductType$(\_,rhs)$ → $\sigma\oplus rhs$
59. | SigDecl$(n,s_p,fld_*,\_)$ → $\mathbf{let}\ \sigma_p\ =\ \sigma\oplus s_p\ \mathbf{in}$
60. $\mathbf{let}\ \sigma_s\ =\ \sigma_p[n\mapsto\mathsf{VarExpr}(n,x)]\ \mathbf{in}$
61. $fold(\oplus,\sigma_s,\mathsf{funs}(x)+fld_*)$
62. | Spec(n,opn_*,sig_*,fun_*) → $fold(\oplus,\sigma,opn_*+fun_*+sig_*)$
63. | → σ

$\mathcal{R}$	: rubyStmt → Store → ($Object$ → Store)	Executes arbitrary Ruby code
τ	: Expr → Type	Type of the given expression
funs	: SigDecl → FunDecl*	Functions where given sig is first param
resolved	: Spec → Bool	Whether all references are resolved

Fig. 8. Helper functions

extracted and defined in the δ function (Fig. 8). It simply maps the input list into a list of VarExpr expressions, each having *name* the same as in the declaration list and *domain* equal to the evaluation of the declared domain against the current store (line 53). It returns that list and the current store extended with those variables (line 54).

Store Extension   The $\oplus$ operator (Fig. 8) is used to extend a store with one or more VarExpr or FunDecl, a Type, and a Spec. If a VarExpr or a FunDecl is given, its name is bound to itself. If a list is given, the operation is folded over the entire list. Extending with a Type reduces to extending with the range of that type. Extending with a SigDecl means recursively adding bindings for its parent signature, adding a binding for the name of that signature, bindings for all the functions that take that signature as the first argument (an auxiliary function funs(x) discovers such functions), and bindings for all its fields. Extending with a Spec adds bindings for all the sigs and functions defined in it, including those from all opened specifications.

5.3   Implementation Considerations

Symbolic Execution   Using the standard Ruby interpreter to symbolically execute αRby programs relieves us from having to keep an explicit representation of the store; instead, the store is implicit in the states of the object in which the execution takes place. Having signatures, fields, and functions represented directly as classes, instance variables, and methods, means having most of the bindings (as defined in Section 5.2) already in place for all sigs and atoms; for all other expressions, missing methods are dynamically forwarded to the signature class corresponding to the expression's type.

One technical challenge is that the semantics of quantifiers requires a new scope to be created, which, for our syntax, Ruby does not already ensure. Consider the following αRby code: **all**(s: S){**some** s}. This is just a hash and a block passed to our domain-specific-language method. When the block is eventually executed (to obtain the symbolic body for this universal quantifier), s must be available as a symbolic variable inside of that block. We do that by first dynamically defining a method with the same name in the context of that block, then calling the block, and, finally, redefining the same method to call super:

```
ctx = block.binding.eval("self")
ctx.define_singleton_method :s, lambda{VarExpr.new(:s, S)}
begin block.call ensure ctx.define_singleton_method(:s) do super() end end
```

Responding to Missing Methods and Constants   To avoid requiring strings instead of identifiers for every new definition (e.g., **sig** :Graph instead of **sig** Graph, where Graph is previously undefined), αRby overrides const_missing and method_missing and instead of failing returns a MissingBuilder instance. Furthermore, MissingBuilder instances also accept a block at creation time, and respond to several operator methods, making constructs like **fun** f[s: S][set S] {} possible. To guard against unintended conversions (e.g., typos), αRby raises a syntax error every time a MissingBuilder is not "consumed" (by certain DSL methods, like **sig** and **fun**) by the end of its scope block.

Online Source Instrumentation   For the purpose of symbolic evaluation, the source code of every αRby function/predicate is instrumented before it is turned into a Ruby method. The need for instrumentation arises because certain operators and control structures, which we would like to treat symbolically, cannot be overridden; examples include all the if-then-else variants, as well as the logic operators. Our instrumentation uses an off-the-self parser and implements a visitor over the generated AST to replace these constructs with appropriate αRby expressions (e.g., x **if** y gets translated to BinExpr.new(IMPLIES, proc{y}, proc{x})). This "traverse and replace" algorithm is far simpler than implementing a full parser for the entire Alloy grammar.

Distinguishing Equivalent Ruby Constructs   Ruby allows different syntactic constructs for the same underlying operation. For example, some built-in infix operators can be written with or without a dot between the left-hand side and the operator (e.g., a∗b is equivalent to a.∗b). Since αRby already performs online source instrumentation, it additionally detects the following syntactic nuances for the purpose of assigning different semantics: (1) in Ruby, "<b2> **if** <b1>" is equivalent to "b1 **and** b2", but our instrumenter always rewrites **and** and **or** to boolean conjunction and disjunction; (2) when prefixed with a dot, operators ∗, < and > are translated to join closure, domain restriction, and range restriction, respectively (.∗, <:, and :> in Alloy).

αRby to Alloy Bridge   All model-finding tasks are delegated to a slightly modified version of the official Alloy Analyzer Java implementation. The main modification we made was adding an extra API method, which additionally accepts a partial instance (represented in a simple textual format independent of the Alloy language). The Alloy Analyzer already has a complex heuristic for computing bounds from the scope specification and certain (automatically detected) idioms; we retain all those features, and on top of them use the αRby-provided partial instance to shrink the bounds further (formalization of which is beyond the scope of this paper). To interoperate between Ruby and Java, we use RJB [14], which conveniently automates most of the process.

6 Related Work

Montaghami and Rayside [11] extended the Alloy language with special syntax for specifying partial instances. They argue convincingly for the importance of having partial instances for Alloy, giving use cases such as test-driven model development, regression testing of models, modeling by example etc. They also provide experimental evidence that staged model finding can lead to better scalability. Their approach is limited to partial instances only, and it does not provide any scripting mechanisms for automating such tasks. Thus to carry out their staged model finding experiments, after obtaining an instance in the first stage, they manually inspected it (e.g., in the visualizer), rewrote it using the new syntax, and then solved in the second stage. Using αRby would automate the whole process, since an αRby instance can provide a set of exact bounds for all included relations, and can handle all the use cases discussed.

A number of tools built on top of Alloy have implemented (often in an ad hoc fashion) one or more features that can now be provided by αRby. Aluminum [13] implements an interesting heuristic for minimizing Alloy instances and by default showing the minimal one first. It also allows the user to augment the current instance by selecting one or more tuples to be included in the next instance. αRby provides a more generic mechanism that lets the user provide an arbitrary formula (possibly involving atoms from the current instance) to be satisfied in the next solution. TACO [5] is a bounded verifier for Java that achieves scalability by relying heavily on the Alloy Analyzer to recognize certain idioms as partial instances; we believe αRby would have made their implementation much simpler.

Our mixed execution was inspired by Rubicon's [12] symbolic evaluator, which also uses the standard Ruby interpreter. Unlike αRby, Rubicon stubs the library code with custom expressions in order to symbolically execute and verify existing web apps.

Many research projects explore the idea of extending a programming language with symbolic constraint-solving features (e.g., [15,10,9,20,19,16]). αRby can be understood as a kind of dual, with the opposite goal. While these efforts aim to bring declarative features in imperative programming, αRby aims to bring imperative features to declarative modeling. Although the basic idea of combining declarative model finding and imperative model finding is shared, the research challenges are very different. As this paper has explained, αRby addresses the challenge of embedding an entire modeling language in a programming language, whereas these related projects instead tend to use a constraint language that is only a modest extension of the programming language's existing expression sublanguage. αRby also addresses the challenge of reconciling two different views of a data structure: one as objects on a heap, and the other as relations (and in this respect is related to work on relational data representation, such as [6]).

7 Conclusion

On the one hand, αRby addresses a collection of very practical problems in the use of a model finding tool. This paper's contribution can thus be regarded as primarily architectural, in demonstrating a different way to build an analysis tool that uses a DSL embedding to allow end-user scripting, rather than a closed compiler-like tool that can be extended only by one of the tool's developers.

On the other hand, αRby suggests a new way to think about a modeling language. The constructs of the language are not treated as functions that generate abstract syntax trees only in a mathematical sense, but are implemented as these functions in a manner that the end user can exploit. This leads us to wonder whether it might be possible to use this style of embedding in the very design of the modeling language. Perhaps, had this approach been available when Alloy was designed, an essential core might have been more cleanly separated from a larger collection of structuring idioms, implemented as functions on top of the core's functions.

Practically speaking, we hope that the developers of tools that use Alloy as a backend will be able to use αRby in their implementations, at the very least making it easier to prototype new functionality. And perhaps the implementors of tools for other declarative languages will find ideas here that they can exploit in similar embeddings.

References

1. Abrial, J., Hoare, A.: The B-Book: Assigning Programs to Meanings. Cambridge University Press (2005)
2. How to think about an Alloy model: 3 levels, http://alloy.mit.edu/alloy/tutorials/online/sidenote-levels-of-understanding.html
3. aRby—An Embedding of Alloy in Ruby, https://github.com/sdg-mit/arby
4. Online collection of aRby examples, https://github.com/sdg-mit/arby/tree/master/lib/arby_models
5. Galeotti, J.P., Rosner, N., López Pombo, C.G., Frias, M.F.: Analysis of invariants for efficient bounded verification. In: ISSTA. ACM (2010)
6. Hawkins, P., Aiken, A., Fisher, K., Rinard, M., Sagiv, M.: Data representation synthesis. ACM SIGPLAN Notices 46 (2011)
7. Jackson, D.: Micromodels of software: Lightweight modelling and analysis with alloy (2002)
8. Jackson, D.: Software Abstractions: Logic, language, and analysis. MIT Press (2006)
9. Köksal, A.S., Kuncak, V., Suter, P.: Constraints as control. ACM SIGPLAN Notices (2012)
10. Milicevic, A., Rayside, D., Yessenov, K., Jackson, D.: Unifying execution of imperative and declarative code. In: ICSE (2011)
11. Montaghami, V., Rayside, D.: Extending alloy with partial instances. In: Derrick, J., Fitzgerald, J., Gnesi, S., Khurshid, S., Leuschel, M., Reeves, S., Riccobene, E. (eds.) ABZ 2012. LNCS, vol. 7316, pp. 122–135. Springer, Heidelberg (2012)
12. Near, J.P., Jackson, D.: Rubicon: bounded verification of web applications. In: FSE. ACM (2012)
13. Nelson, T., Saghafi, S., Dougherty, D.J., Fisler, K., Krishnamurthi, S.: Aluminum: principled scenario exploration through minimality. In: ICSE. IEEE Press (2013)
14. Ruby Java Bridge, http://rjb.rubyforge.org/
15. Samimi, H., Aung, E.D., Millstein, T.: Falling Back on Executable Specifications. In: D'Hondt, T. (ed.) ECOOP 2010. LNCS, vol. 6183, pp. 552–576. Springer, Heidelberg (2010)
16. Siskind, J.M., McAllester, D.A.: Screamer: A portable efficient implementation of nondeterministic common lisp. IRCS Technical Reports Series (1993)
17. Spivey, J.: Understanding Z: a specification language and its formal semantics. Cambridge tracts in theoretical computer science. Cambridge University Press (1988)
18. Torlak, E.: A Constraint Solver for Software Engineering: Finding Models and Cores of Large Relational Specifications. PhD thesis, MIT (2008)
19. Torlak, E., Bodik, R.: Growing solver-aided languages with rosette. In: Proceedings of the 2013 Onward! ACM (2013)
20. Yang, J., Yessenov, K., Solar-Lezama, A.: A language for automatically enforcing privacy policies. ACM SIGPLAN Notices (2012)

MAZE: An Extension of Object-Z for Multi-Agent Systems

Graeme Smith and Qin Li

School of Information Technology and Electrical Engineering
The University of Queensland, Australia

Abstract. The formal development of multi-agent systems (MAS) may involve consideration of system functionality at three distinct levels of abstraction. The *macro* level focusses on the system's overall, global behaviour, independently of how the agents of the system operate and interact. The *meso* level focusses on agent interactions, and the *micro* level on the operation of individual agents. While Object-Z with its high-level support for component-based specifications is well suited to modelling MAS at the macro and meso levels, it can become cumbersome at the micro level where the low-level mechanisms required for dealing with asynchronous communication between agents and timing constraints need to be explicitly defined. In this paper we introduce MAZE, an extension of Object-Z supporting (i) action refinement to facilitate the development process from the macro to micro level, and (ii) a number of syntactic conventions to facilitate micro-level specification. The syntactic conventions are shorthands for existing Object-Z notation and so require no redefinition of Object-Z's semantics.

1 Introduction

A *multi-agent system (MAS)* is a system comprising a number of interacting, autonomous agents. By "autonomous" we mean that the agents can initiate actions without external control. Our notion of an agent includes not only "intelligent" agents, such as those that might be described in terms of their "beliefs" and "desires" [16], but any components that exhibit autonomous behaviour. Components which follow simple protocols such as the sensors in a self-organising sensor network [6], for example, would be regarded as agents, as would the nodes of an *ad-hoc* mobile network which continually adapt their routing patterns to the current network topology [8].

Zambonelli and Omicini [17] argue that the disciplined engineering of MAS should proceed at three distinct levels of abstraction.

1. At the *macro* level the engineer is concerned with the overall system functionality, ignoring the operation and interaction of its agents.
2. At the *meso* level the engineer considers potential agent interactions and interaction paradigms that will lead to the desired system functionality.

Y. Ait Ameur and K.-D. Schewe (Eds.): ABZ 2014, LNCS 8477, pp. 72–85, 2014.

3. At the *micro* level the engineer is concerned with the operation of individual agents, choosing an implementation that results in the required meso-level interactions.

Although Zambonelli and Omicini stop short of proposing a completely formal approach to development, their implied reductionist strategy is, in fact, well suited to an approach based on formal specification and refinement.

In this paper, we introduce an extension of Object-Z [11] called MAZE. Following earlier work on the formal development of MAS [12,13], the extension supports action refinement as a means of developing a specification from the macro level, where operations are coarse-grained, through to the micro level, where the granularity of operations is usually much finer. It also involves a number of syntactic conventions, aimed at facilitating the specification of MAS at the micro level.

We begin in Section 2 by illustrating the use of Object-Z in the macro-level specification of MAS. In Section 3 we motivate the use of action refinement in the formal development of MAS and provide action refinement simulation rules for Object-Z. In Sections 4, 5 and 6 we introduce a number of syntactic conventions for modelling MAS at the micro-level. These conventions allow the specifier to abstract from inter-agent communication mechanisms and associated timing constraints, and instead focus on the functionality of individual agents within a MAS. Together with the action refinement rules they form the extension of Object-Z we call MAZE. We conclude in Section 7.

2 Macro-level Specification

Object-Z [11] is an object-oriented extension of Z [14], in which the notion of a class is introduced to encapsulate a state schema, and associated initial state schema, with all the operation schemas which may change its variables. Classes have been shown useful for modelling the behaviour of agents in MAS [9,13]. As an example, consider the following specification of a simple agent which has an identifier and may become leader of a group of similar agents.

Agent

$id : Identifier$
$leader : \mathbb{B}$

INIT

$\neg\, leader$

BecomeLeader

$\Delta(leader)$

$\neg\, leader \wedge leader'$

The class has two state variables: *id* of a given type *Identifier* and *leader* of type Boolean. Initially, the agent is not a leader and the operation *BecomeLeader* allows it to become a leader. Operations in Object-Z are *guarded*. When the predicate of *BecomeLeader* cannot be satisfied, i.e., because *leader* is already true, then the operation cannot occur. This is in contrast to Z operations which can occur at any time but may have undefined behaviour [14]. The Δ-list in the declaration part of the operation indicates which variables the operation may change; all other variables are implicitly unchanged, i.e., $id' = id$ in *BecomeLeader*.

In this example, we might want our agents to belong to disjoint neighbourhoods and become a leader only when no other agent in their neighbourhood is a leader. Hence, we need to constrain when the *BecomeLeader* action can occur. In the interest of keeping the specification abstract (and hence easy to understand and reason about), we specify such inter-agent constraints (and environmental constraints on the agent in general) in another class describing the entire MAS as shown below. (*Neighbourhood* is a given type and *nh* maps each agent to its neighbourhood.)

$$
\begin{array}{|l}
\underline{\textit{System}} \\
\hline
\begin{array}{l}
agents : \mathbb{F}_1\ Agent \\
nh : Agent \nrightarrow Neighbourhood \\
\hline
\mathrm{dom}\ nh = agents \\
\end{array} \\
\hline
\begin{array}{|l}
\underline{\textit{INIT}} \\
\hline
\forall\, a : agents \bullet a.INIT \\
\end{array} \\
\hline
BecomeLeader \,\widehat{=} \\
\quad [\!]\ a : agents \mid \neg\ (\exists\, b : agents \bullet nh(b) = nh(a) \wedge b.leader) \bullet \\
\qquad a.BecomeLeader \\
\end{array}
$$

The dot notation familiar from object orientation is used to reference variables and the initial condition of class instances, and to apply operations to them. The initial state of *System* states that all agents are in their initial states, i.e., are not leaders. The operation *BecomeLeader* states that a single agent *a* becomes leader provided that none of its neighbours are already leaders. It uses Object-Z's nondeterminstic choice operator $[\!]$ to choose an appropriate agent *a*.

3   Action Refinement

Macro-level specifications such as that of Section 2 abstract from interactions between agents, focussing instead on the outcomes of these interactions. Adding the interactions as we develop the specification to the meso level, and ultimately the micro level, requires the addition of further actions modelling the sending and receiving of messages. Adding actions, however, is not supported by standard data refinement in Object-Z [4]. It is, however, supported by action refinement as defined for action systems [2].

For this reason, Smith and Winter [13] propose adding action refinement to Object-Z. Although Derrick and Boiten [4] define a notion of action refinement for Object-Z, called *non-atomic refinement*, unlike that of action systems it does not allow guards of operations to be strengthened. As shown by Smith and Winter [13], the strengthening of guards is required during refinement to add agent decision-making procedures that determine which of a number of enabled actions occurs (the operations which are not to be performed have their guards strengthened so that they cannot occur). We therefore base our approach on the simulation rules for action refinement in action systems by Back and von Wright [2]. Here we consider the forward simulation rules only, and adapt them for Object-Z. The backwards simulation rules could be similarly adapted.

An action system has a state comprising global, i.e., observable, and local variables, an initialisation condition, and a set of actions. The actions have guards which determine when they are enabled but, unlike Object-Z, the guard being enabled does not guarantee the action's definition can be satisfied. This is instead guaranteed by the action's *precondition*. If an action is enabled in a given state but its precondition is not satisfied, it is said to *abort*. A state in which an action can abort is called an *aborting state*. Action systems behave by repeatedly executing enabled actions until none are enabled, or an enabled action *aborts*. A state in which no actions are enabled is called a *terminating state*.

In order to prove refinement using the simulation rules, the specifier needs to select some of the actions to be *stuttering actions*. A stuttering action must leave the global variables unchanged. All other actions, whether or not they change the global variables, are called *change actions*. For an abstract action system A and a concrete action system C whose states are related by a retrieve relation R, the forward simulation rules are then:

Initialisation: Any initialisation followed by stuttering actions in C simulates (via R) initialisation followed by stuttering actions in A.

Forward simulation: Any change action in C followed by stuttering actions simulates some change action in A followed by stuttering actions, or begins from a state related (by R) to an aborting state of A.

Abort: Any aborting state in C is related only to aborting states in A.

Termination: Any terminating state in C is related only to terminating or aborting states in A.

Infinite stuttering: Any state in C from which infinite stuttering is possible, i.e., an infinite sequence of stuttering actions can occur, is related only to states in A which are either aborting or from which infinite stuttering is possible.

For Object-Z, there are no aborting states (since the guard of an operation guarantees that the operation's definition can be satisfied). Although it has been suggested that Object-Z be extended to include both guards and preconditions [7], in this paper we use standard Object-Z. Hence, the above rules can be defined (in the absence of aborting states) as follows.

For a given Object-Z specification, the change and stuttering actions are particular *occurrences* of the specification's operations, i.e., particular pre-state/

post-state pairs that satisfy an operation's predicate. A single operation can have occurrences which are sometimes change actions, and sometimes stuttering actions, depending on whether or not a certain predicate holds in the post-state of the operation. An example of this will be given when we return to our case study below.

The choice of change and stuttering actions is made by the specifier based on which states they regard as being observable. Note that while we do not require explicit global and local state variables, states can always be re-expressed in terms of such variables to represent any choice of observable states.

Let A be an Object-Z class with state schema $AState$, initial state schema $AInit$, and operation occurrences partitioned into change actions $AChange_0, \ldots,$ $AChange_n$ and stuttering actions $AStutt_0, \ldots, AStutt_m$ for some $n, m : \mathbb{N}$. Similarly, let C be an Object-Z class with state schema $CState$, initial state schema $CInit$, and operation occurrences partitioned into change actions $CChange_0, \ldots,$ $CChange_l$ and stuttering actions $CStutt_0, \ldots, CStutt_k$ for some $l, k : \mathbb{N}$.

Definition 1. *(Action refinement) Let $AStutt = (AStutt_0 \lor \ldots \lor AStutt_m)$ and $CStutt = (CStutt_0 \lor \ldots \lor CStutt_k)$. A is refined by C when there exists a retrieve relation R (modelled by a Z schema as in [4]) which relates the states of C to those of A such that the following hold. ($A \mathbin{\S} B$ is the sequential composition of operations A and B, and A^n is the iteration of operation A n times, e.g., $A^3 = A \mathbin{\S} A \mathbin{\S} A$. The Z notation $\mathrm{pre}\, A$ returns the guard of operation A.)*

Initialisation: *Any initialisation followed by stuttering actions in C simulates initialisation followed by stuttering actions in A. (Schemas are used below as declarations and predicates as in Z [14].)*

$$\forall\, CState;\ CState';\ i : \mathbb{N} \bullet CInit \land CStutt^i \Rightarrow$$
$$(\exists\, AState;\ AState';\ j : \mathbb{N} \bullet AInit \land AStutt^j \land R')$$

Forward Simulation: *Any change action in C followed by stuttering actions simulates some change action in A followed by stuttering actions.*

$$\forall\, AState;\ CState;\ CState';\ c : 0 \mathinner{..} l;\ i : \mathbb{N} \bullet$$
$$R \land CChange_c \mathbin{\S} CStutt^i \Rightarrow$$
$$(\exists\, AState';\ a : 0 \mathinner{..} n;\ j : \mathbb{N} \bullet (AChange_a \mathbin{\S} AStutt^j) \land R')$$

Termination: *Any terminating state in C is related only to terminating states in A.*

$$\forall\, AState;\ CState \bullet$$
$$R \land \neg\ \mathrm{pre}(CChange_0 \lor \ldots \lor CChange_l \lor CStutt) \Rightarrow$$
$$\neg\ \mathrm{pre}(AChange_0 \lor \ldots \lor AChange_n \lor AStutt)$$

Infinite Stuttering: *Any state in C from which infinite stuttering is possible is related only to states in A from which infinite stuttering is possible.*

$$\forall\, AState;\ CState \bullet$$
$$R \land (\forall\, i : \mathbb{N} \bullet \exists\, CState' \bullet CStutt^i \land (\mathrm{pre}\, CStutt)') \Rightarrow$$
$$(\forall\, j : \mathbb{N} \bullet \exists\, AState' \bullet AStutt^j \land (\mathrm{pre}\, AStutt)') \qquad \square$$

As an example of the application of these rules, consider refining the MAS of Section 2 to include an inter-agent strategy for determining the leader. One approach is to assign the node with the minimum identifier amongst all its neighbours as the leader. As pointed out by Gerla et al. [8], this approach can be inefficient, and also requires a *quasi-stationary* assumption that agents are not mobile during leader election and maintenance. They suggest an alternative approach called the "first declaration wins" rule: essentially, the first agent in a neighbourhood to declare itself the leader becomes the leader. Then some means of dealing with contention, i.e., when two agents simultaneously declare themselves leader, is required. That is solved by standard techniques such as the node with the minimum (or maximum) identifier backing off, or both nodes backing off for random times. Such techniques become relevant at the micro-level and can be safely ignored at the meso level.

In the meso-level specification of our case study, an agent has an identifier and a status which indicates whether it is a leader, a follower, or is undecided. Initially, all agents are undecided. An operation *DeclareLeader* models an agent declaring itself leader. It abstractly models an agent sending a message to all its neighbours stating that it is now leader. A second operation *BecomeFollower* abstractly models an agent receiving such a message and hence becoming a follower. This can occur even if the agent is a leader, modelling the agent backing down in the case of contention between prospective leaders.

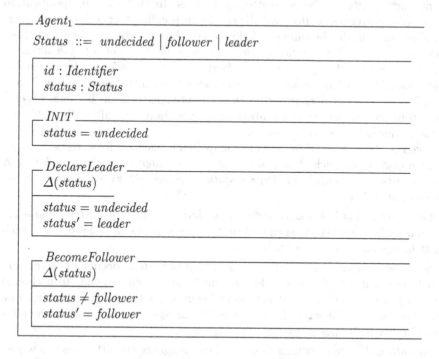

The interaction between agents is again modelled in a class defining the entire MAS. In this case, we specify an operation where a single agent declares itself leader. A second operation models a single agent becoming a follower when one of its neighbours *is* a leader.

$$\boxed{\begin{array}{l} \underline{\ System_1\ } \\[4pt] \quad \boxed{\begin{array}{l} agents : \mathbb{F}_1\ Agent_1 \\ nh : Agent_1 \nrightarrow Neighbourhood \\ \hline \mathrm{dom}\ nh = agents \end{array}} \\[6pt] \quad \underline{\ INIT\ } \\ \quad \forall\, a : agents \bullet a.INIT \\[6pt] \quad DeclareLeader \mathrel{\widehat{=}}\ [\!]\ a : agents \bullet a.DeclareLeader \\ \quad BecomeFollower \mathrel{\widehat{=}} \\ \qquad [\!]\ a : agents \mid (\exists\, b : agents \setminus \{a\} \bullet nh(b) = nh(a) \wedge b.status = leader) \bullet \\ \qquad\quad a.BecomeFollower \end{array}}$$

To show that $System_1$ refines $System$ of Section 2, we first need to partition the operation occurrences into change and stuttering actions. In the abstract specification $System$, all occurrences of the operation $BecomeLeader$ are change actions; there are no stuttering actions. In the concrete specification $System_1$, occurrences of $BecomeFollower$ corresponding to the last neighbouring agent of a leader becoming a follower are change actions. These are those occurrences where the predicate $\exists\, b : agents \bullet nh(b) = nh(a) \wedge b.status = leader \wedge (\forall\, c : agents \setminus \{b\} \mid nh(c) = nh(a) \bullet c.status = follower)$ is true in the post-state. All other operation occurrences are stuttering actions. In other words, the effect of a leader declaration in the concrete specification becomes observable, from the perspective of the abstract specification, only after all neighbours of the leader have become followers.

Given a retrieve relation R which maps each agent a from $System$ to an agent b of $System_1$ such that $a.id = b.id$ and $a.leader \Leftrightarrow (b.status = leader \wedge (\forall\, c : agents \setminus \{b\} \mid nh(c) = nh(b) \bullet c.status = follower))$ we can prove the refinement as follows.

Initialisation. This holds since before any change actions in $System_1$ there are no agents which are leaders such that all their neighbours are followers, and each agent in $System$ is not a leader initially.

Forward Simulation. Whenever a change action of $System_1$ occurs all neighbours of a leader will have become followers which, via R, corresponds to a change action of $System$, i.e., an abstract agent becoming a leader. Furthermore, any stuttering actions following the change action of $System_1$ do not affect the relationship, via R, between $System_1$ and $System$.

Termination. $System_1$ terminates only when all agents are either leaders whose neighbours are not leaders, or followers. In the latter case, the agent will have a neighbour who is a leader (something which can be proved as an invariant).

This corresponds, via R, to all agents in *System* being either a leader or having a neighbour which is a leader. In which case, *System* also terminates.

Infinite stuttering. Each agent can perform *DeclareLeader* and *BecomeFollower* at most once: after *DeclareLeader* the agent never has $status = undecided$ again, and after *BecomeFollower* it never has $status \neq follower$ again. Therefore, with a finite number of agents infinite stuttering is not possible in $System_1$.

4   Micro-level Agent Specification

At the micro level of development we specify the behaviour of agents *locally* without the explicit use of global constraints. The goal is to capture the full behaviour required of the agent in its implementation. While this is possible in standard Object-Z, it can lead to specifications which are awkward to read due to the details of the particular system under development being intermingled with those of the underlying communication mechanisms. In MAZE, we separate these details by capturing the latter implicitly via a number of syntactic conventions.

These conventions are analogous to those used in Z for modelling sequential systems [14]. That is, they are merely a shorthand for what could otherwise be expressed using more basic syntax. For this reason, they do not require an extension to the existing semantics of Object-Z.

The conventions are based on the assumption that all agents interact via asynchronous message-passing. The justification for this is as follows. As there is usually no centralised control in MAS, messages may be sent to an agent at any time, including when it is busy with another message. Hence, messages need to be buffered (since allowing messages to be lost would greatly complicate the simple micro-level protocols we would like to develop). In the implementation of a MAS, the buffering may be part of the communication medium, e.g., when agents are distributed over the Internet, or part of the agent, e.g., when communication is wireless and effectively synchronous between agents.

For specifying agents, a type **message** and two message-related predicates for use in agent specifications are introduced: **send** for modelling messages being sent to the buffer (**send**(m, a) models a message m being sent to agent a, and **send**(m) models a message being broadcast to all connected agents) and **receive** for modelling messages being received from the buffer (**receive**(m, a) models the receipt of a message m from agent a). Each operation in an agent specification may have a single **receive** predicate (as part of its guard) and a single **send** predicate (as part of its postcondition). The type **message** is application-specific and is defined within the agent class.

A predicate **progress**(s, r) is also introduced for use in agent specifications. It is true when the system has progressed to a point where all messages in set s that have been sent by the agent and all messages in set r that have been sent to the agent have been received. This mechanism provides a way of abstracting from the use of timers and timing constraints required in an implementation of the MAS [3,10]. An example is given below.

The formal definitions of **send**, **receive** and **progress** are given in terms of an implicit variable **buffer** in the specification of the MAS. These definitions are provided in Section 5. To see how the predicates are used, we return to our case study.

Assume the mechanism we adopt for handling contention between leader declarations is that the agent with the smallest identifier backs down. Given an infix relation $\_ < \_ : Identifier \leftrightarrow Identifier$ such that $id < rid$ means id is smaller than rid, the required agent class could be specified as follows.

Agent₂ — $Agent_2$

$Status ::= undecided \mid follower \mid declared \mid leader$

message $::= declare \langle\!\langle Identifier \rangle\!\rangle$

$id : Identifier$
$status : Status$

INIT

$status = undecided$

SendDeclaration

$\Delta(status)$

$\textbf{send}(declare(id))$
$status = undecided$
$status' = declared$

BecomeLeader

$\Delta(status)$

$\textbf{progress}(\{declare(id)\}, \{rid : Identifier \setminus \{id\} \bullet declare(rid)\})$
$status = declared$
$status' = leader$

ReceiveDeclaration

$\Delta(status)$
$a : Agent_2$
$rid : Identifier$

$\textbf{receive}(declare(rid), a)$
if $status = undecided \lor status = declared \land id < rid$
then $status' = follower$
else $status' = status$

The type **message** is defined as a Z free type. It comprises one kind of message, *declare*, which carries with it the identifier of the agent making the declaration. The operation *SendDeclaration* broadcasts this message to all neighbouring agents, and sets *status* to *declared*. This status indicates that the agent has declared itself leader, but is waiting for any similar declarations from its neighbours before becoming leader.

The time it needs to wait depends on factors such as any delay caused by the communication medium, and the variance of local clock speeds, and hence response times, between agents [3,10]. These will differ for different implementations. Rather than specify such low-level details, we use a **progress** predicate in operation *BecomeLeader* which ensures that, before the *status* becomes *leader*, the sent *declare* message has been received by all neighbours, and that any *declare* messages sent from neighbours have been received. At that point there can be no further contention.

The final operation *ReceiveDeclaration* models an agent receiving a declaration and either becoming a follower, or ignoring the message. The former happens whenever the agent has status *undecided*, or has status *declared* but has a smaller *id* than that of the agent sending the declaration (whose *id* is included in the received message).

5   Micro-level System Specification

A further syntactic convention is introduced to specify a collection of interacting agents. $\mathbb{T}\,A$ defines a topology of agents of type A in terms of a finite function whose domain is the set of all agents in the topology and which maps each such agent to the agents to which it can send messages, i.e., $\mathbb{T}\,A = (A \nrightarrow \mathbb{F}\,A)$. Note that there are no constraints on the function, allowing unidirectional sending of messages, and agents which are isolated and unable to send or receive messages. Typically, constraints will be added in the specification (e.g., in the predicate of the state schema) to restrict the function as required.

In our case study, for example, we model a topology where agents belong to a unique neighbourhood, and there is bidirectional communication between neighbouring agents. We also require that agents have unique identifiers (to support our chosen contention mechanism).

$System_2$

$agents : \mathbb{T}\,Agent_2$

$\forall\, a : \mathrm{dom}\; agents \bullet \forall\, b : agents(a) \bullet agents(b) = \{a\} \cup agents(a) \setminus \{b\}$
$\forall\, a, b : \mathrm{dom}\; agents \bullet a \neq b \Rightarrow a.id \neq b.id$

Note that the specification $System_2$ does not include explicit initialisation of the agents, nor any operations. This is because as well as introducing a topology of agents of type A, the notation $\mathbb{T}\,A$ implicitly introduces their initialisation (according to the initial state schema of A), and system operations allowing any agent in the topology to perform any of its enabled operations. These operations send messages to and receive messages from an implicit global buffer (similar to that in the Actor specification paradigm of Agha [1]). The buffer is unordered allowing messages to be received by the agent in a different order to which they are sent. This may model the use of different routes through a communication medium such as the Internet, or the ability of an agent to prioritise messages in its internal buffer.

Definition 2. *(Topology) The following two specifications are semantically equivalent where $Op_1, \ldots, Op_n$ are the operations of agent specifiction A.*

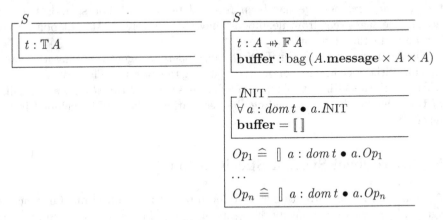

where the implicit variable **buffer** *models the unordered, global buffer as a bag. Each element of the buffer is a tuple* (m, a, b) *where m is a message, a is the agent which sent m, and b is the agent to which the message has been sent.* $\square$

The semantics of the **send** and **receive** predicates used in agent specification is given in terms of the implicit variable **buffer**.

Definition 3. *(Message passing) When we apply the agent operation Op (using the notation a.Op) in a MAS specification with a topology of agents t, the* **receive** *and* **send** *predicates introduce an additional guard G and additional postcondition* **buffer$'$** $=$ (**buffer** $\uplus$ R) $\uplus$ S *to the system operation where*

- G *is* $(m, b, a) \sqsubseteq$ **buffer** *when Op includes* **receive**(m, b)*, and true otherwise.*
- R *is* $[\![(m, b, a)]\!]$ *when Op includes* **receive**(m, b)*, and $\varnothing$ otherwise.*
- S *is* $[\![(m, a, b)]\!]$ *when Op includes* **send**(m, b) *and* $b \in t(a)$*, and, given* $t(a) = \{b_1, \ldots, b_n\}$*,* $[\![(m, a, b_1), \ldots, (m, a, b_n)]\!]$ *when Op includes* **send**(m)*, and $\varnothing$ otherwise.* $\square$

The semantics of **progress** predicates is similarly given in terms of the implicit variable **buffer**.

Definition 4. *(Progress) When we apply the agent operation Op (using a.Op), each predicate of the form* **progress**(s, r) *where s and r are of type* $\mathbb{P}$ **message** *introduces an additional guard:*

$$\forall b : A \bullet (\forall m : s \bullet \neg (m, a, b) \sqsubseteq \textbf{buffer}) \wedge (\forall m : r \bullet \neg (m, b, a) \sqsubseteq \textbf{buffer}) \qquad \square$$

Given these definitions we can show that $System_2$ is an action refinement of $System_1$. We let *all* occurrences of *BecomeLeader* in $System_2$ be change actions. All other actions are stuttering actions.

Let the retrieve relation R map each agent a from $System_1$ to an agent b of $System_2$ such that $a.id = b.id$, and $b.status \in \{undecided, leader\} \Rightarrow a.status = b.status$, and $b.status = declared \Rightarrow a.status = leader$, and $b.status = follower \Rightarrow a.status \in \{leader, follower\}$, and $b.status = follower \Rightarrow a.status = follower$ whenever $\exists c : agents(b) \bullet c.status = leader$. Also, if $nh(a_1) = nh(b_1)$ for any agents a_1 and b_1 in $System_1$, then $b_2 \in agents(a_2)$ for the corresponding agents a_2 and b_2 in $System_2$.

Initialisation. Before a change action in $System_2$, *SendDeclaration* simulates *DeclareLeader* and *ReceiveDeclaration* simulates *BecomeFollower* or *skip* (e.g., when the receiver is a prospective leader with a larger *id*). Hence, the state of $System_2$ is related, via R, to a state of $System_1$ before a change action.

Forward Simulation. An agent a becomes a leader in $System_2$ only when all its neighbours have received its declaration, and all declarations sent by these neighbours have been received by a. Together with the fact that agents have unique identifiers, this ensures that there is never more than one leader in any neighbourhood. Hence, whenever a change action of $System_2$ occurs all neighbours of a leader will have become followers which, via R, corresponds to a change action of $System_1$. As above, subsequent stuttering actions do not affect the relationship, via R, between $System_2$ and $System_1$.

Termination. $System_2$ terminates only when all agents are either leaders whose neighbours will not be leaders (as explained above), or followers. This corresponds, via R, to the states where $System_1$ terminates.

Infinite stuttering. Each agent can perform operations *SendDeclaration* and *BecomeLeader* at most once: since they require the agent to have a particular status to which the agent never returns. Also, since *ReceiveDeclaration* only occurs when a message has been sent, via *SendDeclaration*, with a finite number of agents infinite stuttering in not possible in $System_2$.

6   System Operations at the Micro Level

As well as the agent operations implicitly included in a MAS specification by the notation $\mathbb{T}\,A$, a system class may include explicitly defined operations that change the agents' environment and topology. Such operations may occur either independently (when the environment itself can change) or in response to an agent operation. The latter case requires additional syntax in the form of a tag $< a : \mathrm{dom}\,t \bullet a.Op >$ which appears after the system operation's name indicating that the operation only occurs in conjunction with $a.Op$. (Note that the scope of the variable a in such cases is the entire system operation.)

For example, to include mobility into our case study, we could define a *Move* operation in $Agent_2$ which causes the topology in $System_2$ to change so that an agent a performing a *Move* joins a new neighbourhood.

$\boxed{\begin{array}{l} System_2 \\ \hline agents : \mathbb{T}\ Agent_2 \\ \hline \forall\,a : \mathrm{dom}\ agents \bullet \forall\,b : agents(a) \bullet agents(b) = \{a\} \cup agents(a) \setminus \{b\} \\ \forall\,a, b : \mathrm{dom}\ agents \bullet a \neq b \Rightarrow a.id \neq b.id \\ \hline \boxed{\begin{array}{l} Move < a : \mathrm{dom}\ agents \bullet a.Move > \\ \Delta(agents) \\ \hline \forall\,b : agents(a) \bullet agents'(b) = agents(b) \setminus \{a\} \\ \exists\,s : \mathbb{F}(\mathrm{dom}\ agents \setminus \{a\}) \bullet agents'(a) = s\ \wedge \\ \quad (\forall\,b : s \bullet agents(b) = s \setminus \{b\} \wedge agents'(b) = agents(b) \cup \{a\}) \wedge \\ \quad (\forall\,b : \mathrm{dom}\ agents \setminus (s \cup agents(a) \cup \{a\}) \bullet agents'(b) = agents(b)) \end{array}} \end{array}}$

As with the other syntactic extensions in MAZE, the semantics of system operations is given in terms of equivalent Object-Z syntax.

Definition 5. *(System operations) The following two specifications are semantically equivalent where* $Op_1, \dots, Op_n$ *are the operations of agent specification A.*

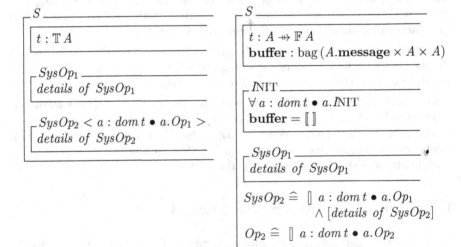

7   Conclusion

This paper has presented MAZE: an extension of Object-Z for the specification and development of multi-agent systems. MAZE supports the development of a multi-agent system from a macro-level specification capturing global functionality to a micro-level specification capturing the local functionality of individual agents. It includes a notion of action refinement as well as a number of syntactic constructs to facilitate micro-level specification.

For presentation purposes, the case study in this paper was necessarily simple. However, MAZE has also been successfully applied by the authors to a complex modular-robotic self-assembly algorithm based on the work of Støy [15].

Future work will look at providing a proof strategy for refinement in MAZE aimed at making formal proof more manageable. One promising direction is the work of Derrick et al. [5] which provides a set of rules for Z action refinement that allow proofs which consider one operation at a time, and that are fully encoded, and proved sound, within the theorem prover KIV.

Acknowledgements This work was supported by Australian Research Council (ARC) Discovery Grant DP110101211.

References

1. Agha, G.: Actors: a model of concurrent computation in distributed systems. MIT Press (1986)
2. Back, R.J.R., von Wright, J.: Trace refinement of action systems. In: Jonsson, B., Parrow, J. (eds.) CONCUR 1994. LNCS, vol. 836, pp. 367–384. Springer, Heidelberg (1994)
3. Chou, C.T., Cidon, I., Gopal, I.S., Zaks, S.: Synchronizing asynchronous bounded delay networks. IEEE Trans. Communications 38(2), 144–147 (1990)
4. Derrick, J., Boiten, E.: Refinement in Z and Object-Z, Foundations and Advanced Applications, 2nd edn. Springer (2014)
5. Derrick, J., Schellhorn, G., Wehrheim, H.: Mechanically verified proof obligations for linearizability. ACM Trans. Program. Lang. Syst. 33(1), 4 (2011)
6. Dressler, F.: Self-organization in sensor and actor networks. Wiley (2007)
7. Fischer, C.: CSP-OZ - a combination of CSP and Object-Z. In: Bowman, H., Derrick, J. (eds.) FMOODS 1997, pp. 423–438. Chapman & Hall (1997)
8. Gerla, M., Kwon, T.J., Pei, G.: On demand routing in large ad hoc wireless networks with passive clustering. In: WCNC 2000, vol. 1, pp. 100–105 (2000)
9. Gruer, P., Hilaire, V., Koukam, A., Cetnarowicz, K.: A formal framework for multi-agent systems analysis and design. Expert System Applications 23(4), 349–355 (2002)
10. Li, Q., Smith, G.: Using bounded fairness to specify and verify ordered asynchronous multi-agent systems. In: ICECCS 2013, pp. 111–120. IEEE Computer Society Press (2013)
11. Smith, G.: The Object-Z Specification Language. Kluwer Academic Publishers (2000)
12. Smith, G., Sanders, J.W.: Formal development of self-organising systems. In: González Nieto, J., Reif, W., Wang, G., Indulska, J. (eds.) ATC 2009. LNCS, vol. 5586, pp. 90–104. Springer, Heidelberg (2009)
13. Smith, G., Winter, K.: Incremental development of multi-agent systems in Object-Z. In: SEW-35. IEEE Computer Society Press (2012)
14. Spivey, J.M.: The Z Notation: a reference manual, 2nd edn. Prentice-Hall (1992)
15. Støy, K.: Using cellular automata and gradients to control self-reconfiguration. Robotics and Autonomous Systems 54, 135–141 (2006)
16. Wooldridge, M.: An Introduction to MultiAgent Systems, 2nd edn. Wiley (2009)
17. Zambonelli, F., Omicini, A.: Challenges and research directions in agent-oriented software engineering. Autonomous Agents and Multi-Agent Systems 9(3), 253–283 (2004)

Quasi-Lexicographic Convergence

Stefan Hallerstede

Department of Engineering
Aarhus University
Denmark
sha@eng.au.dk

Abstract. Anticipation proof obligations for stated variants need to be proved in Event-B even if the variant has no variables in common with anticipated event. This often leads to models that are complicated by additional auxiliary variables and variants that need to take into account these variables. Because of such "encodings" of control flow information in the variants the corresponding proof obligations can usually not be discharged automatically.

We present a new proof obligation for anticipated events that does not have this defect and prove it correct. The proof is fairly intricate due to the nondeterminism of the simulations that link refinements. An informal soundness argument suggests using a lexicographic product in the soundness proof. However, it turns out that a weaker order is required which we call quasi-lexicographic product.

1 Introduction

Event-B provides some flexibility in termination proofs by means of the concept of anticipated events [3]. Anticipated events make it easier to formulate variants for complex models. Ample examples of their use can be found in [1].

The motivation for the work presented in this paper is best illustrated by way of an example. Consider the two fragments of some Event-B machine shown in Fig. 1. To simplify the presentation we have already included an abstract program counter P in the model. Assume

$$P = 3 \Rightarrow w \in 0 .. 4$$

is an invariant of the machine. Concerning the right-hand side only we can prove convergence of event **three** using the variant $4 - w$. The Rodin tool [2] will do this automatically. However, if we also take the left-hand side into account we have to prove anticipation of event **one**

$$(P = 3 \Rightarrow w \in 0 .. 4) \land P = 1 \land y < 2 \Rightarrow 4 - w \in \mathbb{N} \land 4 - w \leq 4 - w . \quad (1)$$

Now, we would fail to prove the first part $4 - w \in \mathbb{N}$ of the conclusion. A possible work-around would be to make the variant "global" using the set

$$(\{P\} \cap \{3\}) \times (0 .. 4 - w) .$$

Y. Ait Ameur and K.-D. Schewe (Eds.): ABZ 2014, LNCS 8477, pp. 86–100, 2014.

```
anticipated event one
when
    P = 1 ∧ y < 2
then
    y := y + 1
end
event two
when
    P = 1 ∧ y = 2
then
    x := y
end
```

```
convergent event three
when
    P = 3 ∧ w < 4
then
    w := w + 1
end
event four
when
    P = 3 ∧ w = 4
then
    v := w
end
```

Fig. 1. Two non-interfering components

This solution is not satisfactory because it complicates the variant and the proof obligations. The proof involves set theory, finite sets and arithmetic and is not done automatically by Rodin tool. We consider tweaking the prover to deal with such instances a bad choice because it would not solve the problem in general and fail occasionally. As our first contribution we show instead that we can drop proof obligations such as (1) for anticipated events entirely. The second contribution concerns the nature of the invariant arising from the use of anticipated events. The soundness proof for the proof obligations requires a generalised form of lexicographic order. This is due to the nondeterminism inherent in the gluing invariant that relates abstract variables to concrete variables. Only special cases such as functional gluing invariants yield lexicographic products. The general case, however, does not.

The informal soundness argument in [3] depends on the abstract variables mentioned in a variant being kept in refinements. Hence, for those variables the refinements are functional; and the claim that termination can be demonstrated by means of a lexicographic product is correct under the given constraint. The only other (semi-) formal soundness proof we are aware of is presented in [8]. However, the proof glosses over a vital fact assuming an equality of abstract sets when only set inclusion is known (see Rem. 7). This way it also achieves to prove that anticipation yields a lexicographic variant.

Overview. We remind the reader of the important properties of well-founded relations in Section 2 and give a short introduction to the used concepts of Event-B in Section 3. The presentation of Event-B is purely set-theoretical and does not discuss Event-B syntax that is used in some examples. Syntax and set-theoretical semantics should be easy to relate though. Details can be found in [1,7]. Section 4 discusses a generalised form of anticipation and convergence based on the concept of quasi-stability. The idea behind quasi-stable relations is to replace the identity relation used in lexicographic products by a more general relation that must not "increase" the first component of the product. Section 5 presents the concept of quasi-lexicographic product that uses such a relation instead of the identity used in the soundness proof of Section 6 In Section 7 we suggest an improvement for the Rodin tool. Section 8 contains the conclusion.

2   Well-Founded Relations

We repeat the main facts about well-founded relations. Most interesting for us is their relationship to transitive closures.

Definition 1. *A predecessor relation m is called* well-founded *if all non-empty subsets z have minima with respect to m,*

$$\forall z \cdot z \neq \varnothing \Rightarrow \exists x \cdot x \in z \wedge \forall y \cdot y \in z \Rightarrow x \mapsto y \notin m \ . \tag{2}$$

Well-foundedness of m is denoted by $\mathbf{wf}\langle m \rangle$.

This is expressed more succinctly using set-theoretic notation (e.g. [1]),

$$\forall z \cdot z \subseteq m^{-1}[z] \Rightarrow z = \varnothing \ . \tag{3}$$

Whereas property (2) is easier to understand, the equivalent set-theoretic statement (3) is easier to apply in proofs.

Later we need well-founded relations that are also transitive. The easiest way to ensure transitivity is to use the transitive closure m^+ of a relation m.

Definition 2. *The* transitive closure m^+ *of a relation m is the smallest relation x satisfying the property $m \cup (m \ ; x) \subseteq x$.*

Clearly, the transitive closure of a relation is a transitive relation.

Lemma 1. $m^+ \ ; m^+ \subseteq m^+$

The proof obligations for anticipation and convergence only require the employed order m to be well-founded. Fortunately, transitive closures of well-founded relations are well-founded. This fact is well-known, e.g. [6].

Lemma 2. $\mathbf{wf}\langle m \rangle \ \Leftrightarrow \ \mathbf{wf}\langle m^+ \rangle$.

Well-foundedness of relations of the shape $c \lhd m$ only concerns subsets of c. This property is sometimes useful in proofs.

Lemma 3. $\mathbf{wf}\langle c \lhd m \rangle \ \Leftrightarrow \ \forall z \cdot z \subseteq c \wedge z \subseteq m^{-1}[z] \Rightarrow z = \varnothing.$

Remark 1. Well-foundedness of $c \lhd m$ implies $c \lhd m$ is irreflexive. If $c \lhd m$ is well-founded and transitive, then $c \lhd m$ is a strict partial order. It is common to require stronger properties of $c \lhd m$ or m like strict partial orders for the loop proof rule in [4]. We aim to keep the number of proof obligations for candidates for m low, hence, we only require well-foundedness of $c \lhd m$, following the approach of [5].

3   Models, Consistency and Refinement

Event-B models are composed of *machines* that are related by *refinement*. A machine consists of a collection of *events* that describe the behaviour of the machine. An event is a relation of the shape $e = g \lhd s$ where g is a set called the *guard* of the event and s a relation called the *action* of the event. A dedicated event with the guard $g = {\sim}\varnothing$ is used for the initialisation of a machine.[1]

A machine has an *invariant* i. The invariant is a set that describes properties of the machine that are preserved by its events. This property is called *consistency* of the event, formally, $cns\langle i, e \rangle$.

Definition 3. $cns\langle i, e \rangle \Leftrightarrow e[i] \subseteq i$.

For event $e = g \lhd s$ we usually also require *feasibility*, that is, $i \cap g \subseteq \mathrm{dom}(s)$, or equivalently, $i \cap g \subseteq \mathrm{dom}(e)$. But we do not make use of it in this article.

A machine N *refines* another machine M if M can simulate the behaviour of N. In this relationship we call N the *concrete* machine and M the *abstract* machine. Machine N is related to machine M by means of a *gluing invariant*. The gluing invariant is a relation j that describes the simulation. Machine N refines M if each event e of M is refined by an event f of N, formally, $ref\langle i, j, e, f \rangle$.

Definition 4. $ref\langle i, j, e, f \rangle \Leftrightarrow (i \lhd j)\,;f \subseteq e\,;j$.

The concrete machine N may also introduce new events that are required to refine *skip*, the event that describes stuttering of M. The event *skip* is the identity relation id. Formally, the introduction of a new event corresponds to $ref\langle i, j, \mathrm{id}, f \rangle$. Note that the invariant of the concrete machine N is $j[i]$.

Lemma 4. $cns\langle i, e \rangle \;\wedge\; ref\langle i, j, e, f \rangle \;\Rightarrow\; cns\langle j[i], f \rangle$.

The relation $i \lhd j$ may also serve as the gluing invariant as implied by the following lemma.

Lemma 5. $cns\langle i, e \rangle \;\wedge\; ref\langle i, j, e, f \rangle \;\Rightarrow\; (i \lhd j)\,;f \subseteq e\,;(i \lhd j)$.

Remark 2. Our presentation of the set-theoretical model of Event-B follows [1] by and large. In [1, Chapter 14] Abrial uses the relation $\rho = (i \lhd j)^{-1}$ in place of i and j as we do in Def. 4. Nonetheless, the formalisations are equivalent as indicated by Lemma 5 and inverting the relation ρ.

A tuple $\langle v, c, m \rangle$ where v is a partial function and $c \lhd m$ is a well-founded relation is called a *variant* and v is called the *variant function*. If we say informally "the variant has changed" refer to differing values of v in consecutive states. Let $\langle v, c, m \rangle$ be a variant. We say that event e is *anticipated* if $ant\langle i, e, v, c, m \rangle$.

Definition 5. $ant\langle i, e, v, c, m \rangle \Leftrightarrow i \lhd e \subseteq v\,;(\mathrm{id} \cup c \lhd m)\,;v^{-1}$.

We say that e is *convergent* if $cvg\langle i, e, v, c, m \rangle$.

Definition 6. $cvg\langle i, e, v, c, m \rangle \Leftrightarrow i \lhd e \subseteq v\,;c \lhd m\,;v^{-1}$.

---

[1] The set complement ${\sim}s$ is defined by $x \in {\sim}s \Leftrightarrow x \notin s$.

Remark 3. The original Event-B proof obligation for anticipated events is

$$i \lhd e \subseteq v \, ; c \lhd (\mathrm{id} \cup m) \, ; v^{-1} \, .$$

The new formulation of the proof obligation does not require the proof of membership in c if the e leaves the variant unchanged. In principle one could further generalise the proof obligation to

$$i \lhd e \subseteq v \, ; \mathrm{id} \cup m \, ; v^{-1}$$

but this would be traded against stronger constraints on m. Our intention is to make finding candidates for m as easy as possible, for instance, allowing a cyclic graph restricted to an acyclic tree. Constraining m would often necessitate the introduction of an auxiliary variable (e.g., for recording the acyclic tree directly). Using the new proof obligation of Def. 5 we have two ways to influence how m is used in case the anticipated event is interfering with the convergent event. One is the choice of the set c for determining a subset of m. The other is the function v. If we map all states that do not need to be considered for the convergence proof to the same element z outside c, then to prove $v(x) \mapsto v(y) \in \mathrm{id}$ we can use $v(x) = z$ and $v(y) = z$ in such cases. However, we would still need to verify $v(x) = z$ and $v(y) = z$.

The challenge of proving soundness of the anticipation and convergence proof obligations is clearly related to dealing with the gluing invariant. Fig. 2 shows

invariant $v \in \mathbb{Z}$	invariant $w \subseteq \{x \mid x < v\} \wedge (v \in \mathbb{N}_1 \Rightarrow 0 \in w)$
variant v	anticipated event **conca**
	any x when
anticipated event **absa**	$w \neq \varnothing \wedge x < \max(w)$
begin	then
skip	$w := w \cup \{x\}$
end	end
convergent event **absc**	convergent event **concc**
any x when	when
$v \in \mathbb{N}_1 \wedge x \in \mathbb{N}_1$	$w \cap \mathbb{N} \neq \varnothing$
then	then
$v := v - x$	$w := w \setminus \{\max(w)\}$
end	end

Fig. 2. A nondeterministic refinement with a simple variant

an abstract and a concrete machine of a refinement. The variant for proving convergence of the abstract event **absc** is v, the only variable of the machine. In terms of our set-theoretical model the variant function is id. The set-theoretic gluing invariant is $\{v \mapsto w \mid w \subseteq \{x \mid x < v\} \wedge (v \in \mathbb{N}_1 \Rightarrow 0 \in w)\}$ establishing a many-to-many relationship between the abstract variable v and the concrete variable w. How to use variable w to express the variant in the concrete machine is not at all obvious. But it is necessary, in order to "forget" about the abstract machine and continue working solely with the concrete machine.

4   Convergence and Anticipation

The proof obligations **ant** and **cvg** stated in Section 3 are not suitable for the induction-based soundness proof of Section 6. We need a more general formulation based on the concept of quasi-stability.

Definition 7. *A relation r is called m-quasi-stable, $qs\langle r, m\rangle$, if it is reflexive, transitive and for all predecessors y of x in r, all predecessors of y in m are also predecessors of x in m, that is, the following three conditions are satisfied*

$$\text{id} \subseteq r \ , \tag{4}$$

$$r \, ; r \subseteq r \ , \tag{5}$$

$$\forall x, y \cdot x \mapsto y \in r \Rightarrow m[\{y\}] \subseteq m[\{x\}] \ . \tag{6}$$

Property (6) can be expressed more concisely using set-theoretic notation

$$(6) \ \Leftrightarrow \ \forall p \cdot (r \, ; m)[p] \subseteq m[p] \ .$$

Whereas the set-theoretic formulation of well-foundedness is favourable for use in proof, this does not hold for (6). Instantiating x and y is usually straightforward. Dealing with the set p above easily leads astray when instantiated during a proof. Set-theoretic formulations are not invariably "better".

An m-quasi-stable relation r is also $c \lhd m$-quasi-stable if the set c is invariant under the inverse of r.

Lemma 6. $qs\langle r, m\rangle \ \wedge \ r^{-1}[c] \subseteq c \ \Rightarrow \ qs\langle r, c \lhd m\rangle.$

Finally, for an m-quasi-stable relation r where m is a transitive relation, the transitive closure of $r \cup m$ is also m-quasi-stable.

Lemma 7. $qs\langle r, m\rangle \ \wedge \ m \, ; m \subseteq m \ \Rightarrow \ qs\langle (r \cup m)^+, m\rangle.$

This is all we need to know about quasi-stable relations for now. We are ready to introduce quasi-variants.

Definition 8. *Let i be a set. A tuple $\langle v, r, m\rangle$ is called an i-quasi-variant iff v is a partial function with $i \subseteq \text{dom}(v)$, m well-founded and r is m-quasi-stable. The i-quasi-variant $\langle v, r, m\rangle$ is called transitive if $m \, ; m \subseteq m$.*

Using Def. 8 we define generalisations **ANT** and **CVG** of **ant** and **cvg**. Let $V = \langle v, r, m\rangle$ be an i-quasi-variant in the following two definitions.

Definition 9. $\textbf{\textit{ANT}}\langle i, e, V\rangle \ \Leftrightarrow \ i \lhd e \subseteq v \, ; (r \cup m) \, ; v^{-1}.$

The generalisation only concerns the replacement of id in **ant** by an m-quasi-stable relation r in **ANT**. The corresponding convergence proof obligation **CVG** has the same shape like **cvg** but uses a quasi-variant.

Definition 10. $\textbf{\textit{CVG}}\langle i, e, V\rangle \ \Leftrightarrow \ i \lhd e \subseteq v \, ; m \, ; v^{-1}.$

Remark 4. We have the obvious equivalences

$$\boldsymbol{ANT}\langle i, e, \langle v, \mathrm{id}, c \lhd m \rangle\rangle \;\Leftrightarrow\; \boldsymbol{ant}\langle i, e, v, c, m \rangle \quad \text{and}$$
$$\boldsymbol{CVG}\langle i, e, \langle v, \mathrm{id}, c \lhd m \rangle\rangle \;\Leftrightarrow\; \boldsymbol{cvg}\langle i, e, v, c, m \rangle \;.$$

A consequence of this is that the proof obligations *ant* and *cvg* can serve as base cases in an inductive soundness proof using $\boldsymbol{ANT}$ and $\boldsymbol{CVG}$.

The transitive closure of a relation preserves quasi-stability. Combined with Lemma 7 this property permits to turn an m-quasi-stable relation r into an m^+-quasi-stable relation $(r \cup m)^+$ that matches the shape of the relation $r \cup m$ in Def. 9 and is transitive.

Lemma 8. $\boldsymbol{qs}\langle r, m \rangle \;\Rightarrow\; \boldsymbol{qs}\langle r, m^+ \rangle.$

Remark 5. Any i-quasi-variant $V = \langle v, r, m \rangle$ has an associated transitive i-quasi-variant $W = \langle v, r, m^+ \rangle$ by Lemmas 2 and 8. Furthermore, $\boldsymbol{ANT}\langle i, e, V \rangle$ implies $\boldsymbol{ANT}\langle i, e, W \rangle$, and $\boldsymbol{CVG}\langle i, e, V \rangle$ implies $\boldsymbol{CVG}\langle i, e, W \rangle$. Thus, using the equivalences of Rem. 4 we can use arbitrary quasi-variants on well-founded sets in specifications but assume that we have transitive quasi-variants available whenever needed.

5   Quasi-Lexicographic Products and Power Orders

The combination of refinement and anticipation produces quasi-lexicographic products on power orders. This complication is caused by the nondeterministic relationship between abstract and concrete states induced by the gluing invariant.

Definition 11. *The r-quasi-lexicographic product of two relations m and n, denoted $m \circledast_r n$, is defined as $(m \,_{||} \sim \varnothing) \cup (r \,_{||} n)$.*[2]

The relation $m \circledast_{\mathrm{id}} n$ is the lexicographic product of m and n. The identity keeps the first component "stable" while the second component changes. It breaks the symmetry of a plain union of m and n and as a result preserves well-foundedness. If we replace the identity by an m-quasi-stable relation we achieve the same.

If we unfold the set-theoretical definition of the r-quasi-lexicographic product, we obtain the more familiar formulation

$$m \circledast_r n = \{(p \mapsto x) \mapsto (q \mapsto y) \mid p \mapsto q \notin m \Rightarrow p \mapsto q \in r \land x \mapsto y \in n\} \;. \quad (7)$$

The following lemma provides the main insight of this section. The r-quasi-lexicographic product with an m-quasi-stable relation r of well-founded relations m and n is well-founded.

---

[2] In the Event-B notation the *parallel product* $r \,_{||} s$ of two relations r and s is defined by $(p \mapsto x) \mapsto (q \mapsto y) \in r \,_{||} s \;\Leftrightarrow\; p \mapsto q \in r \land x \mapsto y \in s$.

Lemma 9. $wf\langle m \rangle \,\wedge\, qs\langle r, m \rangle \,\wedge\, wf\langle n \rangle \,\Rightarrow\, wf\langle m \circledast_r n \rangle$.

The power order of a relation m is a relation over the subsets of its domain and range.

Definition 12. *The* power order *of a relation m, denoted by $\mathbb{O}\, m$, is defined as* $\{p \mapsto q \mid p \subseteq m^{-1}[q] \wedge (p = \varnothing \Rightarrow q = \varnothing)\}$.

Using the power order we could state well-foundedness (3) of a relation m in the form $\forall z \cdot z \mapsto z \in \mathbb{O}\, m \Rightarrow z = \varnothing$. Unfolding the set-theoretical notation the power order $\mathbb{O}\, m$ of a relation m has the following shape

$$\{p \mapsto q \mid (\forall x \cdot x \in p \Rightarrow \exists y \cdot y \in q \wedge x \mapsto y \in m) \wedge (p = \varnothing \Rightarrow q = \varnothing)\} \ . \quad (8)$$

Power orders preserve many important properties of a relation such as transitivity, quasi-stability and well-foundedness on non-empty sets. This permits us to lift known well-founded orders to well-founded power orders.

Lemma 10. $r \,;\, r \subseteq r \,\Rightarrow\, (\mathbb{O}\, r) \,;\, (\mathbb{O}\, r) \subseteq \mathbb{O}\, r$.

If a relation r is m-quasi-stable, then $\mathbb{O}\, r$ is $\mathbb{O}\, m$-quasi-stable.

Lemma 11. $qs\langle r, m \rangle \,\Rightarrow\, qs\langle \mathbb{O}\, r, \mathbb{O}\, m \rangle$.

The empty set occurs in a power order only as the pair $\varnothing \mapsto \varnothing$. Hence, removing the empty set from the range of a power order also removes it from its domain. The following lemma is to be used with Lemma 6 and Lemma 13 below. It permits to remove the empty set from a power order while preserving quasi-stability.

Lemma 12. $(\mathbb{O}\, m)^{-1}[\sim\!\{\varnothing\}] \subseteq \sim\!\{\varnothing\}$.

Well-foundedness is only preserved when the empty set is excluded from the power order. In fact, the empty set is introduced for purely technical reasons in the definition of the power order. Removing it would complicate the definition of quasi-stability, in particular. In the soundness proof below the empty set is easily excluded to occur in all cases where well-foundedness of power orders is required.

Lemma 13. $wf\langle m \rangle \,\Rightarrow\, wf\langle \{\varnothing\} \lhd \mathbb{O}\, m \rangle$.

The following lemma permits to construct a quasi-stable quasi-lexicographic product from quasi-stable components. This construction facilitates the introduction of quasi-lexicographic products in refinements where the pair $\langle r, m \rangle$ is part of a quasi-variant of the abstract model and $\langle s, n \rangle$ is part of a quasi-variant of the concrete model.

Lemma 14. $qs\langle r, m \rangle \,\wedge\, qs\langle s, n \rangle \,\Rightarrow\, qs\langle r \parallel s, m \circledast_r n \rangle$.

6 Soundness

Theorem 1 states the main condition for the termination proof in Event-B to be sound. It says that anticipation and convergence are preserved by refinement, and anticipation may be strengthened to convergence; and the variant function can be expressed in terms of concrete variables only.

Theorem 1. *For sets i, relation j, transitive i-quasi-variant V and $j[i]$-quasi-variant W there is a transitive $j[i]$-quasi-variant U such that*
(N) *for all relations f: if $\mathbf{ref}\langle i, j, \mathrm{id}, f \rangle$, then*
   (1) $\mathbf{ANT}\langle j[i], f, W \rangle$ *implies* $\mathbf{ANT}\langle j[i], f, U \rangle$,
   (2) $\mathbf{CVG}\langle j[i], f, W \rangle$ *implies* $\mathbf{CVG}\langle j[i], f, U \rangle$,
(R) *for all relations e, f: if $\mathbf{cns}\langle i, e \rangle$ and $\mathbf{ref}\langle i, j, e, f \rangle$, then*
   (1) $\mathbf{CVG}\langle i, e, V \rangle$ *implies* $\mathbf{CVG}\langle j[i], f, U \rangle$,
   (2) $\mathbf{ANT}\langle i, e, V \rangle$ *and* $\mathbf{ANT}\langle j[i], f, W \rangle$ *imply* $\mathbf{ANT}\langle j[i], f, U \rangle$,
   (3) $\mathbf{ANT}\langle i, e, V \rangle$ *and* $\mathbf{CVG}\langle j[i], f, W \rangle$ *imply* $\mathbf{CVG}\langle j[i], f, U \rangle$.

Proof. Let $V = \langle v, r, m \rangle$ and $W = \langle w, s, n \rangle$. In a refinement the abstract quasi-variant function v is only accessible by means of the gluing invariant $i \lhd j$; we define $\phi = v^{-1} \, ; \, (i \lhd j)$. Let $U = \langle u, t, o \rangle$ be given by

$$u = (\lambda x \cdot \top \mid \phi^{-1}[\{x\}] \mapsto w[\{x\}])$$
$$t = \mathbb{O}\,(r \cup m)^{+} \,\|\, \mathbb{O}\,s$$
$$o = (((\{\varnothing\} \lhd \mathbb{O}\,m) \circledast_{\mathbb{O}\,(r \cup m)^{+}} (\{\varnothing\} \lhd \mathbb{O}\,n))^{+} \, .$$

It is easy to verify that $j[i] \subseteq \mathrm{dom}(u)$. Furthermore, relation o is well-founded because

$$\top$$
$\Rightarrow$ $\langle$ V is an i-quasi-variant $\rangle$
     $\mathbf{qs}\langle r, m \rangle$
$\Rightarrow$ $\langle$ V is transitive and Lemma 7 $\rangle$
     $\mathbf{qs}\langle (r \cup m)^{+}, m \rangle$
$\Rightarrow$ $\langle$ Lemma 11 $\rangle$
     $\mathbf{qs}\langle \mathbb{O}\,(r \cup m)^{+}, \mathbb{O}\,m \rangle$
$\Rightarrow$ $\langle$ Lemma 12 and Lemma 6 $\rangle$
     $\mathbf{qs}\langle \mathbb{O}\,(r \cup m)^{+}, \{\varnothing\} \lhd \mathbb{O}\,m \rangle$ (9)
$\Rightarrow$ $\langle$ $\mathbf{wf}\langle m \rangle$ because V is an i-quasi-variant, and Lemma 13 $\rangle$
     $\mathbf{qs}\langle \mathbb{O}\,(r \cup m)^{+}, \{\varnothing\} \lhd \mathbb{O}\,m \rangle \wedge \mathbf{wf}\langle \{\varnothing\} \lhd \mathbb{O}\,m \rangle$
$\Rightarrow$ $\langle$ $\mathbf{wf}\langle n \rangle$ because W is a $j[i]$-quasi-variant, and Lemma 13 $\rangle$
     $\mathbf{qs}\langle \mathbb{O}\,(r \cup m)^{+}, \{\varnothing\} \lhd \mathbb{O}\,m \rangle \wedge \mathbf{wf}\langle \{\varnothing\} \lhd \mathbb{O}\,m \rangle \wedge \mathbf{wf}\langle \{\varnothing\} \lhd \mathbb{O}\,n \rangle$
$\Rightarrow$ $\langle$ Lemma 9 $\rangle$

$$\boldsymbol{wf}\langle(\{\varnothing\} \lhd \mathbb{O}\, m) \circledast_{\mathbb{O}\,(r \cup m)^+} (\{\varnothing\} \lhd \mathbb{O}\, n)\rangle$$
$\Rightarrow \quad \langle$ Lemma 2 $\rangle$
$$\boldsymbol{wf}\langle o\rangle \ .\qquad\qquad\qquad\qquad\qquad\qquad\qquad\qquad (10)$$

And, U is a transitive $j[i]$-quasi-variant because

$$\top$$
$\Rightarrow \quad \langle\ \boldsymbol{qs}\langle s,n\rangle$ because W is a $j[i]$-quasi-variant, and Lemma 11 $\rangle$
$$\boldsymbol{qs}\langle \mathbb{O}\, s, \mathbb{O}\, n\rangle$$
$\Rightarrow \quad \langle$ Lemma 12 and Lemma 6 $\rangle$
$$\boldsymbol{qs}\langle \mathbb{O}\, s, \{\varnothing\} \lhd \mathbb{O}\, n\rangle$$
$\Rightarrow \quad \langle$ (9), Lemma 14, Lemma 8, def. of t and o $\rangle$
$$\boldsymbol{qs}\langle t, o\rangle$$
$\Rightarrow \quad \langle$ (10) and Lemma 1, and $j[i] \subseteq \mathrm{dom}(u)$ $\rangle$
$$U \text{ is a transitive } j[i]\text{-quasi-variant .}$$

Moreover, the $j[i]$-quasi-variant U satisfies the two conditions (N) and (R). Now, claims (N1) and (N2) are consequences of claims (R2) and (R3) with $e = \mathrm{id}$ because $\boldsymbol{cns}\langle i, \mathrm{id}\rangle$ and $\boldsymbol{ANT}\langle i, \mathrm{id}, V\rangle$, the latter being a consequence of $\mathrm{id} \subseteq r$. Thus, it only remains to be shown that U satisfies (R).

We begin with the proof of (R1). We have

$$(i \lhd j)\,;f \qquad\qquad \langle\ \boldsymbol{cns}\langle i,e\rangle,\ \boldsymbol{ref}\langle i,j,e,f\rangle \text{ and Lemma 5 }\rangle$$
$$\subseteq (i \lhd e)\,;(i \lhd j) \qquad \langle\ \boldsymbol{CVG}\langle i,e,V\rangle\ \rangle$$
$$\subseteq v\,;m\,;v^{-1}\,;(i \lhd j) \qquad \langle \text{ def. of } \phi\ \rangle$$
$$\subseteq v\,;m\,;\phi\ ,$$

hence,

$$(i \lhd j)\,;f \ \subseteq\ v\,;m\,;\phi\ .\qquad\qquad\qquad\qquad (11)$$

Using this,

$$x \mapsto y \in j[i] \lhd f$$
$\Rightarrow \quad \langle$ Lemma 15 below with "$k := m$" $\rangle$
$$\phi^{-1}[\{x\}] \mapsto \phi^{-1}[\{y\}] \in \mathbb{O}\, m$$
$\Rightarrow \quad \langle\ \mathrm{dom}((i \lhd j)\,;f) \subseteq \mathrm{dom}(v) \text{ by (11) }\rangle$
$$\phi^{-1}[\{x\}] \mapsto \phi^{-1}[\{y\}] \in \mathbb{O}\, m \land \phi^{-1}[\{x\}] \neq \varnothing$$
$\Rightarrow \quad \langle \text{ def. of } \lhd\ \rangle$
$$\phi^{-1}[\{x\}] \mapsto \phi^{-1}[\{y\}] \in \{\varnothing\} \lhd \mathbb{O}\, m$$
$\Rightarrow \quad \langle \text{ def. of } o\ \rangle$

$$(\phi^{-1}[\{x\}] \mapsto w[\{x\}]) \mapsto (\phi^{-1}[\{y\}] \mapsto w[\{y\}]) \in o$$
$$\Leftrightarrow \quad \langle \text{ def. of } u \rangle$$
$$x \mapsto y \in u \,;\, o \,;\, u^{-1} \ .$$

Hence, (R1) holds. Claims (R2) and (R3) both assume $\boldsymbol{ANT}\langle i, e, V \rangle$. Thus, similarly to (11) we have

$$(i \lhd j) \,;\, f \ \subseteq \ v \,;\, (r \cup m) \,;\, \phi \ . \tag{12}$$

Now,

$$x \mapsto y \in j[i] \lhd f$$
$$\Rightarrow \quad \langle \text{ (12) and Lemma 15 below with ``} k := r \cup m \text{''} \rangle$$
$$\phi^{-1}[\{x\}] \mapsto \phi^{-1}[\{y\}] \in \mathbb{O}(r \cup m)$$
$$\Rightarrow \quad \langle \ r \cup m \subseteq (r \cup m)^{+} \text{ by def. of }^{+} \rangle$$
$$\phi^{-1}[\{x\}] \mapsto \phi^{-1}[\{y\}] \in \mathbb{O}\,(r \cup m)^{+} \ . \tag{13}$$

As specified in (R2) and (R3) two cases can be distinguished according to $\boldsymbol{ANT}\langle j[i], f, W \rangle$ and $\boldsymbol{CVG}\langle j[i], f, W \rangle$. The former implies

$$x \mapsto y \in j[i] \lhd f \ \Rightarrow \ w(x) \mapsto w(y) \in s \vee w(x) \mapsto w(y) \in n \ . \tag{14}$$

and the latter

$$x \mapsto y \in j[i] \lhd f \ \Rightarrow \ w(x) \mapsto w(y) \in n \ . \tag{15}$$

Thus, (R3) follows because

$$(13)$$
$$\Rightarrow \quad \langle \ x \mapsto y \in j[i] \lhd f \text{ and (15) } \rangle$$
$$\phi^{-1}[\{x\}] \mapsto \phi^{-1}[\{y\}] \in \mathbb{O}\,(r \cup m)^{+} \wedge w(x) \mapsto w(y) \in n$$
$$\Rightarrow \quad \langle \ x \in \mathrm{dom}(w), \ y \in \mathrm{dom}(w) \text{ and } w \text{ is a function } \rangle$$
$$\phi^{-1}[\{x\}] \mapsto \phi^{-1}[\{y\}] \in \mathbb{O}\,(r \cup m)^{+} \wedge w[\{x\}] \mapsto w[\{y\}] \in \{\varnothing\} \lhd \mathbb{O}\,n$$
$$\Rightarrow \quad \langle \text{ def. of } o \rangle$$
$$(\phi^{-1}[\{x\}] \mapsto w[\{x\}]) \mapsto (\phi^{-1}[\{y\}] \mapsto w[\{y\}]) \in o$$
$$\Leftrightarrow \quad \langle \text{ def. of } u \rangle$$
$$x \mapsto y \in u \,;\, o \,;\, u^{-1} \ .$$

Concerning (R2) observe that the case "$w(x) \mapsto w(y) \in n$" of (14) is already covered by the proof of (R3). With respect to the other case we have

$$(13)$$
$$\Rightarrow \quad \langle\, x \mapsto y \in j[i] \lhd f \text{ and } x \mapsto y \in j[i] \lhd f \Rightarrow w(x) \mapsto w(y) \in s \,\rangle$$
$$\phi^{-1}[\{x\}] \mapsto \phi^{-1}[\{y\}] \in \mathbb{O}\,(r \cup m)^{+} \wedge w(x) \mapsto w(y) \in s$$
$$\Rightarrow \quad \langle\, x \in \mathrm{dom}(w),\, y \in \mathrm{dom}(w) \text{ and } w \text{ is a function} \,\rangle$$
$$\phi^{-1}[\{x\}] \mapsto \phi^{-1}[\{y\}] \in \mathbb{O}\,(r \cup m)^{+} \wedge w[\{x\}] \mapsto w[\{y\}] \in \mathbb{O}\, s$$
$$\Rightarrow \quad \langle\, \text{def. of } t \,\rangle$$
$$(\phi^{-1}[\{x\}] \mapsto w[\{x\}]) \mapsto (\phi^{-1}[\{y\}] \mapsto w[\{y\}]) \in t$$
$$\Leftrightarrow \quad \langle\, \text{def. of } u \,\rangle$$
$$x \mapsto y \in u\,;t\,;u^{-1} \; .$$

Finally, (R2) follows because

$$(13) \hspace{6cm} \langle\, \text{see above} \,\rangle.$$
$$\Rightarrow x \mapsto y \in u\,;t\,;u^{-1} \vee x \mapsto y \in u\,;o\,;u^{-1} \quad \langle\, \text{distributivity of } \cup \text{ and } ; \,\rangle$$
$$\Leftrightarrow x \mapsto y \in u\,;(t \cup o)\,;u^{-1} \; .$$

This concludes the proof of Theorem 1. $\hspace{5cm}$ $\square$

The following lemma shows how a concrete convergence or anticipation condition $(i \lhd j)\,;f \subseteq v\,;k\,;\phi$ induces a power ordering of the concrete event f.

Lemma 15. *Let $\phi = v^{-1}\,;(i \lhd j)$ and $i \subseteq \mathrm{dom}(v)$. Then*

$$(i \lhd j)\,;f \subseteq v\,;k\,;\phi$$
$$\Rightarrow (\forall x,y \cdot x \mapsto y \in j[i] \lhd f \Rightarrow \phi^{-1}[\{x\}] \mapsto \phi^{-1}[\{y\}] \in \mathbb{O}\,k) \; .$$

Proof. Starting from the premise we have

$$(i \lhd j)\,;f \subseteq v\,;k\,;\phi$$
$$\Rightarrow \quad \langle\, \text{def. of } \phi, \text{ set theory} \,\rangle$$
$$\phi\,;f \subseteq v^{-1}\,;v\,;k\,;\phi$$
$$\Rightarrow \quad \langle\, v \text{ is a partial function} \,\rangle$$
$$\phi\,;f \subseteq k\,;\phi$$
$$\Leftrightarrow \quad \langle\, \text{def. of } \phi \,\rangle$$
$$\phi\,;(j[i] \lhd f) \subseteq k\,;\phi$$
$$\Leftrightarrow \quad \langle\, \text{def. of } \cup, \text{ def. of } ; \,\rangle$$
$$(\forall p,y \cdot (\exists x \cdot p \mapsto x \in \phi \wedge x \mapsto y \in j[i] \lhd f) \Rightarrow p \mapsto y \in k\,;\phi)$$
$$\Leftrightarrow \quad \langle\, \text{predicate logic} \,\rangle$$
$$(\forall x,y \cdot x \mapsto y \in j[i] \lhd f \Rightarrow (\forall p \cdot p \mapsto x \in \phi \Rightarrow p \mapsto y \in k\,;\phi)) \; . \hspace{1cm} (16)$$

Now,

$$x \mapsto y \in j[i] \lhd f$$
$\Rightarrow$ $\langle\, (16) \,\rangle$
$$\forall p \cdot p \mapsto x \in \phi \Rightarrow p \mapsto y \in k \,;\phi$$
$\Rightarrow$ $\langle$ def. of $;$, set theory $\rangle$
$$\forall p \cdot p \in \phi^{-1}[\{x\}] \Rightarrow \exists q \cdot q \in \phi^{-1}[\{y\}] \wedge p \mapsto q \in k \qquad\qquad (17)$$
$\Rightarrow$ $\langle$ shape (8) of $\mathbb{O}$, and $\phi^{-1}[\{x\}] \neq \varnothing$ because $i \subseteq \mathrm{dom}(v)$ $\rangle$
$$\phi^{-1}[\{x\}] \mapsto \phi^{-1}[\{y\}] \in \mathbb{O}k \ .$$

Thus, Lemma 15 holds.   □

Remark 6. In a functional refinement $(i \lhd j)^{-1}$ is a partial function, hence, ϕ^{-1} is a function. Now, because ϕ^{-1} is a function and $x \in \mathrm{dom}(\phi^{-1})$ we have

(17)
$$\Rightarrow \phi^{-1}(x) \mapsto \phi^{-1}(y) \in k \quad \text{where } k = \mathrm{id} \text{ or } k = m.$$

For refinements that are not functional we can only assume that ϕ^{-1} is a relation. This leads to the use of power sets and power orders and requires the generalisation to quasi-lexicographical products.

Remark 7. Continuing for relational refinements from (17) with $k = \mathrm{id}$ would yield

(17)
$\Rightarrow$ $\langle\, k = \mathrm{id} \,\rangle$
$$\forall p \cdot p \in \phi^{-1}[\{x\}] \Rightarrow \exists q \cdot q \in \phi^{-1}[\{y\}] \wedge p \mapsto q \in \mathrm{id}$$
$\Rightarrow$ $\langle$ def. of id $\rangle$
$$\forall p \cdot p \in \phi^{-1}[\{x\}] \Rightarrow \exists q \cdot q \in \phi^{-1}[\{y\}] \wedge p = q$$
$\Rightarrow$ $\langle$ one-point rule $\rangle$
$$\forall p \cdot p \in \phi^{-1}[\{x\}] \Rightarrow p \in \phi^{-1}[\{y\}]$$
$\Rightarrow$ $\langle$ def. of $\subseteq$ $\rangle$
$$\phi^{-1}[\{x\}] \subseteq \phi^{-1}[\{y\}]$$

This gives an increasing sequence of sets, a candidate for a quasi-stable relation. Repeating the process with $k = \{p \mapsto q \mid p \subseteq q\}$ and proceeding similarly for the well-founded relation m indicates the need for the constructions presented in this article.

7   An Improved Proof Obligation for Anticipated Events

The current proof obligation for anticipated events could be rewritten in the following shape

$$x \mapsto y \in i \lhd e \wedge x \mapsto y \notin v \,; c \lhd \mathrm{id} \,; v^{-1} \ \Rightarrow \ x \mapsto y \in v \,; c \lhd m \,; v^{-1}$$

Similarly the new proof obligation could be rewritten to

$$x \mapsto y \in i \lhd e \land x \mapsto y \notin v \,; \mathrm{id}\,; v^{-1} \;\Rightarrow\; x \mapsto y \in v \,; c \lhd m \,; v^{-1}$$

and further

$$x \mapsto y \in i \lhd e \land v(x) \neq v(y) \;\Rightarrow\; v(x) \in c \land v(x) \mapsto v(y) \in m \ .$$

And this proof obligation would only need to be generated when $x \neq y$. Following this approach no proof obligation would be generated in the situation described in the introductory example in place of (1).

8   Conclusion

The presented improvement of the anticipation proof obligation should be easy to incorporate into the Rodin tool. Fewer proof obligations need to be generated. The new proof obligation helps to keep models simple: by using the fact that some event is non-interfering on some set of variables we permit variants to be specified "locally" without referring to abstract program counters or similar constructs. This could also be useful for composing models where non-interference is common. (With the current proof rule we would have to change some variant expressions in order for termination claims to remain valid.)

We have also developed the concept of quasi-lexicographic product that is necessary for the soundness proof of anticipation and refinement. All lemmas mentioned in the paper have been proved with the Rodin tool. We are not sure whether a formalisation of Theorem 1 would be possible with reasonable effort in the tool. After all, it was never intended for deeper mathematical work.

Acknowledgement. I am grateful to the anonymous reviewers for their thorough work and constructive suggestions. I am particularly indebted to reviewer 2 who pointed out an error in the original proof of Theorem 1.

References

1. Abrial, J.-R.: Modeling in Event-B: System and Software Engineering. Cambridge University Press (2010)
2. Abrial, J.-R., Butler, M.J., Hallerstede, S., Hoang, T.S., Mehta, F., Voisin, L.: Rodin: an open toolset for modelling and reasoning in event-B. STTT 12(6), 447–466 (2010)
3. Abrial, J.-R., Cansell, D., Méry, D.: Refinement and Reachability in Event_B. In: Treharne, H., King, S., C. Henson, M., Schneider, S. (eds.) ZB 2005. LNCS, vol. 3455, pp. 222–241. Springer, Heidelberg (2005)
4. Apt, K.R., de Boer, F.S., Olderog, E.-R.: Verification of Sequential and Concurrent Programs. Texts in Computer Science. Springer (2009)

5. Dijkstra, E.W., Scholten, C.S.: Predicate Calculus and Program Semantics. Springer, NY (1990)
6. Dijkstra, E.W., van Gasteren, A.J.M.: Well-foundedness and the transitive closure, AvG88/EWD1079 (1990)
7. Hallerstede, S.: On the purpose of event-B proof obligations. Formal Asp. Comput. 23(1), 133–150 (2011)
8. Yilmaz, E.: Tool Support for Qualitative Reasoning in Event-B. Master's thesis, Department of Computer Science, ETH Zurich (2010)

Towards B
as a High-Level Constraint Modelling Language
Solving the Jobs Puzzle Challenge

Michael Leuschel and David Schneider

Institut für Informatik, Universität Düsseldorf*
Universitätsstr. 1, 40225 Düsseldorf
`david.schneider@hhu.de, leuschel@cs.uni-duesseldorf.de`

Abstract. We argue that B is a good language to conveniently express a wide range of constraint satisfaction problems. We also show that some problems can be solved quite effectively by the PROB tool. We illustrate our claim on several examples, such as the *jobs puzzle* - for which we solve the challenge set out by Shapiro. Here we show that the B formalization is both very close to the natural language specification and can still be solved efficiently by PROB. Our approach is particularly interesting when a high assurance of correctness is required. Indeed, compared to other existing approaches and tools, validation and double checking of solutions is available for PROB and formal proof can be applied to establish important properties or provide an unambiguous semantics to the problem specification.

Keywords: B-method, constraint programming, logic programming, Alloy, Kodkod, optimization.

1 Introduction

The B-method [1] is a formal method for specifying safety critical systems, reasoning about those systems and generating code that is correct by construction. The B-Language, part of this method, is a rich, mathematical language based on abstract machines and built around the concepts of first order logic, higher-order relations and set theory. Due to its expressiveness the B-Language allows its users to formalize and express complex problems in a succinct and elegant way on a high level of abstraction.

Initially, the B-method was supported by two tools, BToolkit and Atelier B, which both provided mainly automatic and interactive proving environments, as well as code generators. Later, the PROB validation tool provided automatic animation and model checking. Due to the characteristics of B, PROB gradually evolved into a constraint solving tool for the B language, in order to automatically determine values for parameters and quantified variables. This opens up

---

* Part of this research has been sponsored by the EU FP7 project 287563 (AD-VANCE).

Y. Ait Ameur and K.-D. Schewe (Eds.): ABZ 2014, LNCS 8477, pp. 101–116, 2014.
© Springer-Verlag Berlin Heidelberg 2014

new uses of B beyond developing safety critical systems. Indeed, in this paper we want to show that B is a language well suited to express general constraint satisfaction problems. Moreover, in combination with the constraint solving capabilities of PROB, we can solve a non-trivial class of these problems.

First we present a case study which highlights the expressiveness of B as a constraint modelling language. The case study is based on the *jobs puzzle* [26] and takes on the challenges identified and discussed by Shapiro [21]. One of these challenges is to provide a formalization of the puzzle that follows closely the English text of the puzzle. This aspect makes it particularly interesting, as it allows us to showcase how, using B, these kinds of problems can be expressed very conveniently and still be solved efficiently with PROB. The puzzle, the challenge and our solution to the puzzle are discussed and compared to other solutions in Section 2.

Later in the paper, we establish that PROB as a tool can be used to solve an interesting class of constraint satisfaction problems efficiently, providing a good balance between the ease of expressing a problem on an abstract level using B and the efficiency of solving these problems. In Section 3 we discuss a series of problems that are expressed nicely in B and can be solved by PROB for a wide range of values, such as the *N-Queens* problem, the *Peaceable Armies of Queens* and the *Graph Isomorphism* problem. We compare the results to a selection of different tools to show that PROB gives competitive results, while still having room for improvement as discussed in Section 5. Finally, in Section 4 we present how the constraint solving features of PROB, showcased in this article, are being used in several industrial applications.

Alternate Approaches to Constraint Solving The mathematical language of B is quite close to that of Z and TLA+. As PROB can deal with those formalisms [17,8], the gist of the paper is also valid for those languages. Similarly, VDM and abstract state machines are probably also well suited to express constraint satisfaction problems.

Dedicated Constraint Solving Libraries. Alternate approaches to our high-level formal methods approach are dedicated constraint-solving libraries embedded in general purpose programming languages. Examples are the CLP(FD) library of SICStus Prolog [4] or the ILOG solver. These libraries require a much higher modelling effort and a relatively high level of expertise, but can obviously obtain better performance. Another possible approach is the Zinc modelling language [14]. It provides a higher level encoding than for example CLP(FD), but still cannot deal with higher-order sets or relations. Also, to our knowledge, neither Zinc nor any other tool we are aware of can deal with unbounded constraint satisfaction problems.

SMT-Based Approaches. It would be interesting to see how an expert in the Formula language [10], which maps to the Z3 SMT solver, would encode the problems in this paper, and how the solving times compare with those of PROB. Recently, an Event-B to SMT-LIB converter has become available for the Rodin platform [6]. It is very useful for proof, but as shown in [18] not suitable for constraint solving. For example, it was not possible to solve various simpler problems, such as the *Who killed Agatha* puzzle or a graph colouring problem. As such, we did not attempt to use this B to SMT converter on the examples in this paper.

SAT-Based Approaches. The Alloy language [9] was designed from the outset to be able to effectively translated to SAT problems. This leads to certain expressivity restrictions, e.g., higher-order relations and sets are not allowed as their SAT encoding would become too large to be tractable. PROB follows another principle: it accepts the B language in full with all its consequences, and tries to solve constraints for as many relevant models as possible. Note that PROB also has a backend [18] which translates B constraints into SAT. This uses the same Kodkod library [25] that Alloy employs, and can deal much better with certain relational constraints, but similarly only translates first order sets and relations (the rest are left for the traditional PROB solver). In this paper we discuss and compare several B solutions with Alloy counterparts and also discuss the ProB Kodkod backend.

Finally, one could think of using model checking rather than constraint solving. In fact, we have experimented with various solutions for the puzzles using efficient model checkers such as Spin or TLC. However, for constraint satisfaction problems model checking amounts to naive, brute force search and is rarely able to solve more complicated constraints.

2 On the Expressiveness of B - The Jobs Puzzle

The first and most detailed example we discuss is the *jobs puzzle*. This puzzle was originally published in 1984 by Wos et al. [26] as part of a collection of puzzles for automatic reasoners. A reference implementation of the puzzle, by one of the authors of the book, using OTTER [15] can be found online.[1]

The puzzle consists of eight statements that describe the problem domain and provide some constraints on the elements of the domain. The problem is about a set of people and a set of jobs; the question posed by the puzzle is: who holds which job? The text of the puzzle as presented in [21] is as follows:

- There are four people: Roberta, Thelma, Steve, and Pete.
- Among them, they hold eight different jobs.
- Each holds exactly two jobs.
- The jobs are: chef, guard, nurse, clerk, police officer (gender not implied), teacher, actor, and boxer.
- The job of nurse is held by a male.
- The husband of the chef is the clerk.
- Roberta is not a boxer.
- Pete has no education past the ninth grade.
- Roberta, the chef, and the police officer went golfing together.

---

[1] http://www.mcs.anl.gov/~wos/mathproblems/jobs.txt

What makes this puzzle interesting for automatic reasoners, is that not all the information required to solve the puzzle is provided explicitly in the text. The puzzle can only be solved if certain implicit assumptions about the world are taken into account, such as: the names in the puzzle denote gender or that some of the job names imply the gender of the person that holds it.

2.1 Shapiro's Challenge

Shapiro [21], following the original authors' remarks, that formalizing the puzzle was at times hard and tedious, identified three challenges posed by the puzzle with regard to automatic reasoners. According to Shapiro [21], the challenges posed by the *jobs puzzle* are to:

- formalize it in a non-difficult, non-tedious way
- formalize it in a way that adheres closely to the English statement of the puzzle
- have an automated general-purpose commonsense reasoner that can accept that formalization and solve the puzzle quickly.

Any formalization also needs to encode the implicit knowledge used to solve the puzzle for the automatic reasoners while still trying to satisfy the aspects mentioned above. Addressing this challenge makes this puzzle a good case-study for the expressiveness of B to formalize such a problem.

2.2 A Solution to the Jobs Puzzle Using B

The B encoding of the puzzle uses plain predicate logic, combined with set theory and arithmetic. We will show how this enables a very concise encoding of the problem, staying very close to the natural language requirements. Moreover, the puzzle can be quickly solved using the constraint solving capabilities of PROB. Following the order of the sentences in the puzzle we will discuss one or more possibilities to formalize them using B.

To express *"There are four people: Roberta, Thelma, Steve, and Pete"* we define a set of people, that holds the list of names:

```
PEOPLE={"Roberta", "Thelma", "Steve", "Pete"}
```

We are using strings here to describe the elements of the set. This has the advantage, that the elements of the set are implicitly different.[2] Alternatively, we could use enumerated or deferred sets defined in the SETS section of a B machine.

As stated above we need some additional information that is not included in the puzzle to solve it. The first bit of information is that the names used in the puzzle imply the gender. In order to express this information we create two sets, MALE and FEMALE which are subsets of PEOPLE and contain the corresponding names.

---

[2] This encoding allows us to input the puzzle directly into the PROB console (available at http://stups.hhu.de/ProB/index.php5/ProB_Logic_Calculator).

```
FEMALE={"Roberta", "Thelma"} & MALE={"Steve", "Pete"}
```

The next statement of the puzzle is: "*among them, they hold eight different jobs*". This can be formalized in B using a function that maps from a job to the corresponding person that holds this job using a total surjection from JOBS to PEOPLE:

```
HoldsJob : JOBS -->> PEOPLE
```

Although redundant, as we will see below, to express "*Among them, they hold eight different jobs*" we can add the assertion that the cardinality of HoldsJob is 8. This is possible, because in B functions and relations can be treated as sets of pairs, where each pair consists of an element of the domain and the corresponding element from the range of the relation.

```
card(HoldsJob) = 8
```

Constraining the jobs each person holds, the puzzle states: "*Each holds exactly two jobs*". To express this we use the inverse relation of HoldsJob, it maps a PERSON to the JOBS associated to her. The inverse function or relation is expressed in B using the ~ operator. For readability we assign the inverse of HoldsJob to a variable called JobsOf. JobsOf is in this case is a relation, because, as stated above, each person holds two jobs.

```
JobsOf = HoldsJob~
```

Because JobsOf is a relation and not a function, in order to read the values, we need to use B's relational image operator. This operator maps a subset of the domain to a subset of the range, instead of a single value. To read the jobs Steve holds, the relational image of JobsOf is used as shown below:

```
JobsOf[{"Steve"}]
```

Using the JobsOf relation we can express the third sentence of the puzzle using a universally quantified expression over the set PEOPLE. The Universal quantification operator ($\forall$) is expressed in B using the ! symbol followed by the name of the variable that is quantified. This way of expressing the constraint is close to the original text of the puzzle, saying that the set of jobs each person holds has a cardinality of two.

```
!x.(x : PEOPLE => card(JobsOf[{x}]) = 2)
```

The fourth sentence assigns the set of job names to the identifier JOBS. This statement also constraints the cardinality of HoldsJob to 8.

```
JOBS = {"chef", "guard", "nurse", "clerk", "police", "teacher", "actor", "boxer"}
```

The following statements further constrain the solution. First "*The job of nurse is held by a male*", which we can express using the HoldsJob function and the set MALE by stating that the element of PEOPLE that HoldsJob("nurse") points to is also an element of the set MALE.

$$\text{HoldsJob("nurse") : MALE}$$

Additionally, we add the next bit of implicit information, which is that typically a distinction is made between actress and actor, and therefore the job name actor implies that it is held by a male. This information is formalized, similarly as above.

$$\text{HoldsJob("actor") : MALE}$$

The next sentence: "*The husband of the chef is the clerk*" contains two relevant bits of information, based on another implicit assumption, which is that marriage usually is between one female and one male. With this in mind, we know that the chef is female and the clerk is male. One possibility is to do the inference step manually and encode this as:

$$\text{HoldsJob("chef") : FEMALE \& HoldsJob("clerk") : MALE}$$

Alternatively, and in order to stay closer to the text of the puzzle we can add a function Husband that maps from the set FEMALE to the set MALE as a partial injection. We use a partial function, because we do not assume that all elements of FEMALE map to an element of MALE.

$$\text{Husband : FEMALE >+> MALE}$$

To add the constraint using this function we state that the tuple of the person that holds the job as chef and the person that holds the job as clerk are an element of this function when treated as a set of tuples.

$$\text{(HoldsJob("chef"), HoldsJob("clerk")) : Husband}$$

The next piece of information is that "*Roberta is not a boxer*". Using the JobsOf relation we can express this close to the original sentence, by stating: boxer is not one of Roberta's jobs. This can be expressed using the relational image of the JobsOf relation:

$$\text{"boxer" /: JobsOf[\{"Roberta"\}]}$$

The next sentence provides the information that "*Pete has no education past the ninth grade*". This again needs some contextual information to be useful in order to find a solution for the puzzle [21]. To interpret this sentence we need to know that the jobs of police officer, teacher and nurse require an education of more than 9 years. Hence the information we get is that Pete does not hold any of these jobs. Doing this inference step we could, as above, state something along the lines of HoldsJob("police") /= "Pete", etc. for each of the jobs. The solution used here, tries to avoid doing the manual inference step. Although we still need to provide the information needed to draw the conclusion that Pete does not hold any of these three jobs. We create a set of those jobs that need higher education:

```
QualifiedJobs = {"police", "teacher", "nurse"}
```

Using the relational image operator we can now say that Pete is not among the ones that hold any of these jobs. The relational image can be used to get the set of items in the range of function or relation for all elements of a subset of the domain.

```
"Pete" /: HoldsJob[QualifiedJobs]
```

Finally, the last piece of information is that "*Roberta, the chef, and the police officer went golfing together*", from this we can infer that Roberta, the chef, and the police officer are all different persons. We write this in B stating that the set of Roberta, the person that holds the job as chef, and the person that is the police officer has cardinality 3, using a variable for the set for readability.

```
Golfers = {"Roberta", HoldsJob("chef"), HoldsJob("police")} & card(Golfers) = 3
```

By building the conjunction of all these statements, PROB searches for a valid assignment to the variables introduced that satisfies all constraints, generating a valid solution that answers the question posed by the puzzle "*who holds which job?*" in form of the HoldsJob function. The solution found by PROB is depicted in Fig. 1.[3]

This satisfies, in our eyes, the challenges identified by Shapiro. In the sense that the formalization, is not difficult, although it uses a formal language. The elements of this language are familiar to most programmers or mathematicians and it builds upon well understood and widely known concepts. The brevity of the solution shows that using an expressive high-level language it is possible to encode the puzzle without having tedious tasks in order to be able to solve the puzzle at all.

The encoding of the sentences follows the structure of the English statements very closely. We avoid the use of quantification wherever possible and use set based expressions that relate closely to the puzzle. We are able to encode the additional knowledge needed to solve puzzle in a straight forward way, that is also close to how this would be expressed as statements in English. Lastly it is worth to note that the formalization of "*Each holds exactly two jobs*" is the one furthest away from the English expression, using quantifications and set cardinality expressions.

2.3 Related Work

In his paper Shapiro discusses several formalizations of the puzzle with regard to the identified challenges. A further formalization using controlled natural

---

[3] We used the "Visualize State as Graph" command and then adapted the generated graph using OmniGraffle

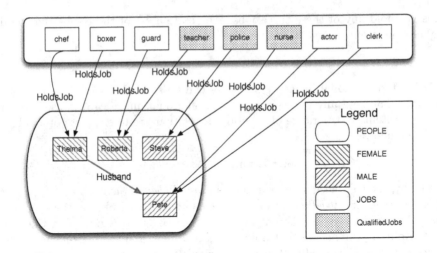

Fig. 1. The solution to the Jobs puzzle, depicted graphically

language and answer set programming (ASP) was presented in [19] by Schwitter et al.

The first of the solutions discussed by Shapiro is a solution from the TPTP website, encoded as a set of clauses and translated to FOL. The main disadvantages of this encoding is that it requires 64 clauses to encode the problem and many of them are needed to define equality among jobs and names. This is in contrast to our B encoding using either enumerated sets or strings, where all elements are implicitly assumed to be different. Thus the user does not have to define the concept of equality for simple atoms.

The second solution discussed by Shapiro uses SNePS [22], a common sense and natural language reasoning system designed with the goal to "have a formal logical language that captured the expressibility of the English language" [21]. The language has a unique name assumption and set arguments making the encoding simpler and less tedious. On the other hand the lack of support for modus tolens requires rewriting some of the statements in order to solve the puzzle.

The last formalization discussed by Shapiro uses Lparse and Smodles [16] which uses stable model semantics with an extended logic programming syntax. According to Shapiro several features of Lparse/Smodels are simmilar to those of SNePS. This formalization also simplifies the encoding of the puzzle, but according to Schwitter et al. both solutions still present a "considerable conceptual gap between the formal notations and the English statements of the puzzle" [19].

Schwitter et al. in their paper "The jobs puzzle: Taking on the challenge via controlled natural language processing" [19] present a solution to the *jobs puzzle* using controlled natural language and a translation to ASP to solve the *jobs puzzle* in a novel way that stays very close to the English statements of the puzzle and satisfying the challenges posed by Shapiro. To avoid the mismatch between natural and controlled natural languages Schwitter et al. describe the use of a development environment that supports the user to input valid statements according to the rules of the controlled language. A solution using a mathematical, but high level language like B avoids this problems by having a formal and, for most, familiar language used to formalize the problem.

3   Solving Constraint Problems with B and PROB

PROB is able to solve our formalization of the *jobs puzzle*, presented in the previous Section in about $10ms$, finding the two possible instantiations of the variables that represent the (same) solution to the puzzle.

In this section we will present four examples, that can be elegantly expressed with B and discuss how these can be solved with the constraint solving features of PROB.[4] There are many more examples, which unfortunately we cannot present here due to space restrictions.

3.1   Subset Sum

The first example is the subset sum problem 7.8.1 from page 340 of "Optimization Models For Decision Making: Volume 1".[5] Expressing this problem just takes this one line of B and PROB can solve it in less than 5 ms:

```
coins = {16,17,23,24,39,40} & stolen : coins --> NATURAL & SIGMA(x).(x:coins|stolen(x)*x)=100
```

The goal is to determine how many bags of coins were lost to amount for 100 missing coins. The bags can have different sizes, specified in the set named `coins`. In order to find the result with B, we create a function that maps from the different coin-bag sizes to a number; this number represents how many bags of that size were stolen. The instantiation of the function `stolen` is constrained by the last expression, which states that the sum of all coins in the missing bags is 100. This is expressed using the `SIGMA` operator, which returns the sum of all values calculated in the associated expression, akin the mathematical Σ operator. An interesting aspect is that we have not explicitly expressed an upper bound on `coins` (`NATURAL` stands for the set of mathematical natural numbers); PROB determines the upper bound itself during the constraint solving process. Finally, we can check that there is only one solution by checking:

```
card({c,s|c = {16,17,23,24,39,40} & s : c --> NATURAL & SIGMA(x).(x:c|s(x)*x)=100})=1
```

---

[4] The source code of the examples can be obtained at the following web site:
http://stups.hhu.de/w/Pub:Constraint_Solving_in_B
[5] http://ioe.engin.umich.edu/people/fac/books/murty/opti_model/

3.2 N-Queens

The well known *N-Queens* problem[6] is a further problem, that can be expressed very succinctly in B by specifying a constant queens, which has to satisfy the following axioms:

```
queens : 1..n >-> 1..n & !(q1,q2).(q1:1..n & q2:2..n & q2>q1
     => queens(q1)+(q2-q1) /= queens(q2) & queens(q1)+(q1-q2) /= queens(q2))
```

The total and injective function queens maps the index of each column to the index of the row where the queen is placed in that column. The formula states that for each pair of columns, the queens placed on those columns are not on the same diagonal. From the set of functions from $1..n$ to $1..n$ PROB discards those candidates that violate the condition on the diagonals and instantiates queens to the first solution that satisfies it.

To get a better impression of how PROB performs for this and hopefully similar problems we compared, as shown in Table 1, the B implementation on PROB 1.3.7-beta9 to a C iterative implementation, a version in Prolog using CLP(FD)[7] running on SWI-Prolog 6.6.4, a version in Prolog taken from [24] Chapter 14[8] , a version written in Alloy using the Minisat and Sat4J SAT-solvers of Alloy 4.2. We ran the examples on a MacBook Pro with a four core Intel i7 Processor with 2.2 Ghz and 8 Gb. RAM for increasing values of n as reported below. All reported times represent the time needed to find a first valid configuration. For each tool we stopped collecting data after the first computation that took more than 10000 ms. to find a solution.

Table 1. Time in *ms.* to find a first solution to the *N-Queens* problem.

n	PROB	C	swipl	swipl CLP(FD)	Alloy Sat4J	Alloy Minisat
8	1	4	1	7	75	81
10	1	4	1	6	190	120
20	28	34	4941	30	3483	10019
30	67	11467	-[9]	49	6296	-
40	122	-	-	77	57992	-
50	250	-	-	483	-	-
60	349	-	-	479	-	-
70	426	-	-	178	-	-
80	623	-	-	278	-	-
90	885	-	-	759	-	-
100	1028	-	-	442	-	-

Table 2. LoC for the different solutions compared

Language	LoC
B (PROB)	2
Alloy	21
Prolog	60
swipl CLP(FD)	122
C	171

There are some pathological cases, not shown here, where PROB, but also other tools perform very badly (such as $n = 88$) but for most inputs of $n < 101$ PROB finds a solution in up to 1.2 seconds.

---

[6] http://en.wikipedia.org/w/index.php?oldid=587668943
[7] http://www.logic.at/prolog/queens/queens.pl
[8] http://bach.istc.kobe-u.ac.jp/llp/bench/queen_p.pl
[9] We canceled this run after 40 minutes without result.

The results show that constraint based solutions to this problem, with increasing board sizes give better results than the brute force versions. Among the constraint based results, the direct encoding in Prolog using CLP(FD) is generally faster than PROB; considering the higher abstraction level of B these results are to be expected. Taking the size of the implementations into account (as reported in Table 2) gives evidence that using B to encode such a problem and PROB to solve it is a good trade-off between the size, the complexity of the implementation and the time required to find a solution.[10]

3.3   Peaceable Armies of Queens

A challenging constraint satisfaction problem, related to the previous, was proposed by Bosch (Optima 1999) and taken up in [23]. It consists of setting up opposing armies of queens of the same size on a $n \times n$ chessboard so that no queen attacks a queen of the opposing colour.

Smith et. al [23] report that the integer linear programming tool CPLEX took 4 hours to find an optimal solution for $n = 8$ (which is 9 black and 9 white queens). Optimal here means placing as many queens as possible, i.e., there is a solution for 9 queens but none for 10 queens. In order to determine the optimal solution for $n = 8$ the ECLiPSe and the ILOG solver are reported to take just over 58 minutes and 27 minutes and 40 seconds respectively (Table 1 in [23]). After applying the new symmetry reduction techniques proposed in [23], the solving time was reduced to 10 minutes and 40 seconds for ECLiPSe and 5 minutes 31 seconds for the ILOG solver.

In a first instance, we have encoded the puzzle as a constraint satisfaction problem, i.e., determining whether for a given board size n and a given number of queens q we can find a correct placement of the queens. We first introduce the following DEFINITION:

```
ORDERED(c,r) == (!i.(i:1..(q-1) => c(i) <= c(i+1)) &
                 !i.(i:1..(q-1) => (c(i)=c(i+1) => r(i) < r(i+1))))
```

The encoding of the problem is now relatively straightforward:

```
blackc : 1..q --> 1..n &  blackr : 1..q --> 1..n &
whitec : 1..q --> 1..n &  whiter : 1..q --> 1..n &
ORDERED(blackc,blackr) & ORDERED(whitec,whiter) &

!(i,j).(i:1..q & j:1..q => blackc(i) /= whitec(j)) &
!(i,j).(i:1..q & j:1..q => blackr(i) /= whiter(j)) &
!(i,j).(i:1..q & j:1..q => blackr(i) /= whiter(j)+(blackc(i)-whitec(j))) &
!(i,j).(i:1..q & j:1..q => blackr(i) /= whiter(j)-(blackc(i)-whitec(j))) &

whitec(1) < blackc(1) /* simple symmetry breaking */
```

The solving time on a MacBook Pro with a four core Intel i7 Processor with 2.2 Ghz and 8 Gb. RAM is as follows using PROB 1.3.7-beta9:

---

[10] We are aware that the different solutions compared might not represent the best possible solution in each formalism.

Table 3. Runtime in seconds to solve the *Peaceable Queens* problem

Board Size n	Queens q	Result	ProB	Alloy Minisat	Alloy Sat4J
7	7	sat	0.29 sec	3.04 sec	14.90 sec
7	8	unsat	17.6 sec	-	-
8	1	sat	0.00 sec	0.01 sec	0.02 sec
8	2	sat	0.00 sec	0.02 sec	0.03 sec
8	3	sat	0.01 sec	0.03 sec	0.05 sec
8	4	sat	0.01 sec	0.09 sec	0.17 sec
8	5	sat	0.02 sec	0.21 sec	0.45 sec
8	6	sat	0.02 sec	0.22 sec	1.29 sec
8	7	sat	0.06 sec	1.29 sec	2.10 sec
8	8	sat	0.51 sec	2.58 sec	6.68 sec
8	9	sat	12.10 sec	122.87 sec	106.74 sec
8	10	unsat	661 sec	-	-

We have also encoded the problem in Alloy. Table 3 shows the results for board sizes of 7 and 8 for ProB and Alloy. We have used Alloy 4.2. For a board size of 8 and 9 queens of each colour the solving time was 107 seconds compared to ProB's 13 seconds. We were unable to confirm the absence of a solution for 10 queens of each colour (Alloy was still running after more than 4 hours). We have also directly used the Kodkod library, through our B to Kodkod translation [18]. Here the runtime was typically a bit slower than running Alloy (e.g., 149 seconds instead of 107 using SAT4J and 123 seconds using MiniSat for 9 queens of each colour).

Note 1. Model checking using TLC basically hopeless; much like N-queens where Spin can only solve the puzzle for relatively small values ProB can also solve the puzzle for $n = 8$, $q = 9$ and an additional two kings (on of each colour; see [7]). The solving time is about one hour.

This example has shown that even problems considered challenging by the constraint programming community can be solved, and that they can be expressed with very little effort. The graphical visualization features of ProB were easy to setup (basically just defining an animation function in B and declaring a few images; see [13]) and were extremely helpful in debugging the model.

3.4   Extended Graph Isomorphism

The *Graph Isomorphism* Problem[11] is the final problem discussed here. To determine if two graphs are isomorphic we search for a bijection of the vertices which preserves the neighbourhood relationship.

Using B, graphs can be represented as relations of nodes, where the vertices are represented by the tuples in the relation, seen as a set. An undirected graph can hence be easily represented as the union of the directed graph relation with the inverse of the relation, basically duplicating all vertices.

Using B we can state the problem using an existential quantification over the formula used to define the *graph isomorphism* problem, following closely the mathematical problem definition. Existential quantification is expressed in

---

[11] http://en.wikipedia.org/w/index.php?oldid=561096064

B using the # operator, which corresponds to the ∃ symbol in mathematical notation. We state that there is a total bijection that maps the vertex set from one graph to the other such that two nodes are adjacent in the domain iff they are adjacent in the range of the bijection. Additionally we only need to encode the entities needed in the quantification in order to solve this problem for two specific graphs.

```
MACHINE CheckGraphIsomorphism
SETS Nodes = {a,b,c,d,e, x,y,z,v,u}
DEFINITIONS
   G1 == {a|->b, a|->c, a|->d, b|->c, b|->d, c|->e, d|->e};
   G2 == {x|->v, x|->u, x|->z, y|->v, y|->u, z|->v, z|->u}
CONSTANTS graph1, graph2, relevant
PROPERTIES
   graph1: Nodes <-> Nodes & graph2: Nodes <-> Nodes &
   graph1 = G1\/G1~ & graph2 = G2 \/ G2~ & /* generate undirected graphs */
   relevant = dom(graph1) \/ dom(graph2) \/ ran(graph1) \/ ran(graph2) &
   #p.(p : relevant >->> relevant &
     !(x, y).(x : relevant & y : relevant =>
       (x|->y : graph1 <=> p(x)|->p(y) : graph2)))
END
```

The above specification can easily be extended with additional constraints. An industrial application of such an extended *graph isomorphism* problem was presented by ClearSy [5]. Here, ClearSy used B and PROB to find graph isomorphisms between high level control flow graphs and control flow graphs extracted from machine code gathered through a black box compiler. In addition, the memory mapping used by the compiler had to be inferred by constraint solving. For the main problem on graphs with 192 nodes each, the solution was found by PROB in 10 seconds. The ability to easily express graph isomorphism and pair it with other domain specific predicates was an important aspect in this application.

4   Industrial Applications

As hinted in the previous section where we describe how ClearSy used B and PROB to decompile machine code, the expressivity of B in combination with the constraint solving capabilities of PROB are being used in several industrial applications, in order to validate inputs for models or to solve problems similar to those shown in this paper.

A further application by Siemens is described in [12]. Siemens use the B-method to develop safety critical software for various driverless train systems. These systems make use of a generic software component, which is then instantiated for a particular line by setting a larger number of parameters (i.e., the topology, the style of trains with their parameters). In order to establish the correctness of the generic software, a large number of properties about these parameters and the system in general are expressed in B.

The data validation problem is then to validate these properties for particular data values. One difficulty is the size of the parameters; the other is the fact that some properties are expressed in terms of abstract values and not in terms of concrete parameter values (which are linked to abstract values via a gluing invariant in the B refinement process). Initially, the data validation was carried out in Atelier-B using custom proof rules [3]. However, many properties could not be checked in this way (due to excessive runtime or memory consumption) and had to be checked manually. As described in [12], Siemens now use the PROB constraint solver to this end and have thus dramatically reduced the time to validate a new parametrisation of their generic software. This success led to this technique also being applied in Alstom and the development of a custom tool DTVT with the company ClearSy [11]. The company Systerel is also using B for data validation for a variety of customers [2]. To this end, Systerel uses a B evaluation engine which is also at the heart of Brama [20] combined with PROB.

It is interesting to note, that data validation is now also being applied in contexts where the system itself was not developed using B; the B language is "just" used to clearly stipulate important safety properties or regulations about data. This shows a shift from using B to formally prove systems or software correct, to using B as an expressive specification language for various constraint satisfaction problems. In the traditional use of B to develop software or systems correct by construction, refinement and the proving process play a central role. In this novel use of B, those aspects of B are almost completely absent. There often is no use of refinement but the properties to be checked or solved become larger and more difficult, making the use of traditional provers nigh impossible.

The PROB tool is now used by various companies for similar data validation tasks, sometimes even in contexts where B itself is not used for the system development process. In those cases, the underlying language of B turns out to be very expressive and efficient to cleanly encode a large class of data properties.

5   Conclusion and Future Work

In the first part of this article we focused on the language B and its power to express constraint satisfaction problems. Using B, we have taken on the challenges identified by Shapiro regarding the *jobs puzzle*. Our primary goal was to show that while B is a formal specification language, it is also very expressive and can be used to encode constraint satisfaction problems in a readable and concise way. Our solution to the *jobs puzzle* addresses all the challenges posed by this puzzle, we were able to create an encoding, that using mainly simple B constructs, creates a simple and straight-forward formalization. We only have to additionally provide an encoding of the implicit information required to solve the puzzle, which can also be achieved in a way that is not complex and close to a translation to English. Our encoding follows the original text closely and PROB is able to solve the puzzle efficiently.

In the second part of this article we focused on solving constraint satisfaction problems written in B using PROB. We outlined on two similar examples that it is possible to efficiently solve problems encoded on a high abstraction level.

The pairing of PROB and B can be a good trade-off between the efficiency of low level constraint solvers that make the encoding of problems very hard and high level systems that have to pay the price of abstraction by the increased amount of computation needed to solve problems. Surprisingly, for some problems (see Section 3.3) we are competitive with low-level modern constraint solving techniques. But obviously, many large scale industrial optimization problems are still out of reach of our approach. Also, compared to Alloy, the standard PROB solver is weak for certain relational operators such as image or transitive closure. But we work on further improving the constraint solving capabilities of PROB and thus reducing the overhead associated with the abstraction level of B, allowing to use PROB on more problems and domains.

As shown in the *Peaceable Queens* example, B and PROB are still awkward for solving optimization problems. The current solution is to setup a problem twice and search for a solution where one problem is solved and the other one not. This is an area we intend to do further research in the future.

Finally we discussed the industrial uses cases of B in combination with the constraint solving features of PROB. All the aspects discussed in this article show the advantages of using a feature rich and high-level language such as B to encode complex problems and at the same time making use of a high-level constraint solver to solve them. Compared to other approaches to constraint solving, ours has the advantage of extensive validation of the tool along with a double chain [11] to cross-check results, and the ability to apply proof to (parts of) B models. This makes B and PROB particularly appealing to solving constraints in safety critical application areas.

Acknowledgements. We thank Daniel Plagge and Dominik Hansen for interesting discussions and feedback. Daniel Plagge also devised the Alloy solutions for the N-Queens puzzle and the peaceable armies of queens. We thank anonymous referees for their useful feedback. We also thank Mats Carlsson and Per Mildner for their support around SICStus Prolog and CLP(FD).

References

1. Abrial, J.-R.: The B-Book. Cambridge University Press (1996)
2. Badeau, F., Doche-Petit, M.: Formal data validation with Event-B. CoRR abs/1210.7039, 2012 (2012) Proceedings of DS-Event-B 2012, Kyoto
3. Boite, O.: Méthode B et validation des invariants ferroviaires. Master's thesis, Université Denis Diderot, Mémoire de DEA de logique et fondements de l'informatique (2000)
4. Carlsson, M., Ottosson, G.: An Open-Ended Finite Domain Constraint Solver. In: Glaser, H., Hartel, P., Kuchen, H. (eds.) PLILP 1997. LNCS, vol. 1292, pp. 191–206. Springer, Heidelberg (1997)
5. ClearSy. Data Validation & Reverse Engineering, http://www.data-validation.fr/data-validation-reverse-engineering/ (June 2013)
6. Déharbe, D., Fontaine, P., Guyot, Y., Voisin, L.: SMT solvers for Rodin. In: Derrick, J., Fitzgerald, J., Gnesi, S., Khurshid, S., Leuschel, M., Reeves, S., Riccobene, E. (eds.) ABZ 2012. LNCS, vol. 7316, pp. 194–207. Springer, Heidelberg (2012)
7. Gent, I.P., Petrie, K.E., Puget, J.-F.: Symmetry in constraint programming. Foundations of Artificial Intelligence 2, 329–376 (2006)

8. Hansen, D., Leuschel, M.: Translating TLA$^+$ to B for validation with PROB. In: Derrick, J., Gnesi, S., Latella, D., Treharne, H. (eds.) IFM 2012. LNCS, vol. 7321, pp. 24–38. Springer, Heidelberg (2012)

9. Jackson, D.: Alloy: A lightweight object modelling notation. ACM Transactions on Software Engineering and Methodology 11, 256–290 (2002)

10. Jackson, E.K., Levendovszky, T., Balasubramanian, D.: Reasoning about meta-modeling with formal specifications and automatic proofs. In: Whittle, J., Clark, T., Kühne, T. (eds.) MODELS 2011. LNCS, vol. 6981, pp. 653–667. Springer, Heidelberg (2011)

11. Lecomte, T., Burdy, L., Leuschel, M.: Formally checking large data sets in the railways. CoRR, abs/1210.6815 (2012) Proceedings of DS-Event-B 2012, Kyoto

12. Leuschel, M., Falampin, J., Fritz, F., Plagge, D.: Automated property verification for large scale B models with ProB. Formal Asp. Comput. 23(6), 683–709 (2011)

13. Leuschel, M., Samia, M., Bendisposto, J., Luo, L.: Easy Graphical Animation and Formula Viewing for Teaching B. In: The B Method: from Research to Teaching, pp. 17–32 (2008)

14. Marriott, K., Nethercote, N., Rafeh, R., Stuckey, P.J., de la Banda, M.G., Wallace, M.: The design of the Zinc modelling language. Constraints 13(3), 229–267 (2008)

15. Mccune, W.: Otter 3.3 reference manual (2003)

16. Niemelä, I., Simons, P., Syrjänen, T.: Smodels: A system for answer set programming. CoRR, cs.AI/0003033 (2000)

17. Plagge, D., Leuschel, M.: Validating Z Specifications Using the PROB Animator and Model Checker. In: Davies, J., Gibbons, J. (eds.) IFM 2007. LNCS, vol. 4591, pp. 480–500. Springer, Heidelberg (2007)

18. Plagge, D., Leuschel, M.: Validating B,Z and TLA$^+$ using PROB and Kodkod. In: Giannakopoulou, D., Méry, D. (eds.) FM 2012. LNCS, vol. 7436, pp. 372–386. Springer, Heidelberg (2012)

19. Schwitter, R.: The jobs puzzle: Taking on the challenge via controlled natural language processing. Theory and Practice of Logic Programming 13, 487–501 (2013)

20. Servat, T.: BRAMA: A new graphic animation tool for B models. In: Julliand, J., Kouchnarenko, O. (eds.) B 2007. LNCS, vol. 4355, pp. 274–276. Springer, Heidelberg (2007)

21. Shapiro, S.C.: The jobs puzzle: A challenge for logical expressibility and automated reasoning. In: AAAI Spring Symposium: Logical Formalizations of Commonsense Reasoning (2011)

22. Shapiro, S.C., The SNePS Implementation Group.: SNePS 2.7.1 User's Manual, Department of Computer Science and Engineering University at Buffalo, The State University of New York (December 2010)

23. Smith, B.M., Petrie, K.E., Gent, I.P.: Models and symmetry breaking for peaceable armies of queens. In: Régin, J.-C., Rueher, M. (eds.) CPAIOR 2004. LNCS, vol. 3011, pp. 271–286. Springer, Heidelberg (2004)

24. Sterling, L., Shapiro, E.: The Art of Prolog. MIT Press (1986)

25. Torlak, E., Jackson, D.: Kodkod: A relational model finder. In: Grumberg, O., Huth, M. (eds.) TACAS 2007. LNCS, vol. 4424, pp. 632–647. Springer, Heidelberg (2007)

26. Wos, L., Overbeek, R., Lusk, E., Boyle, J.: Automated Reasoning: Introduction and Applications. Prentice-Hall, Englewood Cliffs (1984)

Analysis of Self-⋆ and P2P Systems Using Refinement

Manamiary Bruno Andriamiarina[1], Dominique Méry[1,⋆], and Neeraj Kumar Singh[2]

[1] Université de Lorraine, LORIA, BP 239, 54506 Vandœuvre-lès-Nancy, France
{Manamiary.Andriamiarina,Dominique.Mery}@loria.fr
[2] McMaster Centre for Software Certification,
McMaster University, Hamilton, Ontario, Canada
singhn10@mcmaster.ca, Neerajkumar.Singh@loria.fr

Abstract. Distributed systems and applications are becoming increasingly complex, due to factors such as dynamic topology, heterogeneity of components, failure detection. Therefore, they require effective techniques for guaranteeing safety, security and *convergence*. The self-⋆ systems are based on the idea of managing efficiently complex systems and architectures without user interaction. This paper presents a methodology for verifying distributed systems and ensuring safety and *convergence* requirements: *Correct-by-construction* and *service-as-event* paradigms are used for formalizing the system requirements using incremental refinement in EVENT B. Moreover, this paper describes a mechanized proof of correctness of the self-⋆ systems along with a case study related to the P2P-based self-healing protocol.

Keywords: Distributed systems, self-⋆, self-healing, self-stabilization, P2P, EVENT B, liveness, *service-as-event*.

1 Introduction

Nowadays, our daily lives are affected by technologies such as computers, chips, smartphones. These technologies are integrated into large distributed systems that are widely used, which provide required functionalities, (*emergent* [11]) behaviors and properties from interactions between components. Self-⋆ systems and their autonomous properties (e.g, self-stabilizing systems autonomically recovering from faults [5]) tend to take a growing importance in the development of distributed systems. In this study, we use the *correct by construction* approach [7] for modelling the distributed self-⋆ systems. Moreover, we emphasize on the *service-as-event* [2] paradigm, that identifies the phases of *self-stabilization* mechanism.

We consider that a system is characterized by events modifying the states of a system, and modelling abstract phases/procedures or basic actions according to the abstraction level. We define a self-stabilizing system S

Fig. 1. Diagram for a Self-Stabilizing System S

⋆ This work was supported by grant ANR-13-INSE-0001 (The IMPEX Project http://impex.loria.fr) from the Agence Nationale de la Recherche (ANR).

Y. Ait Ameur and K.-D. Schewe (Eds.): ABZ 2014, LNCS 8477, pp. 117–123, 2014.

with three states (see in Fig.1): *legal states* (*correct states* satisfying a safety property *P*), *illegal states* (violating the property *P*) and *recovery states* (states leading from *illegal* to *legal states*). The system S is represented by a set of events $\mathcal{M} = \mathcal{CL} \cup \mathcal{ST} \cup \mathcal{F}$. The subset $\mathcal{CL}$ models the computation steps of the system and introduces the notion of *closure* [4] : any computation starting from a *legal state* leads to another *legal state*. The occurence of a fault, modelled by an event $f \in \mathcal{F}$ (dotted transition in Fig.1), leads the system S into an *illegal state*. When a fault occurs, we assume that some procedures identify the current *illegal states* and simulate the stabilization (recovery ($r \in \mathcal{ST}$) and convergence ($r \in \mathcal{CV}$, with $\mathcal{CV} \subseteq \mathcal{ST}$)) procedure to legal states.

This paper is organised as follows. Section 2 introduces the formal verification approach including *service-as-event* paradigm and illustrates the proposed methodology with the study of the self-healing P2P-based protocol [8]. Section 3 finally concludes the paper along with future work.

2   Stepwise Design of the Self-healing Approach

In this section, we propose a formal methodology for self-⋆ systems that integrates the EVENT B method, the related toolbox RODIN platform and elements of temporal logics, such as traces properties (liveness). Using refinement, we gradually build models of self-⋆ systems in the EVENT B framework [1]. Moreover, we use the *service-as-event* paradigm to describe the *stabilization* and *convergence* from *illegal* states to *legal* ones. The concept of *refinement diagrams* [2,9] intends to capture the intuition of the designer for deriving progressively the target self-⋆ system.

2.1   Introduction to the Self-healing P2P-Based Approach

Fig. 2. Architecture

The development of *self-healing P2P-based approach* is proposed by Marquezan et al. [8], where the *reliability* of a P2P-system is the main concern. The self-healing process ensures that if a management service (a *task* executed by peers) of the system enters a *faulty/failed* state, then a self-healing/recovery procedure guarantees that the service switches back to a *legal* state. The self-healing is as follows: **(1) Self-detection** identifies failed instances (peers) of a management service. **(2) Self-activation** is started, whenever a management service is detected as failed. A failed service does not trigger recovery if there are still enough instances for running the service; otherwise, **(3) Self-configuration** repairs the service: new peers running the service are instantiated, and the service is returned into a *legal* state. We illustrate the use of *service-as-event* paradigm and *refinement diagrams* with the formal design of *self-healing approach*.

2.2   The Formal Design

Figure 2 depicts the formal design of *self-healing P2P-based approach*. The model M0 abstracts the approach. The refinements M1, M2, M3 introduce the *self-detection*, *self-activation* and *self-configuration*. Models from M4 to M20 are used for localising the self-healing algorithm. The last refinement M21 presents a local model that describes procedures for recovering process of P2P system.

Abstracting the Self-healing Approach (M0). We use the *service-as-event* paradigm to describe the main *functionality (i.e. recovery) offered* by the self-healing protocol. Each service (s) is described by two states: RUN (*legal/running* state) and $FAIL$ (*illegal/faulty* state). A property $P \mathrel{\widehat{=}} (s \mapsto RUN \in serviceState)$ expresses that a service (s) is in a *legal running* (RUN) state. An event FAILURE leads service (s) into a faulty state ($FAIL$), satisfying $\neg P$. The *self-healing* of service (s) is expressed by a liveness (*leads to*) property as follows : $(\neg P) \rightsquigarrow P$, meaning that each faulty state will *eventually* be followed by a legal one. The procedure is stated by an abstract event HEAL, where service (s) recovers from a *faulty* state to a *legal running* one. The refinement diagram[1] (see Fig.3) and events sum up the abstraction of a *recovery* procedure.

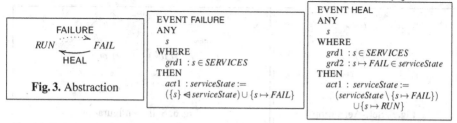

Fig. 3. Abstraction

```
EVENT FAILURE
ANY
    s
WHERE
    grd1  : s ∈ SERVICES
THEN
    act1  : serviceState :=
        ({s} ⩤ serviceState) ∪ {s ↦ FAIL}
```

```
EVENT HEAL
ANY
    s
WHERE
    grd1  : s ∈ SERVICES
    grd2  : s ↦ FAIL ∈ serviceState
THEN
    act1  : serviceState :=
        (serviceState \ {s ↦ FAIL})
            ∪ {s ↦ RUN}
```

This *macro/abstract view* of the *self-healing* is detailed by refinement[2], using intermediate steps guided by the three phases : *Self-detection*, *Self-activation* and *Self-configuration*. New variables denoted by $NAME\_\{Refinement\ Level\}$ are introduced.

Introducing the Self-detection (M1). A new state (FL_DT_1) defines the *detection of failures* : a service (s) can **suspect** and **identify** a failure ($FAIL\_1$) before triggering recovery (HEAL). We introduce a new property $R_0 \mathrel{\widehat{=}} (s \mapsto FL\_DT\_1 \in serviceState\_1)$ and a new event FAIL_DETECT. The steps of self-detection are introduced, using the inference rules [6] related to the operator *leads to* ($\rightsquigarrow$), as illustrated by refinement diagram 4 and proof tree. The event FAIL_DETECT expresses the *self-detection*: the failed state ($FAIL\_1$) of a service (s) is detected (state FL_DT_1). The property $(\neg P) \rightsquigarrow R_0$ is expressed by the event FAIL_DETECT. $R_0 \rightsquigarrow P$ is defined by the event HEAL, where the service (s) is restored to a *legal running* state after failure detection. The same method is applied to identify all the phases of *self-healing* algorithm. Due to   limited   space,

Fig. 4. Self-Detection

$$\frac{(\neg P) \rightsquigarrow R_0 \qquad R_0 \rightsquigarrow P}{(\neg P) \rightsquigarrow P}\ trans$$

we focus on the interesting parts of models and liveness properties. The complete development can be downloaded from web[3] and details can be found in the companion paper [3].

---

[1] The assertions ($s \mapsto st \in serviceState$), describing the state (st) of a service (s), are shorten into (st), in the nodes of the refinement diagrams, for practical purposes.

[2] ⊕: to add elements to a model, ⊖: to remove elements from a model.

[3] http://eb2all.loria.fr/html_files/files/selfhealing/self-healing.zip

EVENT FAILURE REFINES FAILURE ... WHERE $\oplus s \mapsto RUN\_1 \in serviceState\_1$ THEN $\ominus act1 : ...$ $\oplus serviceState\_1 :=$ $(serviceState\_1 \setminus \{s \mapsto RUN\_1\})$ $\cup \{s \mapsto FAIL\_1\}$	EVENT FAIL_DETECT ANY s WHERE $grd1 : s \in SERVICES$ $grd2 : s \mapsto FAIL\_1 \in serviceState\_1$ THEN $act1 : serviceState\_1 :=$ $(serviceState\_1 \setminus \{s \mapsto FAIL\_1\})$ $\cup \{s \mapsto FL\_DT\_1\}$	EVENT HEAL REFINES HEAL ... WHERE $\ominus grd2 ...$ $\oplus s \mapsto FL\_DT\_1 \in serviceState\_1$ THEN $\ominus act1 ...$ $\oplus serviceState\_1 :=$ $(serviceState\_1 \setminus \{s \mapsto FL\_DT\_1\})$ $\cup \{s \mapsto RUN\_1\}$

Introducing the Self-activation (M2) and Self-configuration (M3). The *self-activation* is introduced in M2 (see Fig. 5), where a failure of a service (s) is evaluated as critical or non-critical using a new state FL_ACT_2 and an event FAIL_ACTIV. The *self-configuration* step is introduced in M3 (see Fig.6): if the failure of service (s) is critical, then *self-configuration* for a service (s) is triggered (state FL_CONF_3), otherwise the failure is ignored (state FL_IGN_3).

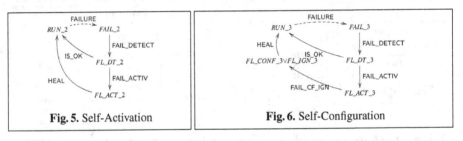

Fig. 5. Self-Activation   **Fig. 6.** Self-Configuration

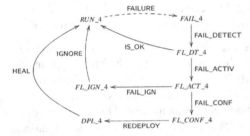

Fig. 7. Self-Healing steps

The Global Behaviour (M4). The models are refined and decomposed into several steps (see Fig.7) [8]. **(1)** *Self-Detection* phase is used to detect any failure in the autonomous system. The event FAIL_DETECT models the failure detection; and the event IS_OK states that if a detected failure of a service (s) is a *false alarm*, then the service (s) returns to a *legal* state (RUN_4). **(2)** *Self-Activation* evaluates detected failures which are actual. The events FAIL_IGN and IGNORE are used to ignore the failure of service (s) when it is not critical (FL_IGN_4). The event FAIL_CONF triggers the reconfiguration of service (s) when failure is critical (FL_CONF_4). **(3)** *Self-Configuration* presents the healing procedure of a *failed* service using an event REDEPLOY.

The refinements M5, M6, M7 introduce gradually the running ($run\_peers(s)$), faulty ($fail\_peers[\{s\}]$), suspicious ($susp\_peers(s)$) and deployed instances ($dep\_inst[\{s\}]$) for a service (s). Each service (s) is associated with the minimal number of instances required for running service (s): during the *self-activation* phase, if the number of running instances of service (s) is below than minimum, failure is critical. Models from M8 to M10 detail the *self-detection* and *self-configuration* phases to introduce the *token owners* for the services. Models from M11 to M20 localise gradually the events (we switch from a *service* point of view to the point of view of peers). Due to limited space[3], in the next section, we present only M21.

```
MACHINE 21 ...
  EVENT SUSPECT_INST
    ANY
      s, susp
    WHERE
      grd1 : s ∈ SERVICES
      grd2 : susp ⊆ PEERS
      grd3 : susp = run_inst(token_owner(s) ↦ s) ∩ unav_peers
      grd4 : suspc_inst(token_owner(s) ↦ s) = ∅
      grd5 : inst_state(token_owner(s) ↦ s) = RUN_4
      grd6 : susp ≠ ∅
    THEN
      act1 : suspc_inst(token_owner(s) ↦ s) := susp
    END
  EVENT FAILURE ...
  EVENT RECONTACT_INST_OK ...
  EVENT RECONTACT_INST_KO ...
  EVENT FAIL_DETECT ...
  EVENT IS_OK ...
  EVENT FAIL_ACTIV ...
  EVENT FAIL_IGNORE ...
  EVENT IGNORE ...
  EVENT FAIL_CONFIGURE ...
  EVENT REDEPLOY_INSTC ...
  EVENT REDEPLOY_INSTS ...
  EVENT REDEPLOY ...
  EVENT HEAL ...
  EVENT MAKE_PEER_UNAVAIL ...
  EVENT UNFAIL_PEER ...
  EVENT MAKE_PEER_AVAIL ...
```

The Local Model (M21). This model details locally the *self-healing* procedure of a service (s). The notion of *token owner* is more detailed: the *token owner* is a peer instance of service (s) that is marked as a *token owner* for the Management Peer Group (MPG), i.e. the set of peers instantiating service (s). It controls *self-healing* by applying *self-detection*, *self-activation*, and *self-configuration* steps. **(1) Self-Detection** introduces an event SUSPECT_INST that states that the *token owner* is able to *suspect* a set (*susp*) of unavailable instances of service (s). Events RECONTACT_INST_OK and RECON-TACT_INST_KO are used to specify the successful and failed recontact, respectively, of the unavailable instances for ensuring failures. Moreover, the *token owner* is able to monitor the status of service (s) using two events FAIL_DETECT, and IS_OK. If instances remain unavailable after the recontacting procedure, the *token owner* informs the safe members of MPG of failed instances (FAIL_DETECT); otherwise, the *token owner* indicates that service (s) is running properly (IS_OK). **(2) Self-Activation** introduces an event FAIL_ACTIV where the *token owner* evaluates if a failure is critical. Event FAIL_IGNORE specifies that the failure is not critical. It is ignored (event IGNORE), if several instances (more than minimum) are running correctly. Otherwise, the failure will be declared critical, and *self-configuration* will be triggered using an event FAIL_CONFIGURE. **(3) Self-Configuration** introduces three events REDEPLOY_INSTC, REDEPLOY_INSTS and REDEPLOY that specify that new instances of running service (s) are deployed until the minimal number of instances is reached. And after, the event HEAL can be triggered, corresponding to the *convergence* of the self-healing process.

Moreover, in this model, we have formulated *hypotheses* for ensuring the correct functioning of the self-healing process: *(1) If the token owner of a service (s) becomes unavailable, at least one peer, with the same characteristics as the disabled token owner (state, local informations about running, failed peers, etc.) can become the new token owner; (2) There is always a sufficient number of available peers that can be deployed to reach the legal running state of a service (s).* In a nutshell, we say that our methodology allows users to understand the self-⋆ mechanisms, to gain insight into their architectures (components, coordination, etc.); and gives evidences of their correctness under some assumptions/hypotheses.

3   Discussion, Conclusion and Future Work

We present a methodology based on liveness properties and *refinement diagrams* for modelling the self-⋆ systems using EVENT B. The key ideas are to characterize the self-stabilizing systems by modes : 1) *legal (correct)* state, 2) *illegal (faulty)* state, and 3) *recovery* state (see Fig.1); then identify the required abstract steps between modes, for ensuring *convergence*; and enrich abstract models using refinement. We have illustrated our methodology with the *self-healing approach* [8]. The complexity of the development is measured by the number of proof obligations (POs) which are automatically/manually discharged (see Table 1). A large majority ($\sim$ 70%) of the 1177 manual proofs is solved by simply running the provers from the Atelier B. The actual summary of POs is given by Table 2. Manually discharged POs require analysis and skills, whereas *quasi*-automatically discharged POs would only need a *tuning* of RODIN (e.g. provers run automatically).

Table 1. Summary of Proof Obligations

Model	Total	Auto		Interactive	
CONTEXTS	30	26	86.67%	4	13.33%
M0	3	3	100%	0	0%
M1	21	15	71.4%	6	28.6%
M2	46	39	84.8%	7	15.2%
M3	68	0	0%	68	100%
M4	142	16	11.27%	126	88.75%
OTHER MACHINES	1111	158	14.22%	953	85.78%
M21	13	0	0%	13	100%
TOTAL	1434	257	17.9%	1177	82.1%

Table 2. Synthesis of POs

Total	Auto		Quasi-Auto		Manual	
1434	257	17.9%	850	59.3%	327	22.8%

Furthermore, our refinement-based formalization produces local models close to the *source code*. Our future works include the generation of applications from the resulting model extending tools like EB2ALL [10]. Moreover, further case studies will help us to discover new patterns that could be implemented in the RODIN platform. Finally, another point would be to take into account dependability properties and concurrency.

References

1. Abrial, J.-R.: Modeling in Event-B: System and Software Engineering. Cambridge University Press (2010)
2. Andriamiarina, M.B., Méry, D., Singh, N.K.: Integrating Proved State-Based Models for Constructing Correct Distributed Algorithms. In: Johnsen, E.B., Petre, L. (eds.) IFM 2013. LNCS, vol. 7940, pp. 268–284. Springer, Heidelberg (2013)
3. Andriamiarina, M.B., Méry, D., Singh, N.K.: Analysis of Self-$\star$ and P2P Systems using Refinement (Full Report). Technical Report, LORIA, Nancy, France (2014)
4. Berns, A., Ghosh, S.: Dissecting self-* properties. In: Proceedings of the 2009 Third IEEE International Conference on Self-Adaptive and Self-Organizing Systems, SASO 2009, pp. 10–19. EEE Computer Society, Washington, DC (2009)
5. Dolev, S.: Self-Stabilization. MIT Press (2000)
6. Lamport, L.: A temporal logic of actions. ACM Trans. Prog. Lang. Syst. 16(3), 872–923 (1994)
7. Leavens, G.T., Abrial, J.-R., Batory, D.S., Butler, M.J., Coglio, A., Fisler, K., Hehner, E.C.R., Jones, C.B., Miller, D., Jones, S.L.P., Sitaraman, M., Smith, D.R., Stump, A.: Roadmap for enhanced languages and methods to aid verification. In: Jarzabek, S., Schmidt, D.C., Veldhuizen, T.L. (eds.) GPCE, pp. 221–236. ACM (2006)

8. Marquezan, C.C., Granville, L.Z.: Self-* and P2P for Network Management - Design Principles and Case Studies. Springer Briefs in Computer Science. Springer (2012)
9. Méry, D.: Refinement-based guidelines for algorithmic systems. International Journal of Software and Informatics 3(2-3), 197–239 (2009)
10. Méry, D., Singh, N.K.: Automatic code generation from event-b models. In: Proceedings of the Second Symposium on Information and Communication Technology, SoICT 2011, pp. 179–188. ACM, New York (2011)
11. Smith, G., Sanders, J.W.: Formal development of self-organising systems. In: González Nieto, J., Reif, W., Wang, G., Indulska, J. (eds.) ATC 2009. LNCS, vol. 5586, pp. 90–104. Springer, Heidelberg (2009)

B Formal Validation of ERTMS/ETCS Railway Operating Rules

Rahma Ben Ayed[1], Simon Collart-Dutilleul[1], Philippe Bon[1],
Akram Idani[2], and Yves Ledru[2]

[1] IFSTTAR-Lille, 20 Rue Elisée Reclus BP 70317,
59666 Villeneuve d'Ascq Cedex, France
{rahma.ben-ayed,simon.collart-dutilleul,philippe.bon}@ifsttar.fr
http://www.ifsttar.fr
[2] UJF-Grenoble 1/Grenoble-INP/UPMF-Grenoble 2/CNRS, LIG UMR 5217, 38041
Grenoble, France
{akram.idani,yves.ledru}@imag.fr
http://www.liglab.fr

Abstract. The *B* method is a formal specification method and a means of formal verification and validation of safety-critical systems such as railway systems. In this short paper, we use the B4MSecure tool to transform the UML models, fulfilling requirements of European Railway Traffic Management System (ERTMS) operating rules, into *B* specifications in order to formally validate them.

Keywords: Railway operating rules, UML models, Role Based Access Control, *B* method, formal validation.

1 Introduction

ERTMS [6] is the European Railway Traffic Management System which is designed to replace the national on-board railway systems in Europe in order to make rail transport safer and more competitive, and to improve cross-border connections. ERTMS includes the European Train Control System (ETCS) which specifies the on-board equipment and its communication with the trackside.

The aim of our work is to confront the European specifications with the national operating rules, as well as the use of formal models to validate whether a given scenario fulfills the specification regarding the functional and safety requirements. We propose to model a nominal scenario of *Movement Authority (MA)*, extracted from ERTMS operating rules, and to translate it into *B* specifications in order to validate it.

In the following section, an overview of the nominal scenario *MA* is given and its UML models are described. Section 3 highlights the *B* formal validation after an automatic translation of these models into *B* specifications using B4MSecure. Finally, section 4 concludes this paper.

Y. Ait Ameur and K.-D. Schewe (Eds.): ABZ 2014, LNCS 8477, pp. 124–129, 2014.
© Springer-Verlag Berlin Heidelberg 2014

2 *Movement Authority* Overview

Movement Authority (MA) is an authorization given to a train to move to a given point as a supervised movement. Some features can be used to define an *MA*, such as sections subdividing it, the time-out value attached to each section, etc. The *MA* function unfolds with interactions between the *OnboardSafetyManagement* (the on-board computer-based machine), the *TracksideSafetyManagement* (the trackside computer-based machine), and the *Driver*, as follows:

MA.1 The *OnboardSafetyManagement* requests an *MA* to the *TracksideSystem*.

MA.2 The *TracksideSafetyManagement* receives the *MA* request from the *TracksideSystem*.

MA.3 The *TracksideSafetyManagement* proposes an *MA* to the *TracksideSystem* after creating it. It can also modify and/or delete the *MA*.

MA.4 The *OnboardSafetyManagement* receives the proposed *MA* from the *TracksideSystem*, authorizes it and processes the *MA* authorization in order to be displayed in the *Driver Machine Interface (DMI)*.

MA.5 The *Driver* reads the authorized *MA*.

Each step of this scenario represents a permission to do an action on an entity by a role. On this basis, 3 roles (*OnboardSafetyManagement, TracksideSafetyManagement, Driver*), 3 system entities (*TracksideSystem, MA, DMI*) and 10 possible permissions (underlined actions) can be extracted.

Our approach consists in, on the one hand, the modeling of ERTMS operating rules in semi-formal UML notations with their graphical views and dedicated profiles extensions taking into account various aspects (structural, dynamic, behavioural, etc.), and on the other hand, their validation and verification with a formal *B* method with its mathematical notations and automated proof. The combination of these two notations has been studied and several approaches of UML to *B* translation have been proposed, cited in [3]. In order to model the scenario above, we use B4MSecure platform supporting the UML/B modeling process and lying within the scope of Model Driven Engineering (MDE).

For the sake of concision, the B4MSecure platform [7] is briefly presented. As an Eclipse platform, it is dedicated to formally validate a functional UML model enhanced by an access control policy. It uses a Role Based Acces Control (RBAC) profile inspired from SecureUML profile [4]. This profile aims at specifying information related to access control in order to model roles and their permisssions. This platform acts in 3 steps: a functional UML class diagram specifying system entities, security UML models with an access control policy and the translation of both models into *B* specifications.

Following the three-stepped approach of B4MSecure, a functional UML class diagram containing all system entities as classes and the relationships between them is built. Then, security UML class diagrams enhance the functional model by expressing which role has the permission to perform a given action in the railway system: a class diagram dedicated to the roles and others dedicated to the access control policies which are based on permissions linking the roles to the entities, such as the access control of the *MA* in Fig. 1. A permission is modeled as a UML association class, between a role and a class of the functional model, with a stereotype *Permission*. For instance, **MA.4** is modeled in Fig. 1 by the permission of the *OnboardSafetyManagement* to authorize the *MA*. All these diagrams are translated to *B* specifications.

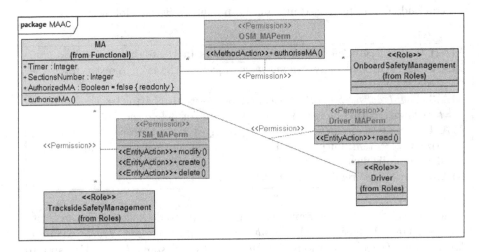

Fig. 1. Roles and permissions associated with MA

3 *B* Formal Validation of *Movement Authority*

The functional model is translated into a unique *B* machine, named *Functional*, and permissions are translated into a *B* machine, named *RBAC_Model*. As shown in Fig. 2, the functional formal model follows a classical translation scheme similar to [5]. The *RBAC_Model* adds variables about permissions and roles. For example, *PermissionAssignement* is a total function from *PERMISSIONS* to the cartesian product *(ROLES * ENTITIES)*, and *isPermitted* is a relation between *ROLES* and *Operations* sets. *PERMISSIONS*, *ENTITIES* and *Operations* are the sets defined in *RBAC_Model*, while *ROLES* is a set defined in the included *UserAssignments* machine. Initialization of these variables is conformant to the SecureUML model. Then, initialization proof obligation, produced by the AtelierB prover for these variables, allows to verify whether the SecureUML model respects RBAC well-formedness rules such as no cycles in role hierarchy, etc. The operations of the security formal model encapsulate the operational part of the functional formal model. Each functional operation is associated with an operation in the security model verifying that a user has permission to call the functional operation. For instance, *secure_MA__authorizeMA* operation of *RBAC_Model* checks the permissions associated with the functional operation *MA__authorizeMA*. *Secured* operations add a statement in the postcondition

```
Machine                          Machine
 Functional                       RBAC_Model
                                  INCLUDES
                                   Functional, UserAssignments
SETS                              SEES
 MA_AS; ...                        ContextMachine
                                  SETS
ABSTRACT_VARIABLES                 ENTITIES = {MA_Label, ...};
 MA, ...                           Attributes = {MA_AuthorizedMA_Label, ...};
                                   Operations ={MA_authorizeMA_Label, ...} ...
INVARIANT                         VARIABLES
 MA <: MA_AS & ...                 PermissionAssignement, isPermitted, ...
                                  INVARIANT
INITIALISATION                     PermissionAssignement: PERMISSIONS --> (ROLES * ENTITIES)
 MA := {}  || ...                  & isPermitted: ROLES <-> Operations ...
                                  INITIAISATION
OPERATIONS                         PermissionAssignement :=
                                   {(OSM_MAPerm|->(OnboardSafetyManagement|->MA_Label)),...}
MA__authorizeMA(Instance)=        OPERATIONS
 PRE                              secure_MA__authorizeMA(Instance)=  PRE
 Instance : MA &                   Instance: MA & MA__AuthorizedMA(Instance) = FALSE THEN
 MA__AuthorizedMA(Instance) = FALSE SELECT MA__authorizeMA_Label : isPermitted[currentRole]
 THEN                               THEN MA__authorizeMA(Instance)
 MA__AuthorizedMA(Instance) := TRUE  END
END; ...                          END; ...

END                              END
```

Fig. 2. *Functional* and *RBAC_Model* machines

e.g *SELECT MA__authorizeMA_Label: isPermitted[currentRole]* in order to verify whether *MA__authorizeMA_Label* is allowed to the connected user using a
particular role. Indeed, *isPermitted* computes, from the initial state, the set of
authorized functional operations for each role.

UML models of extracted ERTMS/ETCS operating rules containing 7 functional classes, 5 roles and 17 permissions are transformed into 830 lines of functional formal model and 1545 lines of security formal model. We use the ProB
animator in order to validate these specifications. A first animation checks the
nominal behaviour of *Movement Authority*. Then variants of this animation check
that the given permissions forbid the execution of secure actions by unautho-
rized roles, since a secure action can be performed only with a permission given
to a role. The ability of the system specified by the class diagram to play the
ERTMS scenarios is checked through animations of the corresponding transformed B model. These animations validate the permissions assigned to each
role. But they don't check that the sequence of actions models the MA protocol.
Actually, the sequence of actions is defined in the animation by the user, but
it is not embedded in the UML/B model. This can be resolved by adding some
contraints as preconditions in secured operations. Nevertheless, adding these
conditions breaks the consistency between the UML model and the B machine.
Owing to the lack of dynamic aspects in UML class diagrams, we intend to explore more UML diagrams as future work. Then we will focus on enriching UML
class diagrams with, for instance, sequence diagrams which model the ordered

interactions in scenarios and deriving B specifications from them in order to validate system's behavior.

At this stage, safety requirements have not yet been integrated to B specifications. As further work, we will consider enriching the B specifications with safety properties stemming from safety requirements of the ERTMS operating rules in order to formally verify them using the B prover. Moreover, SysML requirement diagrams combined with our UML diagrams may guarantee the traceability aspects of system requirements when they will be translated to B specifications.

4 Conclusion

In this short paper, we have presented a *Movement Authority* function extracted from the ERTMS/ETCS operating rules. This function was modeled using UML graphical notations and then translated automatically, via the B4MSecure platform, into B specifications which were checked successfully using the ProB animator. The combination of UML/B aims to ease the understanding of the system with the graphical notations of UML and formally validate system requirements with B formal notations. Research works done in the Selkis project [1], [2] and [3] show the efficiency of this platform and its different steps leading to the formal validation of scenarios in the healthcare Information Systems by seeking for malicious sequences of operations. However, in this paper, we show the use of this existant platform in another context related to distributed railway systems and their operating rules.

Acknowledgements This research is funded by the PERFECT project (ANR-12-VPTT-0010) and is partly supported by the SELKIS project (ANR-08-SEGI-018).

References

1. Ledru, Y., Idani, A., Milhau, J., Qamar, N., Laleau, R., Richier, J.-L., Labiadh, M.-A.: Taking into Account Functional Models in the Validation of IS Security Policies. In: Salinesi, C., Pastor, O. (eds.) CAiSE 2011 Workshops. LNBIP, vol. 83, pp. 592–606. Springer, Heidelberg (2011)
2. Milhau, J., Idani, A., Laleau, R., Labiadh, M.A., Ledru, Y., Frappier, M.: Combining UML, ASTD and B for the formal specification of an access control filter. In: Innovations in Systems and Software Engineering, vol. 7, pp. 303–313. Springer (2011)
3. Idani, A., Labiadh, M.A., Ledru, Y.: Infrastructure dirigée par les modèles pour une intégration adaptable et évolutive de UML et B. Ingénierie des Systèmes d'Information Journal 15, 87–112 (2010)
4. Lodderstedt, T., Basin, D., Doser, J.: SecureUML: A UML-Based Modeling Language for Model-Driven Security. In: Jézéquel, J.-M., Hussmann, H., Cook, S. (eds.) UML 2002. LNCS, vol. 2460, pp. 426–441. Springer, Heidelberg (2002)

5. Laleau, R., Mammar, A.: An overview of a method and its support tool for generating B specifications from UML notations. In: Proceedings of the 15th IEEE International Conference on Automated Software Engieering, ASE 2000, pp. 269–272. IEEE Computer Society, Washington (2000)
6. ERTMS, http://www.ertms.net
7. B4MSecure, http://b4msecure.forge.imag.fr

Modelling Energy Consumption
in Embedded Systems with VDM-RT

José Antonio Esparza Isasa, Peter Würtz Vinther Jørgensen,
Claus Ballegård Nielsen, and Stefan Hallerstede

Department of Engineering, Aarhus University,
Finlandsgade 22, Aarhus N 8200, Denmark
{jaei,pvj,clausbn,sha}@eng.au.dk

Abstract. Today's embedded systems are typically battery powered devices with
a limited energy budget, which makes optimal usage of energy a critical task
faced by embedded system developers. One approach to making an embedded
system energy-efficient is by putting its processing units (CPUs) into less energy
consuming modes when their computational resources are not needed. Physical
prototyping enables exploration of different strategies for energy usage by controlling the power modes of the system CPUs. This may lead to construction of
many different hardware and software systems and thus impose high costs. In
this paper we propose language extensions for VDM-RT that enable modelling
and execution of different strategies for energy usage in embedded systems by
controlling the hardware architecture and power modes of the system CPUs.

Keywords: Energy-aware design, embedded systems, VDM-RT.

1 Introduction

This work presents a formal modelling approach for performing analysis and evaluation
of the energy consumption in embedded systems. The approach is based on the VDM-
RT [4] dialect and specifically focuses on including CPU sleep mode in the simulation.
This mode is an operational state that most of the modern CPUs used in today's micro-
controllers implement. In this mode the system enters a low power consumption mode
in which the CPU is deactivated and waiting for an event to wake up.

We proposed the representation of sleeping states in VDM-RT by applying a design
pattern structure at the modelling level [1]. This initial approach, while effective, introduced an additional level of complexity at the modelling stage. In this paper we present
the extensions to the language to incorporate the notion of a sleeping CPU state. The
preliminary application of this modelling approach to a simple case study has shown
that it is possible to produce estimates of approximately 95% accuracy without having
to prototype each solution under consideration.

The reminder of this paper presents: The language and tool modifications in sections 2 and 3 respectively. An example of how this approach can be used is presented
in section 4. Finally, section 5 elaborates on future work and section 6 concludes this
paper.

Y. Ait Ameur and K.-D. Schewe (Eds.): ABZ 2014, LNCS 8477, pp. 130–135, 2014.

2 Language Modifications

2.1 CPUs in VDM-RT

CPUs in VDM-RT are computational environments where parts of a model can be deployed so they will be executed. They include a real time operating system and can be configured in terms of frequency and scheduling policy. CPUs are configured in a **system** class together with the bus abstraction, that represents connectivity between them. An example of this is shown in Listing 1, where CPUs are declared and model components mapped to them.

```
instance variables
  public static mcu : CPU := new CPU(<FP>, 20E6);
  public static ctrl : Controller := new Controller();
  public static irq : CPU := new CPU(<FP>,20E6);
  public static wu : WakeUpL := new WakeUpL(); ...
operations
  public System : () ==> System
   System () == (
     mcu.deploy(appLogic,"ApplicationLogic");
     irq.deploy(wu,"Interrupt Provider"); ...
```

Listing 1: Main body of the System class showing two CPUs and logic being deployed

VDM-RT CPUs are always in the same operational state, active, and therefore they do not incorporate any representation of low power state.

2.2 Language Extensions

Extending VDM-RT with the sleep operation makes it possible to represent embedded software that sleeps the CPU responsible for executing the application logic. This is shown in Listing 2. In this example the ApplicationLogic executes in APP_TIME time units and then the CPU (microcontroller unit, mcu in Listing 2) is put to sleep. Once woken up execution resumes by invoking the PostWULogic operation.

```
duration (APP_TIME) ApplicationLogic();
System'mcu.sleep(); -- Blocks until activated externally
duration (POST_WAKE_UP) PostWULogic();
```

Listing 2: High-level representation of the embedded software sleeping the CPU

The operation active can be invoked by the parts of the model that represent internal wake-up sources such as the sleep timer or external sources such as interrupts (see Listing 2). This allows the sleeping CPU to resume execution (see Listing 3).

```
System'mcu.active();
```

Listing 3: Static call waking up (activating) the mcu

2.3 Modelling Different Wake-up Policies

The language extensions presented above, and more specifically the way the `active` operation is used, can model different wake-up policies. In order to represent an external interrupt based wake-up policy, the model needs to incorporate additional logic to represent external events that can be fed to the system. These events will trigger the wake-up logic at a certain time as shown in Listing 4.

```
if time > TM1 and time < TM2 then System'FeedEvent();
```

Listing 4: Event fed to the model of the system between TM1 and TM2.

The system model uses a periodic thread deployed on a non-sleeping CPU to represent a hardware block that waits for the external interrupt. This is shown in Listing 5. Once the event has been recognized by the model representing the system, it can invoke the `active` operation and resume application processing.

```
thread periodic (CHECK_EVENT_PERIOD,0,0,0) (monitorEvent);
```

Listing 5: Periodic execution of the monitor event logic

The second wake-up policy is based on sleep timer expiration. From the modelling perspective it can be treated in a manner analogous to the external event modelling approach presented above. A periodic thread runs the sleep timer logic (see Listing 6) that periodically checks if the `overflowOn` time mark has elapsed. If this condition is satisfied the CPU is activated and resumes execution of application logic. It is the responsibility of the application logic to set up the sleep timer again before going to sleep the next time.

```
duration(0)( if (time > overflowOn) then (
   System'mcu.active();) );
```

Listing 6: Sleep Timer logic

3   Tool Modifications

Implementing support for CPU power modes in VDM-RT involved changing both the interpretation of VDM-RT models and the scheduling of the system resources, i.e. the CPUs and buses comprising the VDM-RT system. In this section we highlight the most important changes made to the Overture tool in order to implement the proposed VDM-RT language extensions.

3.1   Interpreter Modifications

The sleep and active operations were added to the existing VDM-RT specification of the CPU class operations. Listing 7 shows the specification of the operations added as part of the tool extensions we have made. Since the sleep and active operations are specified in VDM-RT there was no need to extend the lexer and parser. The type checker, however, needs to decorate the nodes of the abstract syntax tree representing the operation invocations of sleep and activate with appropriate type information.

```
public sleep : () ==> ()
   sleep() == is not yet specified;
public active : () ==> ()
   active() == is not yet specified
```

Listing 7: Specification of the existing sleep and active operations in VDM-RT

Our implementation further introduces a flag called _sleeping that indicates whether a CPU is asleep or active. When the sleep and active operations are invoked in the VDM-RT model the VDM-RT interpreter invokes the **is not yet specified** statement, which results in the VDM-RT execution being delegated to a Java execution environment where the _sleeping flag is set. The implementation allows consecutive invocations of sleep or active although this has no effect.

3.2   Scheduler Modifications

The VDM-RT interpreter of Overture [3] uses a *resource scheduler* for coordinating the activity of resources. The resource scheduler is responsible for controlling the execution on the resources according to their scheduling policy (e.g. fixed priority), which determines the order at which threads execute (on the CPU on which they are allocated) and what message is to be sent on a bus. In addition, the scheduling policy is used for calculating the amount of time a thread is allowed to execute (its time slice). In particular, a periodic thread with period p_0 that is scheduled to run will have to wait p_0 time units before it will execute again.

Active threads can request the resource scheduler to advance the global time by some time step, (e.g. the end of a statement is reached) and the resource scheduler now stops including it in the scheduling process. When all resources have reached this state the resource scheduler will advance the global time by the smallest time step. All waiting threads that are satisfied by this time step are allowed to run (they are considered for scheduling again), while the other threads will have to wait for the time remaining of their time step.

The _sleeping flag is used in the scheduling process to indicate to the resource scheduler that the CPU is asleep. When the scheduler executes the scheduling process it will invoke the reschedule method on every resource. For CPUs that are asleep reschedule will always return **false** to inform the scheduler that the CPU cannot be rescheduled.

4 Application Example

We proposed a concrete modelling structure to use the language extensions presented above. This structure is shown in the class diagram in Fig. 1. This is a generic structure that uses the two energy saving policies presented in section 2.3. This structure uses two VDM-RT CPUs that are connected through a VDM-RT bus (esBUS):

WakeUpHWSource: represents a hardware block that features the necessary components to wake up the CPU from a sleeping state.

MCU: represents a microcontroller unit containing the CPU that will be operating in active and sleep mode depending on the logic under study.

These CPUs run the following model components:

WakeUpSource: defines a generic template that must be realized by the components that wake up the CPU. It specifies the use of the operation mcu.active().

SleepTimer: is a concrete realization of the WakeUpSource that models a wake up configuration based on a resettable timer. This is modelled following the structure proposed above.

WakeUpInterrupt: is a concrete realization of the WakeUpSource that models a wake up configuration based on an interrupt generated on the occurrence of an external event. This is modelled following the structure proposed above.

Application Logic: models the application logic running in the main CPU. This application is able to sleep the CPU through the invocation of the mcu.sleep() operation. This logic will time mark the transitions between the active and sleep states and logs them in a file.

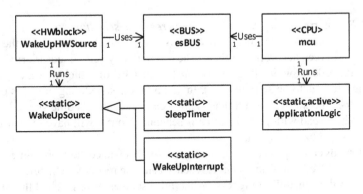

Fig. 1. Generic structure for models using the sleep extension

The Log produced by the application logic can be analysed once the simulation has completed in order to study both power and energy consumption. Based on this log and taking into consideration the manufacturers specification for the CPU under study one can plot directly the power consumption of the CPU over time. In order to determine the energy consumption this curve should be integrated. This analysis is considerably easier than a study based on prototypes and measurements and provides a high level of accuracy ($\approx 95\%$).

5   Future Work

The work presented in this paper is focused on the analysis of CPU energy consumption in embedded systems. However, in some cases communication with other systems also has large impact on energy consumption. VDM-RT incorporates the bus abstraction to represent communication between CPUs. Part of our future work is to evaluate the capability of this abstraction to study energy consumption aspects during the design of communication subsystems. This could require the creation of libraries or language modifications, following a similar approach to the one presented in this paper. The future analysis of energy consumption in communication combined with the work presented in this paper can be beneficial when design trade-offs between computation and communication exists. This is common in design of communication protocols for low power embedded systems. We frame communication, computation and other factors that impact energy consumption under a concrete design approach in [2].

6   Conclusions

Exploring different strategies for efficient energy usage in embedded system development by controlling the power modes of CPUs has most traditionally been done through the construction of different physical prototypes. In this paper we have enabled design space exploration of energy usage in embedded systems using the VDM-RT notation. This enables making design decisions based on accurate energy consumption predictions without having to realise multiple system architectures. We have implemented language extensions for VDM-RT in the Overture tool, which enable controlling power modes of system CPUs at the modelling level. Combining our language extensions with existing support for describing hardware architectures enables reasoning about energy usage in embedded systems using VDM-RT.

References

1. Isasa, J.A.E., Larsen, P.G.: Modelling Different CPU Power States in VDM-RT. In: Proceedings of the 11th Overture Workshop 2013, Aarhus University (June 2013)
2. Isasa, J.A.E., Larsen, P.G., Hansen, F.O.: Energy-Aware Design of Embedded Software through Modelling and Simulation. In: Proceedings of the SYRCoSE 2014 Symposium (May 2014)
3. Lausdahl, K., Larsen, P.G., Battle, N.: A Deterministic Interpreter Simulating a Distributed Real Time System Using VDM. In: Qin, S., Qiu, Z. (eds.) ICFEM 2011. LNCS, vol. 6991, pp. 179–194. Springer, Heidelberg (2011)
4. Verhoef, M.: Modeling and Validating Distributed Embedded Real-Time Control Systems. PhD thesis, Radboud University Nijmegen (2009)

Sealed Containers in Z

Eerke Boiten[1] and Jeremy Jacob[2]

[1] School of Computing, University of Kent, UK
[2] Department of Computer Science, University of York, UK

Abstract. Physical means of securing information, such as sealed envelopes and scratch cards, can be used to achieve cryptographic objectives. Reasoning about this has so far been informal.

We give a model of distinguishable sealed envelopes in Z, exploring design decisions and further analysis and development of such models.

1 Introduction

Physical mechanisms for securing information such as sealed envelopes and scratch cards have powerful properties. They contain information, which remains hidden until an explicitly decided moment. Up to then, anyone who has not constructed the item can plausibly argue ignorance. Reasoning about such mechanisms so far has been done informally, which is insufficient for complex applications like voting and polling protocols [9], and most fundamentally: general cryptographic schemes [8]. In such schemes, scratch cards and sealed envelopes play an intricate role, and the notions of security are sophisticated, starting from possibilistic and extending to probabilistic and complexity aspects.

This paper explores the formal modelling and analysis of sealed distinguishable *envelopes* containing a single *bit*, applied in protocols for *bit commitment*[8]. Our emphasis will be on constructing the model and protocols, looking ahead to semantic requirements for systematic analysis and formal development.

2 Modelling Sealed Envelopes in Z

This is a story about *Agent*s who pass *Envelope*s about:

[*Agent, Envelope*]

Envelopes contain bits, and may be uncreated, closed or open. The value in a closed envelope may only be known by its creator; the value in an open envelope is known by anyone who possesses it. A created envelope is in the possession of exactly one agent in any state. We model an agent knowing the content of an envelope by relations *zero* and *one*. The predicate $a \mapsto e \in zero$ encodes that agent a has direct evidence (it created the envelope, or has seen it when open) that e contains a 0-bit; and similarly for *one*. The predicates below state that all open envelopes are held by some agent, the content of every created envelope is known by some agent, and all agents who know the content

Y. Ait Ameur and K.-D. Schewe (Eds.): ABZ 2014, LNCS 8477, pp. 136–141, 2014.

of an envelope agree on it. Operations on this state need to satisfy the criteria that open envelopes cannot be closed, and agents' knowledge never decreases.

$$
\begin{array}{l}
\underline{\;S\;}\\
holder : Envelope \nrightarrow Agent\\
open : \mathbb{F}\, Envelope\\
zero, one : Agent \leftrightarrow Envelope\\
\hline
open \subseteq \operatorname{dom} holder\\
\operatorname{ran}(zero \cup one) = \operatorname{dom} holder\\
\operatorname{ran} zero \cap \operatorname{ran} one = \varnothing
\end{array}
\qquad
\begin{array}{l}
\underline{\;OpS\;}\\
\Delta S\\
\hline
open \subseteq open'\\
\;\\
zero \subseteq zero'\\
\;\\
one \subseteq one'
\end{array}
$$

There are three operations, to create, move and open sets of envelopes. When an agent $a?$ creates envelopes it does so for a set of envelopes to store 0-bits, $zs?$, and for a set of 1-bits, $os?$ – either may be singleton or empty. New envelopes are held by their creator, are closed, and their values are known to their creator only.

$$
\begin{array}{l}
\underline{\;Create\;}\\
OpS;\; a? : Agent;\; zs?, os? : \mathbb{F}\, Envelope\\
\hline
(zs? \cup os?) \cap \operatorname{dom} holder = \varnothing \;\wedge\; zs? \cap os? = \varnothing\\
holder' = holder \cup ((zs? \cup os?) \times \{a?\}) \;\wedge\; open' = open\\
zero' = zero \cup (\{a?\} \times zs?) \;\wedge\; one' = one \cup (\{a?\} \times os?)
\end{array}
$$

A set of envelopes may be moved to a named agent as long as they have all been created, and are held by one agent. The receiving agent learns the values of any open envelopes. No envelope is opened by this operation.

$$
\begin{array}{l}
\underline{\;Move\;}\\
OpS;\; b? : Agent;\; es? : \mathbb{F}\, Envelope\\
\hline
es? \subseteq \operatorname{dom} holder \wedge \exists a : Agent \bullet holder(\!|\, es?\, |\!) = \{a\}\\
holder' = holder \oplus (es? \times \{b?\}) \wedge open' = open\\
zero' = zero \cup (\{b?\} \times (es? \cap open \cap \operatorname{ran} zero))\\
one' = one \cup (\{b?\} \times (es? \cap open \cap \operatorname{ran} one))
\end{array}
$$

The holder of a set of envelopes can open them. The holder learns their values, they do not change hands.

$$
\begin{array}{l}
\underline{\;Open\;}\\
OpS;\; es? : \mathbb{F}\, Envelope\\
\hline
es? \subseteq \operatorname{dom} holder \;\wedge\; es? \cap open = \varnothing\\
holder' = holder \;\wedge\; open' = open \cup es?\\
\exists a : Agent \bullet holder(\!|\, es?\, |\!) = \{a\} \;\wedge\\
\quad zero' = zero \cup (\{a\} \times (es? \cap \operatorname{ran} zero)) \;\wedge\\
\quad one' = one \cup (\{a\} \times (es? \cap \operatorname{ran} one))
\end{array}
$$

We define a schema that reports an agent's view of a state. This is a *finalisation* operation, not commonly used in Z states-and-operations models, but a good way of encoding non-standard observations of abstract data types. The variables with 0-subscripts are those which represent $b?$'s view of (an instance of) state S.

___

View

$S; S_0; b? : Agent$

___

$holder_0 = holder \rhd \{b?\} \wedge open_0 = open \cap \text{dom } holder_0$
$zero_0 = \{b?\} \lhd zero \wedge one_0 = \{b?\} \lhd one$

___

3   Bit Commitment: A Challenge

Commitment is an essential cryptographic primitive between two parties. In the simplest case, where the value committed to is a single bit, it works as follows.

One party, the *sender*, executes an action $Commit(b)$ for given value b. From then, the *receiver* knows the sender has committed to a bit, but not which. This is the *hiding* property, which a cheating receiver may try to break. The situation now set up is typically exploited in other protocols, e.g. authentication, for further actions. After that, the sender can execute an *Open* action, upon which the receiver finds out b. The value learned should be the same value committed to – this is called the *binding* property, which a cheating sender may try to break. A surrounding protocol would typically be aborted on discovery of foul play.

The obvious but flawed implementation attempt using envelopes just has the sender (here $a?$) passing an envelope with the bit in to the receiver (here $b?$):

$$Commit \mathrel{\widehat{=}} [Create; es? : \mathbb{F}\, Envelope \mid es? = zs? \cup os? \wedge \#es? = 1] \mathbin{\mathring{,}} Move$$

Now we can specialise to the case of committing to a zero-bit, and the recipient's view afterwards (for one-bit it's the analogous *CommitOneView*), and state and prove that the two cases are indistinguishable ("hiding").

$$CommitZero \mathrel{\widehat{=}} [Commit \mid os? = \varnothing]$$
$$CommitZeroView \mathrel{\widehat{=}} \exists\, es?, zs?, os? : \mathbb{F}\, Envelope \bullet CommitZero \mathbin{\mathring{,}} View$$

$$\forall a?, b? : Agent \mid a? \neq b? \bullet \theta CommitZeroView = \theta CommitOneView$$

The recipient, $b?$, can open the envelope to discover the value of the bit in it. We introduce a view for the receiver once that has happened, and a correctness statement for "binding" at this stage:

$$CommitOpenView \mathrel{\widehat{=}} Commit \mathbin{\mathring{,}} Open \mathbin{\mathring{,}} View$$
$$CommitZeroOpenView \mathrel{\widehat{=}} CommitZero \mathbin{\mathring{,}} Open \mathbin{\mathring{,}} View$$

$$\forall\, CommitOpenView \bullet (es? \subseteq \text{ran } zero_0 \setminus \text{ran } one_0 \Rightarrow CommitZeroOpenView)$$
$$\wedge \ (es? \subseteq \text{ran } one_0 \setminus \text{ran } zero_0 \Rightarrow CommitOneOpenView)$$

However, this is not yet a correct implementation of commitment: *Open* is enabled as soon as the commitment has been made. Thus, it does not allow the sender to control *when* the receiver may learn the bit value. Removing the *Move* action from the commitment step also would not solve it, as this would break the *binding* property: the sender could then postpone the choice of bit. In the physical world, as in the cryptographic one, a sledgehammer solution to this is to assume a trusted third party, to which we can hand the envelope (or the bit) for safekeeping between committing and opening.

The exploration of bit commitment as a two party protocol, with the normal assumptions of a fixed number of messages of bounded size, has led to multiple negative results. First, it is impossible to achieve both perfect hiding and perfect binding. Approximate, computational, notions of security have been achieved with practical schemes, but the desirable compositionality property of "Universal Composability" is provably unattainable without further assumptions [5]. All this makes commitment a challenge for formal methods, requiring approximate notions of correctness and having an unsatisfiable "obvious ideal" specification.

4 Envelope Based Commitment Protocols

An envelope-based protocol that is more resistant against cheating needs three additional enhancements. First, the bit in the envelope needs to be masked with a "random" bit r. (In this paper, uniform probabilistic choice is abstracted to non-deterministic choice.) If the sender puts r **xor** v? in the envelope, opening the envelope will not reveal v?. The sender will transmit r to open the commitment, which will allow the receiver to learn v? by cancellation of **xor**. However, this allows the sender to cheat against binding, by transmitting $\neg r$ on opening.

The way out of this is for the *receiver* to prepare envelopes with random bits for the sender. With just two of these, cheating attempts always succeed, but with four it can be detected often enough. The receiver creates these, two with each bit value, and sends them to the sender. He opens three, and expects to find two zeros and a one, or vice versa – if not, the receiver's cheating has been detected. If the receiver biases the choice by sending (say) three zeroes plus a one, this is detected with a chance of one in four which seems small, but can be amplified by repeating the scheme to achieve any required level of security.

Thus, the protocol consists of three communications of envelopes between the parties, with typically a time gap between the second and the third in which the overarching protocol does its job. In traditional protocol notation it is given below, using two bits of extra notation:

- values listed between $[\![\ ]\!]$ brackets are *received* in a non-deterministic order;
- $[x]$ denotes a newly created closed envelope containing the bit x, a new open envelope with x is represented by $\langle x \rangle$.

Preparation: $B \rightarrow A : [\![[1], [1], [0], [0]]\!]$
   A receives these as $E1 \ldots E4$, opens $E1 \ldots E3$ and takes the exclusive-or of their values returning b.
   If $E1 \ldots E3$ all had the same value, *A* finds *B* has cheated and aborts.
Commitment: $A \rightarrow B : \langle vv \rangle$
   The value vv is computed as the exclusive-or of b and the value v? that *A* wants to commit to. It is sent in an open envelope called e? below.

Opening: $A \to B : E4$

B receives and opens this last closed envelope, the value found should be b,
and computes the exclusive-or of b and vv which should be $v?$.
If the envelope received isn't one of the original four created by B, A has
cheated and B aborts the protocol (and any surrounding ones).

The Preparation Phase The receiver creates four envelopes and sends them to the
sender. An honest receiver balances the bits to be sent: two of each. A smart dishonest
receiver sends three plus one (four the same would always be found out). The sender
will detect this case with probability $\frac{1}{4}$. With $a?$ as the sender, and $b?$ as the receiver, we
need shorthands for envelopes being created and moving in opposite directions from
before:

$CreateB \mathrel{\widehat{=}} Create[b?/a?]$ $\qquad$ $MoveBA \mathrel{\widehat{=}} Move[a?/b?]$
$SendFour \mathrel{\widehat{=}} [CreateB; es? : \mathbb{F}\,Envelope \mid es? = zs? \cup os? \land \#es? = 4] \;\mathbin{\raise0.5ex\hbox{\scriptsize{9}}}\; MoveBA$
$Honest \mathrel{\widehat{=}} [SendFour \mid \#zs? = 2]$
$Dishonest \mathrel{\widehat{=}} [SendFour \mid \#zs? = 1 \lor \#os? = 1]$

We introduce $fs?$ for the set of envelopes to be opened, through renaming. With an
honest receiver the sender knows the value in the unopened envelope:

$OpenThree \mathrel{\widehat{=}} [Open[fs?/es?]; \; es? : \mathbb{F}\,Envelope \mid \#fs? = 3 \land fs? \subseteq es?]$
$HonestOpenThree \mathrel{\widehat{=}} Honest \mathbin{\raise0.5ex\hbox{\scriptsize{9}}} OpenThree$
$\forall\,HonestOpenThree \bullet (\#(\mathrm{ran}\,zero \cap fs?) = 1 \Rightarrow es? \setminus fs? \subseteq \mathrm{ran}\,zero)$
$\qquad\qquad\qquad\land (\#(\mathrm{ran}\,one \cap fs?) = 1 \Rightarrow es? \setminus fs? \subseteq \mathrm{ran}\,one)$

The Commitment Step. The relevant definitions (for a zero bit) for this phase are as
follows. Details are in the full paper [4].

$CreateOne \mathrel{\widehat{=}} \exists\,zs? : \mathbb{F}\,Envelope \bullet [Create[e?/os?] \mid \#e? = 1 \land \#zs? = 0]$
$SendOne \mathrel{\widehat{=}} CreateOne \mathbin{\raise0.5ex\hbox{\scriptsize{9}}} Open[e?/es?] \mathbin{\raise0.5ex\hbox{\scriptsize{9}}} Move[e?/es?]$
$CommitToOne \mathrel{\widehat{=}} OpenThree \mathbin{\raise0.5ex\hbox{\scriptsize{9}}} ([SendOne \mid \#(\mathrm{ran}\,zero \cap fs?) = 1]$
$\qquad\qquad\qquad\qquad\qquad \lor [SendZero \mid \#(\mathrm{ran}\,zero \cap fs?) = 2])$
$CommitOView \mathrel{\widehat{=}} \exists\,e?, es?, fs?, zs?, os? \bullet Honest \mathbin{\raise0.5ex\hbox{\scriptsize{9}}} CommitToOne \mathbin{\raise0.5ex\hbox{\scriptsize{9}}} View$

Opening the Commitment. The relevant definitions for this phase are as below, with
analogues for the one-bit:

$MoveLast \mathrel{\widehat{=}} [es?, fs?, ef? : \mathbb{F}\,Envelope \mid Move[ef?/es?] \land ef? = es?/fs?]$
$Same \mathrel{\widehat{=}} [S'; \; e?, ef? : \mathbb{F}\,Envelope \mid e? \cup ef? \subseteq \mathrm{ran}\,zero' \lor e? \cup ef? \subseteq \mathrm{ran}\,one']$
$OpenZero \mathrel{\widehat{=}} MoveLast \mathbin{\raise0.5ex\hbox{\scriptsize{9}}} ([Open[ef?/es?] \land Same)$

5 Discussion and Further Work

We have not considered refinement for these specifications. Rather, we took an approach
that is more common in security: using security properties over an abstract implementation. Better would be to state the security properties abstractly, and produce the implementation as a (gradual, stepwise) refinement of them, for a suitable refinement notion.

Some of our properties even refer directly to the specification internals, which is less abstract. The standard Z style is a natural choice for this work, as it considers state hidden, and these sealed containers are all about subtle information hiding varying over time. The use of finalisations ("views") is a first step towards providing a more abstract observational semantics, in line with previous work by both authors [3,1].

A number of additional aspects need to be considered in the refinement relation. Information flow from hidden to visible variables also needs to be incorporated, for example based on Morgan's shadow semantics [10] or the fog semantics [2] by Banks and Jacob. A promising theoretical framework for integrating the missing probabilistic aspect is the theory by McIver, Morgan and others [7,6]. Integrating also the computational complexity aspects inherent from modern cryptography remains an open problem.

Acknowledgement. This work is supported by the EPSRC CryptoForma Network, on formal methods and cryptography (www.cryptoforma.org.uk). The specification was developed with support from the Z-Eves proof tool [11].

References

1. Banks, M.J., Jacob, J.L.: On modelling user observations in the UTP. In: Qin, S. (ed.) UTP 2010. LNCS, vol. 6445, pp. 101–119. Springer, Heidelberg (2010)
2. Banks, M.J., Jacob, J.L.: On integrating confidentiality and functionality in a formal method. Formal Asp. Comput. (2013), http://link.springer.com/article/10.1007%2Fs00165-013-0285-4
3. Boiten, E., Derrick, J., Schellhorn, G.: Relational concurrent refinement part II: Internal operations and outputs. Formal Asp. Comput. 21(1-2), 65–102 (2009)
4. Boiten, E., Jacob, J.: Modelling sealed envelopes in Z (2014), http://www.cs.kent.ac.uk/people/staff/eab2/papers/envlop.pdf
5. Canetti, R., Fischlin, M.: Universally composable commitments. In: Kilian, J. (ed.) CRYPTO 2001. LNCS, vol. 2139, pp. 19–40. Springer, Heidelberg (2001)
6. Hoang, T.S., McIver, A.K., Meinicke, L., Morgan, C.C., Sloane, A., Susatyo, E.: Abstractions of non-interference security: probabilistic versus possibilistic. Formal Asp. Comput. 26(1), 169–194 (2014)
7. McIver, A., Morgan, C.: Abstraction, Refinement and Proof for Probabilistic Systems. Springer (2004)
8. Moran, T., Naor, M.: Basing cryptographic protocols on tamper-evident seals. In: Caires, L., Italiano, G.F., Monteiro, L., Palamidessi, C., Yung, M. (eds.) ICALP 2005. LNCS, vol. 3580, pp. 285–297. Springer, Heidelberg (2005)
9. Moran, T., Naor, M.: Polling with physical envelopes: A rigorous analysis of a human-centric protocol. In: Vaudenay, S. (ed.) EUROCRYPT 2006. LNCS, vol. 4004, pp. 88–108. Springer, Heidelberg (2006)
10. Morgan, C.: The shadow knows: Refinement of ignorance in sequential programs. Sci. Comput. Program. 74(8), 629–653 (2009)
11. Saaltink, M.: The Z/EVES system. In: Bowen, J.P., Hinchey, M.G., Till, D. (eds.) ZUM 1997. LNCS, vol. 1212, pp. 72–85. Springer, Heidelberg (1997)

Specifying Transaction Control
to Serialize Concurrent Program Executions

Egon Börger[1] and Klaus-Dieter Schewe[2]

[1] Università di Pisa, Dipartimento di Informatica, 56125 Pisa, Italy
boerger@di.unipi.it
[2] Software Competence Centre Hagenberg, 4232 Hagenberg, Austria
klaus-dieter.schewe@scch.at

Abstract. We define a programming language independent transaction
controller and an operator which when applied to concurrent programs
with shared locations turns their behavior with respect to some abstract
termination criterion into a transactional behavior. We prove the correctness property that concurrent runs under the transaction controller
are serialisable. We specify the transaction controller TACTL and the
operator *TA* in terms of Abstract State Machines. This makes TACTL
applicable to a wide range of programs and in particular provides the
possibility to use it as a plug-in when specifying concurrent system components in terms of Abstract State Machines.[1]

1 Introduction

This paper is about the use of transactions as a common means to control concurrent access of programs to shared locations and to avoid that values stored
at these locations are changed almost randomly. A *transaction controller* interacts with concurrently running programs (read: sequential components of an
asynchronous system) to control whether access to a shared location can be
granted or not, thus ensuring a certain form of consistency for these locations. A
commonly accepted consistency criterion is that the joint behavior of all transactions (read: programs running under transactional control) with respect to the
shared locations is equivalent to a serial execution of those programs. Serialisability guarantees that each transaction can be specified independently from the
transaction controller, as if it had exclusive access to the shared locations.

It is expensive and cumbersome to specify transactional behavior and prove
its correctness again and again for components of the great number of concurrent
systems. Our goal is to define once and for all an abstract (i.e. programming language independent) transaction controller TACTL which can simply be "plugged
in" to turn the behavior of concurrent programs (read: components M of any
given asynchronous system $\mathcal{M}$) into a transactional one. This involves to also

---

[1] The first author, Humboldt research prize awardee 2007/08, gratefully acknowledges
partial support by a renewed research grant from the Alexander von Humboldt
Foundation.

Y. Ait Ameur and K.-D. Schewe (Eds.): ABZ 2014, LNCS 8477, pp. 142–157, 2014.
© Springer-Verlag Berlin Heidelberg 2014

define an operator $TA(M, \text{TACTL})$ which forces the programs M to listen to the controller TACTL when trying to access shared locations.

For the sake of generality we define the operator and the controller in terms of Abstract State Machines (ASMs) which can be read and understood as pseudocode so that TACTL and the operator TA can be applied to code written in any programming language (to be precise: whose programs come with a notion of single step, the level where our controller imposes shared memory access constraints to guarantee transactional code behavior). On the other side, the precise semantics underlying ASMs (for which we refer the reader to [5]) allows us to mathematically prove the correctness of our controller and operator.

We concentrate here on transaction controllers that employ locking strategies such as the common two-phase locking protocol (2PL). That is, each transaction first has to acquire a (read- or write-) lock for a shared location, before the access is granted. Locks are released after the transaction has successfully committed and no more access to the shared locations is necessary. There are of course other approaches to transaction handling, see e.g. [6,13,14,16] and the extensive literature there covering classical transaction control for flat transactions, timestamp-based, optimistic and hybrid transaction control protocols, as well as non-flat transaction models such as sagas and multi-level transactions.

We define TACTL and the operator TA in Sect. 2 and the TACTL components in Sect. 3. In Sect. 4 we prove the correctness of these definitions.

2   The Transaction Operator $TA(M, \text{TACTL})$

As explained above, a transaction controller performs the lock handling, the deadlock detection and handling, the recovery mechanism (for partial recovery) and the commit of single machines. Thus we define it as consisting of four components specified in Sect. 3.

> TACTL =
> > LOCKHANDLER
> > DEADLOCKHANDLER
> > RECOVERY
> > COMMIT

The operator $TA(M, \text{TACTL})$ transforms the components M of any concurrent system (read: asynchronous ASM) $\mathcal{M} = (M_i)_{i \in I}$ into components of a concurrent system $TA(\mathcal{M}, \text{TACTL})$ where each $TA(M_i, \text{TACTL})$ runs as transaction under the control of TACTL:

$$TA(\mathcal{M}, \text{TACTL}) = ((TA(M_i, \text{TACTL}))_{i \in I}, \text{TACTL})$$

TACTL keeps a dynamic set *TransAct* of those machines M whose runs it currently has to supervise to perform in a transactional manner until M has *Terminated* its transactional behavior (so that it can COMMIT it).[2] To turn the

---

[2] In this paper we deliberately keep the termination criterion abstract so that it can be refined in different ways for different transaction instances.

behavior of a machine M into a transactional one, first of all M has to register itself with the controller TACTL, read: to be inserted into the set of currently to be handled *TransActions*. To UNDO as part of a recovery some steps M made already during the given transactional run segment of M, a last-in first-out queue *history*(M) is needed which keeps track of the states the transactional run goes through; when M enters the set *TransAct* the *history*(M) has to be initialized (to the empty queue).

The crucial transactional feature is that each non private (i.e. shared or monitored or output) location l a machine M needs to read or write for performing a step has to be *LockedBy*(M) for this purpose; M tries to obtain such locks by calling the LOCKHANDLER. In case no *newLocks* are needed by M in its *currState* or the needed *newLocks* can be *Granted* by the LOCKHANDLER, M performs its next step; in addition, for a possible future recovery, the machine has to RECORD in its *history*(M) the current values of those locations which are (possibly over-) written by this M-step together with the obtained *newLocks*. Then M continues its transactional behavior until it is *Terminated*. In case the needed *newLocks* are *Refused*, namely because another machine N in *TransAct* for some needed l has *W-Locked*(l, N) or (in case M wants a W-(rite)Lock) has *R-Locked*(l, N), M has to *Wait* for N; in fact it continues its transactional behavior by calling again the LOCKHANDLER for the needed *newLocks*—until the needed locked locations are unlocked when N's transactional behavior is COMMITed, whereafter a new request for these locks this time may be *Granted* to M.[3]

As a consequence deadlocks may occur, namely when a cycle occurs in the transitive closure *Wait** of the *Wait* relation. To resolve such deadlocks the DEADLOCKHANDLER component of TACTL chooses some machines as *Victims* for a recovery.[4] After a victimized machine M is *Recovered* by the RECOVERY component of TACTL, so that M can exit its *waitForRecovery* state, it continues its transactional behavior.

This explains the following definition of $TA(M, \text{TACTL})$ as a control state ASM, i.e. an ASM with a top level Finite State Machine control structure. We formulate it by the flowchart diagram of Fig. 1, which has a precise control state ASM semantics (see the definition in [5, Ch.2.2.6]). The components for the recovery feature are highlighted in the flowchart by a colouring that differs from that of the other components. The macros which appear in Fig. 1 and the components of TACTL are defined below.

---

[3] As suggested by a reviewer, a refinement (in fact a desirable optimization) consists in replacing such a waiting cycle by suspending M until the needed locks are released. Such a refinement can be obtained in various ways, a simple one consisting in letting M simply stay in *waitForLocks* until the *newLocks* *CanBeGranted* and refining LOCKHANDLER to only choose pairs $(M, L) \in \text{LockRequest}$ where it can GRANTREQUESTEDLOCKS(M, L) and doing nothing otherwise (i.e. defining REFUSEREQUESTEDLOCKS$(M, L) = \text{skip}$). See Sect. 3.

[4] To simplify the serializability proof in Sect.3 and without loss of generality we define a reaction of machines M to their victimization only when they are in *ctl_state*$(M) = $ TA-*ctl* (not in *ctl_state*$(M) = \text{waitForLocks}$). This is to guarantee that no locks are *Granted* to a machine as long as it does *waitForRecovery*.

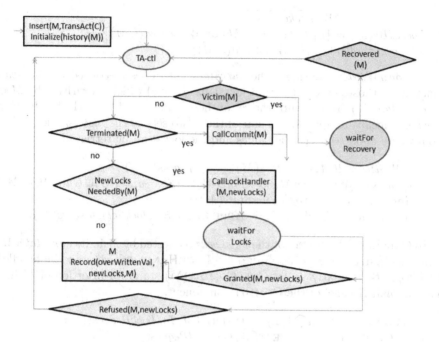

Fig. 1. TA(M,C)

The predicate $NewLocksNeededBy(M)$ holds if in the current state of M at least one of two cases happens:[5] either M to perform its step in this state reads some shared or monitored location which is not yet $LockedBy(M)$ or M writes some shared or output location which is not yet $LockedBy(M)$ for writing. A location can be $LockedBy(M)$ for reading ($R\text{-}Locked(l, M)$) or for writing ($W\text{-}Locked(l, M)$). Formally:

$NewLocksNeededBy(M) =$
$\quad newLocks(M, currState(M))^6 \neq (\emptyset, \emptyset)$
$newLocks(M, currState(M))^7 = (R\text{-}Loc, W\text{-}Loc)$
$\quad$ **where**
$\quad R\text{-}Loc = ReadLoc(M, currState(M)) \cap (SharedLoc(M) \cup MonitoredLoc(M))$
$\qquad \cap \overline{LockedBy(M)}^8$
$\quad W\text{-}Loc = WriteLoc(M, currState(M)) \cap (SharedLoc(M) \cup OutputLoc(M))$

---

[5] See [5, Ch.2.2.3] for the classification of locations and functions.

[6] For layout reasons we omit in Fig.1 the arguments of the functions $newLocks$ and $overWrittenVal$.

[7] By the second argument $currState(M)$ of $newLocks$ (and below of $overWrittenVal$) we indicate that this function of M is a dynamic function which is evaluated in each state of M, namely by computing in this state the sets $ReadLoc(M)$ and $WriteLoc(M)$; see Sect. 4 for the detailed definition.

[8] By $\overline{X}$ we denote the complement of X.

$$\cap \overline{W\text{-}LockedBy(M)}$$
$$LockedBy(M) = \{l \mid R\text{-}Locked(l, M) \text{ or } W\text{-}Locked(l, M)\}$$
$$W\text{-}LockedBy(M) = \{l \mid W\text{-}Locked(l, M)\}$$

The *overWrittenValues* are the *currState(M)*-values (retrieved by the *eval*-function) of those shared or output locations $(f, args)$ which are written by M in its *currState(M)*. To RECORD the set of these values together with the obtained *newLocks* means to append the pair of these two sets to the *history* queue of M from where upon recovery the values and the locks can be retrieved.

$$overWrittenVal(M, currState(M)) = \{((f, args), val) \mid$$
$$(f, args) \in WriteLoc(M, currState(M)) \cap (SharedLoc(M) \cup OutputLoc(M))$$
$$\textbf{and } val = eval(f(args), currState(M))\}$$
$$\text{RECORD}(valSet, lockSet, M) = \text{APPEND}((valSet, lockSet), history(M))$$

To CALLLOCKHANDLER for the *newLocks* requested by M in its *currState(M)* means to INSERT$(M, newLocks)$ into the LOCKHANDLER's set of to be handled *LockRequests*. Similarly we let CALLCOMMIT(M) stand for insertion of M into a set *CommitRequest* of the COMMIT component.

$$\text{CALLLOCKHANDLER}(M, L) = \text{INSERT}((M, L), LockRequest)$$
$$\text{CALLCOMMIT}(M) = \text{INSERT}(M, CommitRequest)$$

3   The Transaction Controller Components

A CALLCOMMIT(M) by machine M enables the COMMIT component. Using the **choose** operator we leave the order in which the *CommitRequest*s are handled refinable by different instantiations of TACTL.

COMMITing M means to UNLOCK all locations l that are *LockedBy(M)*. Note that each lock obtained by M remains with M until the end of M's transactional behavior. Since M performs a CALLCOMMIT(M) when it has *Terminated* its transactional computation, nothing more has to be done to COMMIT M besides deleting M from the sets of *CommitRequest*s and still to be handled *TransActions*.[9]

Note that the locations $R\text{-}Locked(l, M)$ and $W\text{-}Locked(l, M)$ are shared by the COMMIT, LOCKHANDLER and RECOVERY components, but these components never have the same M simultaneously in their request resp. *Victim* set since when machine M has performed a CALLCOMMIT(M), it has *Terminated* its transactional computation and does not participate any more in any $(M, L) \in$ *LockRequest* or *Victimization*.

COMMIT =
    if *CommitRequest* $\neq \emptyset$ **then**
        choose $M \in CommitRequest$ COMMIT(M)

---

[9] We omit clearing the *history(M)* queue since it is initialized when M is inserted into *TransAct*(TACTL).

where
    COMMIT(M) =
      forall $l \in LockedBy(M)$ UNLOCK(l, M)
      DELETE($M, CommitRequest$)
      DELETE($M, TransAct$)
    UNLOCK(l, M) =
      if $R\text{-}Locked(l, M)$ **then** $R\text{-}Locked(l, M) := false$
      if $W\text{-}Locked(l, M)$ **then** $W\text{-}Locked(l, M) := false$

As for COMMIT also for the LOCKHANDLER we use the **choose** operator to leave the order in which the *LockRequests* are handled refinable by different instantiations of TACTL.

The strategy we adopt for lock handling is to refuse all locks for locations requested by M if at least one of the following two cases happens:

- some of the requested locations is *W-Locked* by another transactional machine $N \in TransAct$,
- some of the requested locations is a *WriteLocation* that is *R-Locked* by another transactional machine $N \in TransAct$.

This definition, which is specified below by the predicate *CannotBeGranted*, implies that multiple transactions may simultaneously have a *R-Lock* on some location. To REFUSEREQUESTEDLOCKS it suffices to set the communication interface *Refused* of $TA(M, TACTL)$; this makes M *Wait* for each location l that is $W\text{-}Locked(l, N)$ and for each *WriteLocation* that is $R\text{-}Locked(l, N)$ by some other transactional component machine $N \in TransAct$.

LOCKHANDLER =
  if $LockRequest \neq \emptyset$ **then**
    choose $(M, L) \in LockRequest$
      HANDLELOCKREQUEST(M, L)
  where
    HANDLELOCKREQUEST(M, L) =
      if $CannotBeGranted(M, L)$
        then REFUSEREQUESTEDLOCKS(M, L)
        else GRANTREQUESTEDLOCKS(M, L)
      DELETE($(M, L), LockRequest$)
    $CannotBeGranted(M, L)$ =
      let $L = (R\text{-}Loc, W\text{-}Loc), Loc = R\text{-}Loc \cup W\text{-}Loc$
        forsome $l \in Loc$ **forsome** $N \in TransAct \setminus \{M\}$
          $W\text{-}Locked(l, N)$ **or**
            $(l \in W\text{-}Loc$ **and** $R\text{-}Locked(l, N))$
    REFUSEREQUESTEDLOCKS(M, L) = ($Refused(M, L) := true$)
    GRANTREQUESTEDLOCKS(M, L) =
      let $L = (R\text{-}Loc, W\text{-}Loc)$
        forall $l \in R\text{-}Loc$ ($R\text{-}Locked(l, M) := true$)
        forall $l \in W\text{-}Loc$ ($W\text{-}Locked(l, M) := true$)
      $Granted(M, L) := true$

A *Deadlock* originates if two machines are in a *Wait* cycle, otherwise stated if for some (not yet *Victim*ized) machine M the pair (M, M) is in the transitive (not reflexive) closure *Wait** of *Wait*. In this case the DEADLOCKHANDLER selects for recovery a (typically minimal) subset of *Deadlocked* transactions *toResolve*—they are *Victim*ized to *waitForRecovery*, in which mode (control state) they are backtracked until they become *Recovered*. The selection criteria are intrinsically specific for particular transaction controllers, driving a usually rather complex selection algorithm in terms of number of conflict partners, priorities, waiting time, etc. In this paper we leave their specification for TACTL abstract (read: refinable in different directions) by using the **choose** operator.

DEADLOCKHANDLER =
  if *Deadlocked* $\cap$ $\overline{Victim}$ $\neq$ $\emptyset$ **then** // there is a Wait cycle
    choose *toResolve* $\subseteq$ *Deadlocked* $\cap$ $\overline{Victim}$
      forall $M \in$ *toResolve* *Victim*$(M) :=$ *true*
  where
    Deadlocked $= \{M \mid (M, M) \in M^*\}$
    $M^* =$ TransitiveClosure(*Wait*)
    Wait$(M, N) =$ **forsome** l *Wait*(M, l, N)
    Wait$(M, l, N) =$
      $l \in$ *newLocks*$(M,$ *currState*$(M))$ **and** $N \in$ *TransAct* $\setminus \{M\}$ **and**
      W-Locked(l, N) **or** $(l \in$ *W-Loc* **and** *R-Locked*$(l, N))$
        where *newLocks*$(M,$ *currState*$(M)) = ($*R-Loc*, *W-Loc*$)$

Also for the RECOVERY component we use the **choose** operator to leave the order in which the *Victim*s are chosen for recovery refinable by different instantiations of TACTL. To be *Recovered* a machine M is backtracked by UNDO(M) steps until M is not *Deadlocked* any more, in which case it is deleted from the set of *Victim*s, so that be definition it is *Recovered*. This happens at the latest when *history*(M) has become empty.

RECOVERY =
  if *Victim* $\neq$ $\emptyset$ **then**
    choose $M \in$ *Victim* TRYTORECOVER(M)
  where
    TRYTORECOVER$(M) =$
      if $M \notin$ *Deadlocked* **then** *Victim*$(M) :=$ *false*
        else UNDO(M)
    Recovered =
      $\{M \mid$ *ctl-state*$(M) =$ *waitForRecovery* **and** $M \notin$ *Victim*$\}$
    UNDO$(M) =$
      let $($*ValSet*, *LockSet*$) =$ *youngest*(*history*$(M))$
        RESTORE$($*ValSet*$)$
        RELEASE$($*LockSet*$)$
        DELETE$(($*ValSet*, *LockSet*$),$ *history*$(M))$
    where

$\text{RESTORE}(V) =$
  forall $((f, args), v) \in V\ f(args) := v$
$\text{RELEASE}(L) =$
  let $L = (R\text{-}Loc, W\text{-}Loc)$
  forall $l \in Loc = R\text{-}Loc \cup W\text{-}Loc$ $\text{UNLOCK}(l, M)$

Note that in our description of the DEADLOCKHANDLER and the (partial) RECOVERY we deliberately left the strategy for victim seclection and UNDO abstract leaving fairness considerations to be discussed elsewhere. It is clear that if always the same victim is selected for partial recovery, the same deadlocks may be created again and again. However, it is well known that fairness can be achieved by choosing an appropriate victim selection strategy.

4   Correctness Theorem

In this section we show the desired correctness property: if all monitored or shared locations of any M_i are output or controlled locations of some other M_j and all output locations of any M_i are monitored or shared locations of some other M_j (closed system assumption)[10], each run of $TA(\mathcal{M}, \text{TACTL})$ is equivalent to a serialization of the terminating M_i-runs, namely the M_{i_1}-run followed by the M_{i_2}-run etc., where M_{i_j} is the j-th machine of $\mathcal{M}$ which performs a commit in the $TA(\mathcal{M}, \text{TACTL})$ run. To simplify the exposition (i.e. the formulation of statement and proof of the theorem) we only consider machine steps which take place under the transaction control, in other words we abstract from any step M_i makes before being INSERTed into or after being DELETed from the set $TransAct$ of machines which currently run under the control of TACTL.

First of all we have to make precise what a *serial* multi-agent ASM run is and what *equivalence* of $TA(\mathcal{M}, \text{TACTL})$ runs means in the general multi-agent ASM framework.

Definition of Run Equivalence. Let $S_0, S_1, S_2, \ldots$ be a (finite or infinite) run of $TA(\mathcal{M}, \text{TACTL})$. In general we may assume that TACTL runs forever, whereas each machine $M \in \mathcal{M}$ running as transaction will be terminated at some time – at least after commit M will only change values of non-shared and non-output locations[11]. For $i = 0, 1, 2, \ldots$ let Δ_i denote the unique, consistent update set defining the transition from S_i to S_{i+1}. By definition of $TA(\mathcal{M}, \text{TACTL})$ the update set is the union of the update sets of the agents executing $M \in \mathcal{M}$ resp. TACTL:

$$\Delta_i = \bigcup_{M \in \mathcal{M}} \Delta_i(M) \cup \Delta_i(\text{TACTL}).$$

---

[10] This assumption means that the environment is assumed to be one of the component machines.

[11] It is possible that one ASM M enters several times as a transaction controlled by TACTL. However, in this case each of these registrations will be counted as a separate transaction, i.e. as different ASMs in $\mathcal{M}$.

$\Delta_i(M)$ contains the updates defined by the ASM $TA(M, \text{TACTL})$ in state S_i[12] and $\Delta_i(\text{TACTL})$ contains the updates by the transaction controller in this state. The sequence of update sets $\Delta_0(M)$, $\Delta_1(M)$, $\Delta_2(M)$, ... will be called the *schedule* of M (for the given transactional run).

To generalise for transactional ASM runs the equivalence of transaction schedules known from database systems [6, p.621ff.] we now define two *cleansing operations* for ASM schedules. By the first one (i) we eliminate all (in particular unsuccessful-lock-request) computation segments which are without proper M-updates; by the second one (ii) we eliminate all M-steps which are related to a later $\text{UNDO}(M)$ step by the RECOVERY component:

(i) Delete from the schedule of M each $\Delta_i(M)$ where one of the following two properties holds:

- $\Delta_i(M) = \emptyset$ (M contributes no update to S_i),
- $\Delta_i(M)$ belongs to a step of an M-computation segment where M in its *ctl_state(M)* =TA-*ctl* does $\text{CALLLOCKHANDLER}(M, newLocks)$ and in its next step moves from *waitForLocks* back to control state TA$-ctl$ because the LOCKHANDLER *Refused(M, newLocks)*.[13]

In such computation steps M makes no proper update.

(ii) Repeat choosing from the schedule of M a pair $\Delta_j(M)$ with later $\Delta_{j'}(M)$ ($j < j'$) which belong to the first resp. second of two consecutive M-Recovery steps defined as follows:

- a (say M-RecoveryEntry) step whereby M in state S_j moves from TA-*ctl* to *waitForRecovery* because it became a *Victim*,
- the next M-step (say M-RecoveryExit) whereby M in state $S_{j'}$ moves back to control state TA-*ctl* because it has been *Recovered*.

In these two M-Recovery steps M makes no proper update. Delete:

(a) $\Delta_j(M)$ and $\Delta_{j'}(M)$,
(b) the $((Victim, M), true)$ update from the corresponding $\Delta_t(\text{TACTL})$ ($t < j$) which in state S_j triggered the M-RecoveryEntry,
(c) $\text{TRYTORECOVER}(M)$-updates in any $\Delta_{i+k}(\text{TACTL})$ between the considered M-RecoveryEntry and M-RecoveryExit step ($i < j < i + k < j'$),
(d) each $\Delta_{i'}(M)$ belonging to the M-computation segment from TA-*ctl* back to TA-*ctl* which contains the proper M-step in S_i that is UNDOne in S_{i+k} by the considered $\text{TRYTORECOVER}(M)$ step; besides control state and RECORD updates these $\Delta_{i'}(M)$ contain updates (ℓ, v) with

---

[12] We use the shorthand notation $\Delta_i(M)$ to denote $\Delta_i(TA(M, \text{TACTL}))$; in other words we speak about steps and updates of M also when they really are done by $TA(M, \text{TACTL})$. Mainly this is about transitions between the control states, namely TA-*ctl*, *waitForLocks*, *waitForRecovery* (see Fig.1), which are performed during the run of M under the control of the transaction controller TACTL. When we want to name an original update of M (not one of the updates of *ctl_state(M)* or of the RECORD component) we call it a proper M-update.

[13] Note that by eliminating this $\text{CALLLOCKHANDLER}(M, L)$ step also the corresponding LOCKHANDLER step $\text{HANDLELOCKREQUEST}(M, L)$ disappears in the run.

$\ell = (f, (val_{S_i}(t_1), \ldots, val_{S_i}(t_n)))$ where the corresponding UNDO updates are $(\ell, val_{S_i}(f(t_1, \ldots, t_n))) \in \Delta_{i+k}(\text{TACTL})$,

(e) the HANDLELOCKREQUEST$(M, newLocks)$-updates in $\Delta_{l'}(\text{TACTL})$ corresponding to M's CALLLOCKHANDLER step (if any: in case $newLocks$ are needed for the proper M-step in S_i) in state S_l ($l < l' < i$).

The sequence $\Delta_{i_1}(M), \Delta_{i_2}(M), \ldots$ with $i_1 < i_2 < \ldots$ resulting from the application of the two cleansing operations as long as possible – note that confluence is obvious, so the sequence is uniquely defined – will be called the *cleansed schedule* of M (for the given run).

Before defining the equivalence of transactional ASM runs let us remark that $TA(\mathcal{M}, \text{TACTL})$ has indeed several runs, even for the same initial state S_0. This is due to the fact that a lot of non-determinism is involved in the definition of this ASM. First, the submachines of TACTL are non-deterministic:

- In case several machines $M, M' \in \mathcal{M}$ request conflicting locks at the same time, the LOCKHANDLER can only grant the requested locks for one of these machines.
- Commit requests are executed in random order by the COMMIT submachine.
- The submachine DEADLOCKHANDLER chooses a set of victims, and this selection has been deliberately left abstract.
- The RECOVERY submachine chooses in each step a victim M, for which the last step will be undone by restoring previous values at updated locations and releasing corresponding locks.

Second, the specification of $TA(\mathcal{M}, \text{TACTL})$ leaves deliberately open, when a machine $M \in \mathcal{M}$ will be started, i.e., register as a transaction in *TransAct* to be controlled by TACTL. This is in line with the common view that transactions $M \in \mathcal{M}$ can register at any time to the transaction controller TACTL and will remain under its control until they commit.

Definition 1. Two runs $S_0, S_1, S_2, \ldots$ and $S'_0, S'_1, S'_2, \ldots$ of $TA(\mathcal{M}, \text{TACTL})$ are *equivalent* iff for each $M \in \mathcal{M}$ the cleansed schedules $\Delta_{i_1}(M), \Delta_{i_2}(M), \ldots$ and $\Delta'_{j_1}(M), \Delta'_{j_2}(M), \ldots$ for the two runs are the same and the read locations and the values read by M in S_{i_k} and S'_{j_k} are the same.

That is, we consider runs to be equivalent, if all transactions $M \in \mathcal{M}$ read the same locations and see there the same values and perform the same updates in the same order disregarding waiting times and updates that are undone.

Definition of Serializability. Next we have to clarify our generalised notion of a serial run, for which we concentrate on committed transactions – transactions that have not yet committed can still undo their updates, so they must be left out of consideration[14]. We need a definition of the read- and write-locations of

---

[14] Alternatively, we could concentrate on complete, infinite runs, in which only committed transactions occur, as eventually every transaction will commit – provided that fairness can be achieved.

M in a state S, i.e. $ReadLoc(M, S)$ and $WriteLoc(M, S)$ as used in the definition of $newLocks(M, S)$.

The definition of $Read/WriteLoc$ depends on the locking level, whether locks are provided for variables, pages, blocks, etc. To provide a definite definition, in this paper we give the definition at the level of abstraction of the locations of the underlying class $\mathcal{M}$ of component machines (ASMs) M. Refining this definition (and that of $newLocks$) appropriately for other locking levels does not innvalidate the main result of this paper.

We define $ReadLoc(M, S) = ReadLoc(r, S)$ and analogously $WriteLoc(M, S) = WriteLoc(r, S)$, where r is the defining rule of the ASM M. Then we use structural induction according to the definition of ASM rules in [5, Table 2.2]. As an auxiliary concept we need to define inductively the read and write locations of terms and formulae. The definitions use an interpretation I of free variables which we suppress notationally (unless otherwise stated) and assume to be given with (as environment of) the state S. This allows us to write $ReadLoc(M, S)$, $WriteLoc(M, S)$ instead of $ReadLoc(M, S, I)$, $ReadLoc(M, S, I)$ respectively.

Read/Write Locations of Terms and Formulae. For state S let I be the given interpretation of the variables which may occur freely (in given terms or formulae). We write $val_S(construct)$ for the evaluation of $construct$ (a term or a formula) in state S (under the given interpretation I of free variables).

$ReadLoc(x, S) = WriteLoc(x, S) = \emptyset$ for variables x
$ReadLoc(f(t_1, \ldots, t_n), S) =$
$\quad \{(f, (val_S(t_1), \ldots, val_S(t_n)))\} \cup \bigcup_{1 \leq i \leq n} ReadLoc(t_i, S)$
$WriteLoc(f(t_1, \ldots, t_n), S) = \{(f, (val_S(t_1), \ldots, val_S(t_n)))\}$

Note that logical variables are not locations: they cannot be written and their values are not stored in a location but in the given interpretation I from where they can be retrieved.

We define $WriteLoc(\alpha, S) = \emptyset$ for every formula α because formulae are not locations one could write into. $ReadLoc(\alpha, S)$ for atomic formulae $P(t_1, \ldots, t_n)$ has to be defined as for terms with P playing the same role as a function symbol f. For propositional formulae one reads the locations of their subformulae. In the inductive step for quantified formulae $domain(S)$ denotes the superuniverse of S minus the Reserve set [5, Ch.2.4.4] and I_x^d the extension (or modification) of I where x is interpreted by a domain element d.

$ReadLoc(P(t_1, \ldots, t_n), S) =$
$\quad \{(P, (val_S(t_1), \ldots, val_S(t_n)))\} \cup \bigcup_{1 \leq i \leq n} ReadLoc(t_i, S)$
$ReadLoc(\neg \alpha) = ReadLoc(\alpha)$
$ReadLoc(\alpha_1 \wedge \alpha_2) = ReadLoc(\alpha_1) \cup ReadLoc(\alpha_2)$
$ReadLoc(\forall x \alpha, S, I) = \bigcup_{d \in domain(S)} ReadLoc(\alpha, S, I_x^d)$

Note that the values of the logical variables are not read from a location but from the modified state environment function I_x^d.

Read/Write Locations of ASM Rules

$ReadLoc(\textbf{skip}, S) = WriteLoc(\textbf{skip}, S) = \emptyset$
$ReadLoc(t_1 := t_2, S) = ReadLoc(t_1, S) \cup ReadLoc(t_2, S)$
$WriteLoc(t_1 := t_2, S) = WriteLoc(t_1, S)$
$ReadLoc(\textbf{if } \alpha \textbf{ then } r_1 \textbf{ else } r_2, S) =$
$\quad ReadLoc(\alpha, S) \cup \begin{cases} ReadLoc(r_1, S) \textbf{ if } val_S(\alpha) = true \\ ReadLoc(r_2, S) \textbf{ else} \end{cases}$
$WriteLoc(\textbf{if } \alpha \textbf{ then } r_1 \textbf{ else } r_2, S) = \begin{cases} WriteLoc(r_1, S) \textbf{ if } val_S(\alpha) = true \\ WriteLoc(r_2, S) \textbf{ else} \end{cases}$
$ReadLoc(\textbf{let } x = t \textbf{ in } r, S, I) = ReadLoc(t, S, I) \cup ReadLoc(r, S, I_x^{val_S(t)})$
$WriteLoc(\textbf{let } x = t \textbf{ in } r, S, I) = WriteLoc(r, S, I_x^{val_S(t)})$ // call by value
$ReadLoc(\textbf{forall } x \textbf{ with } \alpha \textbf{ do } r, S, I) =$
$\quad ReadLoc(\forall x\alpha, S, I) \cup \bigcup_{a \in range(x,\alpha,S,I)} ReadLoc(r, S, I_x^a)$
$\qquad \textbf{where } range(x, \alpha, S, I) = \{d \in domain(S) \mid val_{S, I_x^d}(\alpha) = true\}$
$WriteLoc(\textbf{forall } x \textbf{ with } \alpha \textbf{ do } r, S, I) = \bigcup_{a \in range(x,\alpha,S,I)} WriteLoc(r, S, I_x^a)$

In the following cases the same scheme applies to read and write locations:[15]

$Read[Write]Loc(r_1 \textbf{ par } r_2, S) =$
$\quad Read[Write]Loc(r_1, S) \cup Read[Write]Loc(r_2, S)$
$Read[Write]Loc(r(t_1, \ldots, t_n), S) = Read[Write]Loc(P(x_1/t_1, \ldots, x_n/t_n), S)$
$\quad \textbf{where } r(x_1, \ldots, x_n) = P$ // call by reference
$Read[Write]Loc(r_1 \textbf{ seq } r_2, S, I) = Read[Write]Loc(r_1, S, I) \cup$
$\quad \begin{cases} Read[Write]Loc(r_2, S + U, I) \textbf{ if } yields(r_1, S, I, U) \textbf{ and } Consistent(U) \\ \emptyset \qquad\qquad\qquad\qquad\qquad\qquad \textbf{else} \end{cases}$

For **choose** rules we have to define the read and write locations simultaneously to guarantee that the same instance satisfying the selection condition is chosen for defining the read and write locations of the rule body r:

if $range(x, \alpha, S, I) = \emptyset$ **then**
$\quad ReadLoc(\textbf{choose } x \textbf{ with } \alpha \textbf{ do } r, S, I) = ReadLoc(\exists x\alpha, S, I)$
$\quad WriteLoc(\textbf{choose } x \textbf{ with } \alpha \textbf{ do } r, S, I) = \emptyset$ // empty action
else choose $a \in range(x, \alpha, S, I)$
$\quad ReadLoc(\textbf{choose } x \textbf{ with } \alpha \textbf{ do } r, S, I) =$
$\qquad ReadLoc(\exists x\alpha, S, I) \cup ReadLoc(r, S, I_x^a)$
$\quad WriteLoc(\textbf{choose } x \textbf{ with } \alpha \textbf{ do } r, S, I) = WriteLoc(r, S, I_x^a)$

We say that M has or is committed (in state S_i, denoted $Committed(M, S_i)$) if step COMMIT(M) has been performed (in state S_i).

Definition 2. A run of $TA(\mathcal{M}, \text{TACTL})$ is *serial* iff there is a total order $<$ on $\mathcal{M}$ such that the following two conditions are satisfied:

(i) If in a state M has committed, but M' has not, then $M < M'$ holds.

---

[15] In $yields(r_1, S, I, U)$ U denotes the update set produced by rule r_1 in state S under I.

(ii) If M has committed in state S_i and $M < M'$ holds, then the cleansed schedule $\Delta_{j_1}(M'), \Delta_{j_2}(M'), \ldots$ of M' satisfies $i < j_1$.

That is, in a serial run all committed transactions are executed in a total order and are followed by the updates of transactions that did not yet commit.

Definition 3. A run of $TA(\mathcal{M}, \text{TACTL})$ is *serialisable* iff it is equivalent to a serial run of $TA(\mathcal{M}, \text{TACTL})$.[16]

Theorem 1. *Each run of $TA(\mathcal{M}, \text{TACTL})$ is serialisable.*

Proof. Let $S_0, S_1, S_2, \ldots$ be a run of $TA(\mathcal{M}, \text{TACTL})$. To construct an equivalent serial run let $M_1 \in \mathcal{M}$ be a machine that commits first in this run, i.e. $Committed(M, S_i)$ holds for some i and whenever $Committed(M, S_j)$ holds for some $M \in \mathcal{M}$, then $i \leq j$ holds. If there is more than one machine M_1 with this property, we randomly choose one of them.

Take the run of $TA(\{M_1\}, \text{TACTL})$ starting in state S_0, say $S_0, S_1', S_2', \ldots, S_n'$. As M_1 commits, this run is finite. M_1 has been DELETEd from *TransAct* and none of the TACTL components is triggered any more: neither COMMIT nor LOCKHANDLER because *CommitRequest* resp. *LockRequest* remain empty; not DEADLOCKHANDLER because *Deadlock* remains false since M_1 never *Waits* for any machine; not RECOVERY because *Victim* remains empty. Note that in this run the schedule for M_1 is already cleansed.

We now define a run $S_0'', S_1'', S_2'', \ldots$ (of $TA(\mathcal{M} - \{M_1\}, \text{TACTL})$, as has to be shown) which starts in the final state $S_n'' = S_0''$ of the $TA(\{M_1\}, \text{TACTL})$ run and where we remove from the run defined by the cleansed schedules $\Delta_i(M)$ for the originally given run all updates made by steps of M_1 and all updates in TACTL steps which concern M_1. Let

$$\Delta_i'' = \bigcup_{M \in \mathcal{M} - \{M_1\}} \Delta_i(M) \cup \{(\ell, v) \in \Delta_i(\text{TACTL}) \mid (\ell, v) \text{ does not concern } M_1\}.$$

That is, in Δ_i'' all updates are removed from the original run which are done by M_1—their effect is reflected already in the initial run segment from S_0 to S_n'—or are LOCKHANDLER updates involving a *LockRequest*(M_1, L) or are *Victim*$(M_1) := true$ updates of the DEADLOCKHANDLER or are updates involving a TRYTORECOVER(M_1) step or are done by a step involving a COMMIT(M_1).

Lemma 1. $S_0'', S_1'', S_2'', \ldots$ is a run of $TA(\mathcal{M} - \{M_1\}, \text{TACTL})$.

Lemma 2. The run $S_0, S_1', S_2', \ldots, S_n', S_1'', S_2'', \ldots$ of $TA(\mathcal{M}, \text{TACTL})$ is equivalent to the original run $S_0, S_1, S_2, \ldots$.

By induction hypothesis $S_0'', S_1'', S_2'', \ldots$ is serialisable, so $S_0, S_1', S_2', \ldots$ and thereby also $S_0, S_1, S_2, \ldots$ is serialisable with $M_1 < M$ for all $M \in \mathcal{M} - \{M_1\}$. □

---

[16] Modulo the fact that ASM steps permit simultaneous updates of multiple locations, this definition of serializability is equivalent to Lamport's sequential consistency concept [15].

Proof.(Lemma 1). We first show that omitting in Δ_i'' every update from $\Delta_i(\text{TACTL})$ which concerns M_1 does not affect updates by TACTL in S_i'' concerning $M \neq M_1$. In fact starting in the final M_1-state S_0'', $TA(\mathcal{M} - \{M_1\}, \text{TACTL})$ makes no move with a $Victim(M_1) := true$ update and no move of $\text{COMMIT}(M_1)$ or $\text{HANDLELOCKREQUEST}(M_1, L)$ or $\text{TRYTORECOVER}(M_1)$

It remains to show that every M-step defined by $\Delta_i''(M)$ is a possible M-step in a $TA(\mathcal{M} - \{M_1\}, \text{TACTL})$ run starting in S_0''. Since the considered M-schedule $\Delta_i(M)$ is cleansed, we only have to consider any proper update step of M in state S_i'' (together with its preceding lock request step, if any). If in S_i'' M uses $newLocks$, in the run by the cleansed schedules for the original run the locks must have been granted after the first COMMIT, which is done for M_1 before S_0''. Thus these locks are granted also in S_i'' as part of a $TA(\mathcal{M} - \{M_1\}, \text{TACTL})$ run step. If no $newLocks$ are needed, that proper M-step depends only on steps computed after S_0'' and thus is part of a $TA(\mathcal{M} - \{M_1\}, \text{TACTL})$ run step. $\quad\square$

Proof.(Lemma 2) The cleansed machine schedules in the two runs, the read locations and the values read there have to be shown to be the same. First consider any $M \neq M_1$. Since in the initial segment $S_0, S_1', S_2', \ldots, S_n'$ no such M makes any move so that its update sets in this computation segment are empty, in the cleansed schedule of M for the run $S_0, S_1', S_2', \ldots, S_n', S_1'', S_2'', \ldots$ all these empty update sets disappear. Thus this cleansed schedule is the same as the cleansed schedule of M for the run $S_n', S_1'', S_2'', \ldots$ and therefore by definition of $\Delta_i''(M) = \Delta_i(M)$ also for the original run $S_0, S_1, S_2, \ldots$ with same read locations and same values read there.

Now consider M_1, its schedule $\Delta_0(M_1), \Delta_1(M_1), \ldots$ for the run $S_0, S_1, S_2, \ldots$ and the corresponding cleansed schedule $\Delta_{i_0}(M_1), \Delta_{i_1}(M_1), \Delta_{i_2}(M_1), \ldots$. We proceed by induction on the cleansed schedule steps of M_1. When M_1 makes its first step using the $\Delta_{i_0}(M_1)$-updates, this can only be a proper M_1-step together with the corresponding RECORD updates (or a lock request directly preceding such a $\Delta_{i_1}(M_1)$-step) because in the computation with cleansed schedule each lock request of M_1 is granted and M_1 is not $Victimized$. The values M_1 reads or writes in this step (in private or locked locations) have not been affected by a preceding step of any $M \neq M_1$—otherwise M would have locked before the non-private locations and keep the locks until it commits (since cleansed schedules are without UNDO steps), preventing M_1 from getting these locks which contradicts the fact that M_1 is the first machine to commit and thus the first one to get the locks. Therefore the values M_1 reads or writes in the step defined by $\Delta_{i_0}(M_1)$ (resp. also $\Delta_{i_1}(M_1)$) coincide with the corresponding location values in the first (resp. also second) step of M_1 following the cleansed schedule to pass from S_0 to S_1' (case without request of $newLocks$) resp. from S_0 to S_1' to S_2' (otherwise). The same argument applies in the inductive step which establishes the claim. $\quad\square$

5 Conclusion

In this article we specified (in terms of Abstract State Machines) a transaction controller TaCtl and a transaction operator which turn the behaviour of a set of concurrent programs into a transactional one under the control of TaCtl. In this way the locations shared by the programs are accessed in a well-defined manner. For this we proved that all concurrent transactional runs are serialisable.

The relevance of the transaction operator is that it permits to concentrate on the specification of program behavior ignoring any problems resulting from the use of shared locations. That is, specifications can be written in a way that shared locations are treated as if they were exclusively used by a single program. This is valuable for numerous applications, as shared locations (in particular, locations in a database) are common, and random access to them is hardly ever permitted.

Furthermore, by shifting transaction control into the rigorous framework of Abstract State Machines we made several extensions to transaction control as known from the area of databases [6]. In the classical theory schedules are sequences containing read- and write-operations of the transactions plus the corresponding read- and write-lock and commit events, i.e., only one such operation or event is treated at a time. In our case we exploited the inherent parallelism in ASM runs, so we always considered an arbitrary update set with usually many updates at the same time. Under these circumstances we generalised the notion of schedule and serialisability in terms of the synchronous parallelism of ASMs. In this way we stimulate also more parallelism in transactional systems.

Among further work we would like to be undertaken is to provide a (proven to be correct) implementation of our transaction controller and the *TA* operator, in particular as plug-in for the CoreASM [7,8] or Asmeta [4] simulation engines. We would also like to see refinements or adaptations of our transaction controller model for different approaches to serialisability [13], see also the ASM-based treatment of multi-level transaction control in [14]. Last but not least we would like to see further detailings of our correctness proof to a mechanically verified one, e.g. using the ASM theories developed in KIV (see [1] for an extensive list of relevant publications) and PVS [9,12,11] or the (Event- [3]) B [2] theorem prover for an (Event-) B transformation of $TA(\mathcal{M}, \text{TaCtl})$ (as suggested in [10]).

Acknowledgement. We thank Andrea Canciani and some of our referees for useful comments to improve the paper.

References

1. The KIV system, http://www.informatik.uni-augsburg.de/lehrstuehle/swt/se/kiv/
2. Abrial, J.-R.: The B-Book. Cambridge University Press, Cambridge (1996)
3. Abrial, J.-R.: Modeling in Event-B. Cambridge University Press (2010)
4. The Abstract State Machine Metamodel website (2006), http://asmeta.sourceforge.net

5. Börger, E., Stärk, R.F.: Abstract State Machines. A Method for High-Level System Design and Analysis. Springer (2003)
6. Elmasri, R., Navathe, S.B.: Fundamentals of Database Systems. Addison Wesley (2006)
7. Farahbod, R., et al.: The CoreASM Project, http://www.coreasm.org
8. Farahbod, R., Gervasi, V., Glässer, U.: CoreASM: An Extensible ASM Execution Engine. Fundamenta Informaticae XXI (2006)
9. Gargantini, A., Riccobene, E.: Encoding Abstract State Machines in PVS. In: Gurevich, Y., Kutter, P.W., Odersky, M., Thiele, L. (eds.) ASM 2000. LNCS, vol. 1912, pp. 303–322. Springer, Heidelberg (2000)
10. Glässer, U., Hallerstede, S., Leuschel, M., Riccobene, E.: Integration of Tools for Rigorous Software Construction and Analysis (Dagstuhl Seminar 13372). Dagstuhl Reports 3(9), 74–105 (2014)
11. Goerigk, W., Dold, A., Gaul, T., Goos, G., Heberle, A., von Henke, F.W., Hoffmann, U., Langmaack, H., Pfeifer, H., Ruess, H., Zimmermann, W.: Compiler correctness and implementation verification: The verifix approach. In: Fritzson, P. (ed.) Int. Conf. on Compiler Construction, Proc. Poster Session of CC 1996, Linköping, Sweden, IDA Technical Report LiTH-IDA-R-96-12 (1996)
12. Goos, G., von Henke, H., Langmaack, H.: Project Verifix, http://www.info.uni-karlsruhe.de/projects.php/id=28&lang=en
13. Gray, J., Reuter, A.: Transaction Processing: Concepts and Techniques. Morgan Kaufmann (1993)
14. Kirchberg, M., Schewe, K.-D., Zhao, J.: Using Abstract State Machines for the design of multi-level transaction schedulers. In: Abrial, J.-R., Glässer, U. (eds.) Rigorous Methods for Software Construction and Analysis. LNCS, vol. 5115, pp. 65–77. Springer, Heidelberg (2009)
15. Lamport, L.: How to make a multiprocessor computer that correctly executes multiprocess programs. IEEE Trans. on Computers 28(9) (September 1979)
16. Schewe, K.-D., Ripke, T., Drechsler, S.: Hybrid concurrency control and recovery for multi-level transactions. Acta Cybernetica 14(3), 419–453 (2000)

Distributed Situation Analysis
A Formal Semantic Framework

Narek Nalbandyan[1], Uwe Glässer[1],
Hamed Yaghoubi Shahir[1], and Hans Wehn[2,*]

[1] Software Technology Lab, Simon Fraser University, B.C., Canada
[2] MDA Systems Ltd, Research & Development, B.C., Canada
{nnalband,glaesser,syaghoub}@sfu.ca,hw@mdacorporation.com

Abstract. Situation Analysis is critical for dynamic decision-making in responding to real-world situations. The complex and intricate nature of situation analysis processes calls for evolutionary modeling and formal engineering methods that facilitate experimental validation of abstract mathematical descriptions to link the essential design aspects with rapid prototyping in early development phases. For the transition from abstract concepts and requirements to precise specifications to high level design of situation analysis systems, we derive here a generic ASM ground model as a framework for defining the precise meaning of fundamental situation analysis concepts applicable to different application domain models.

Keywords: Abstract state machines, ASM ground model, ASM refinement, Model-driven engineering, Maritime situation awareness.

1 Introduction

Situation Analysis (SA) is generally defined as a *process*, the examination of a situation, its elements, and their relations, to provide and maintain a state of situation awareness [1], which is critical for situation assessment in dynamic decision-making to respond to security threats and disaster situations. Situation awareness refers to the perception of the elements in the environment within a space-time continuum, the comprehension of their meaning, and the projection of their status in the near future [2]. Computational situation analysis models are important in situations that routinely call for intelligent coordination and management of multiple mobile surveillance platforms (satellites, surveillance aircrafts, drones, radar, etc.) operating in vast geographical areas. Situational evidence is collected for maritime security and military surveillance to protect critical infrastructure against a variety of threats and illegal activities, and also for emergency management to coordinate rescue and relief operations after a disaster such as a major flooding, oil spill, earthquake, or tsunami. Increasing complexity of decision-making problems and elaborate contextual situations have

---

* This research has been funded by NSERC under the Discovery Grant program and the Collaborative Research & Development program jointly with MDA Systems Ltd.

Y. Ait Ameur and K.-D. Schewe (Eds.): ABZ 2014, LNCS 8477, pp. 158–173, 2014.

altered the nature of decision support. Interactive decision-making, to be facilitated by decision support systems, has become a complex multifaceted process
that requires "solving various semi- to ill-structured problems involving multiple
attributes, objectives and goals" [3]. A comprehensive design approach based on
evolutionary and iterative processes is needed to address the non-deterministic
complexities arising from today's real-world decision-making requirements [4].

In this paper we propose a formal semantic framework for engineering and
prototyping of Distributed Situation Analysis (DSA) models using the Abstract
State Machine (ASM) method [5] for the construction and analysis of *ground
models* and their refinements [6]. Progressive real-world SA system design calls
for evolutionary modeling and formal methods for experimental validation of
mathematical models to link essential design aspects with rapid prototyping
in early development phases to facilitate: *i*) analysis of the problem space, *ii*)
reasoning about complex design decisions, and *iii*) systematic derivation of conformance criteria for checking the validity of domain models. A formal approach
is justified and unavoidable if one strives for robustness, reliability and repro-
ducibility of results, satisfaction of constraints, and a language to represent and
reason about complex dynamic situations [7]. Formulating SA models in abstract
computational terms turns out to be challenging since the underlying concepts
lack semantic clarity (as will be explained), and faces two notoriously problematic questions: 1) *How can one formulate the problem in a concise but precise and
yet understandable form?* and 2) *How can one describe and evaluate a solution
prior to building it into a system?*

Various SA models have been proposed and studied by different communities
and in different application fields (see Sections 2-3). The two novel contributions
of this paper are the fully decentralized organization of the analysis framework
based on an asynchronous computation model with multiple mobile observers,
and a clear separation of generic system concepts from domain specific analysis
methods and fusion algorithms needed for processing *time series* (sequences of
consecutive data points collected over time) to detect and identify objects of
interest, analyze and assess their behavior. Observers cooperate in gathering situational evidence by monitoring discrete events, distributed in time and space,
to develop a coherent and consistent global picture of a dynamic situation as
it unfolds. Our goal is to derive a robust and extensible, generic framework as
a semantic foundation for defining the meaning of fundamental SA concepts in
abstract computational terms as distributed ASM ground model, the validity of
which can be established by analyzing and reasoning about its key attributes,
consistency, correctness and *completeness* [6], and by observation and experimentation with realistic scenarios, maritime security in our case. To the best of
our knowledge the approach presented here is new and original.

Section 2 discusses related work. Section 3 defines the problem scope. Next,
Section 4 explains common situation analysis concepts and illustrates the DSA
system model. Section 5 then introduces the ASM ground model and refinements
for various core components of the DSA framework, and Section 6 summarizes
lessons learned. Section 7 concludes the paper.

2   Related Work

We discuss here related work on formal analysis and design of situation analysis models and on distributed problem solving frameworks.

Formal Approaches to Situation Analysis: The three major approaches to formalizing SA concepts can be characterized as using: 1) general purpose formal logics (e.g., [8]), 2) methods based on machine learning and ontologies (e.g., [9]), and 3) distinct SA methods and models like Dynamic Case-Based Reasoning [10] and State Transition Data Fusion (STDF) [11]. STDF, an extension of the JDL Data Fusion Model [1], the dominant sensor fusion paradigm, provides a unification of both sensor and higher-level fusion across three principal levels [11]: Object Assessment, Situation Assessment, and Impact Assessment. Like the JDL model, STDF is a functional model, although less abstract than the JDL model, and seeks to demonstrate a unifying framework across the three levels.

Some of the above works make specific commitments prior to rigorous requirements analysis and design validation. The approach proposed in [8] attempts to make their decision support model operational by employing reasoners that are associated with the specific logics. A situation analysis system must necessarily be able to reason about data over time, an aspect that is not supported by the applied reasoners; hence, the authors had to design their framework so as to work around this issue.

Although the above approaches may not lend themselves well to SA and decision support systems design as addressed here, they each have their own strengths and benefits. Approaches specifically focusing on early phases of problem formulation and design are explored in [12] and [7,13].

There is a necessity for integrated approaches to design SA decision support systems for complex domains [4]. "Large, complex systems are hard to evolve without undermining their dependability. Often change is disproportionately costly, ... system architectures are pivotal in meeting the above challenge. ... First, dependability properties tend to be emergent, and are much more readily modeled and controlled at an architectural level." [4,14].

Formal frameworks proposed for SA models mainly focus on specific and mostly theoretical aspects. Practical needs call for adaptive and evolutionary design methods that encompass iterative modeling, interactive simulation and experimental validation for the purpose of linking essential aspects of design with rapid prototyping of executable SA models in early phases to facilitate: 1) analysis of the problem space, 2) reasoning about design decisions, and 3) deriving conformance criteria for checking the validity of SA domain models.

Distributed Problem Solving: Distributed problem solving is defined as the cooperative solution of problems by a decentralized and loosely coupled collection of problem solvers [15]. This approach provides increased performance through scalability, robustness, and fault tolerance, as such a system can easily respond to dynamic changes by reconfiguration as needed to maintain its function.

The Multi-agent systems paradigm facilitates understanding and building distributed systems [16]. Multi-agent systems are "composed of multiple interacting computing elements, known as agents", where an agent is "a computer system situated in some environment and capable of autonomous action in this environment in order to meet its objectives" [16]. Wooldridge and Jennings [17] suggest that, in order to meet their design objectives, agents need to be *proactive*, i.e., able to take initiatives in performing goal-directed actions; *reactive*, i.e., able to perceive the environment and react to its changes, and *social*, i.e., able to interact with other agents and humans.

According to [16], a multi-agent system can be conceptually divided into two components: the agents and the environment (viewed as a special agent). Each agent has a local state at any time, which represents all the information the agent possesses. The state of the environment represents information about the outside world. Agents may have partial views of the world, which may or may not be accurate. The behavior of environment is typically nondeterministic, for it is not always possible to accurately predict the behavior of environment ahead of time. The global state of the system is composed of the combination of local states of its constituent agents and the environment.

3 Problem Description

The three main, partially overlapping, views on situation analysis across different communities are: the cognitive science perspective [18], which comes from the human factors community and focuses on measuring *team situation awareness* by "taking the system itself as the unit of analysis and focusing on the interactions between the parts of the system and resultant emerging behavior rather than study its parts in isolation" [18]; the computer science perspective represented by agent-based systems [12], which views SA systems as systems operated by multiple intelligent autonomous agents; and the information fusion perspective, which provides a framework for SA from the viewpoint of data fusion, defined as "the process of utilizing one or more data sources over time to assemble a representation of aspects of interest in an environment" [11].

While these perceptions of processes for which we need computational models appropriately define situation analysis from their respective viewpoints, there are noticeable semantic gaps between them. In particular, the cognitive science approach is more at the abstraction level of human interactions and thus does not provide an explicit representation of a system model and its entities. On the other hand, existing agent-based approaches, while providing clear system models, do not provide experimental platforms for testing their models and do not explicitly define the underlying SA concepts. Lastly, existing data fusion approaches lack intuitive formalizations for building systems. To the best of our knowledge, currently there is no formal SA framework that provides notational means for turning abstract requirements into precise specifications of relevant concepts, and subsequently using these specifications for high level system design. The goal of this paper is to unify the three viewpoints mentioned above through a novel Distributed Situation Analysis framework, filling in the major semantic gaps and coming up with a comprehensive definition of the framework.

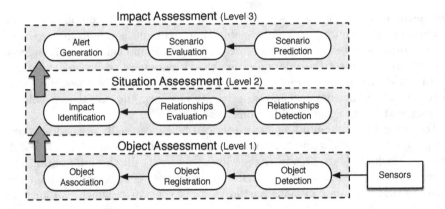

Fig. 1. Fusion Model: Adopted from Simplified JDL Model

Figure 1 illustrates our interpretation of the common ground in terms of an abstract fusion model adopted from the Simplified JDL model [1]. In this paper, we focus on the Object Assessment and Situation Assessment layers.

SA systems inevitably deal with complex situations: processing complex data, resolving conflicting observations, reasoning about uncertainty, adjusting system parameters like sensor characteristics and positions, etc. While the main role of the system is not (and cannot be) to make decisions, it assists and guides human operators in making decisions. Accuracy, performance and reliability are vital factors. This calls for explorative modeling to identify and accurately define the key concepts so as to gradually refine complex models; likewise, this entails formal approaches to properly define semantics in abstract mathematical terms.

4   Key Concepts in DSA

In this section, we identify common concepts associated with SA system models and define a DSA architecture which is generic and potentially suitable in diverse application domains, including maritime safety and security, air traffic control, road traffic management, search and rescue operations, and more. We start with two notorious challenges that any such system needs to address.

Observation Uncertainty: Realistic DSA system models need to take into account adversarial impact factors of the physical environment in which a system operates. Observers use mobile and stationary sensor platforms for monitoring areas of interest to gather situational evidence of events distributed in time and space. This results in a series of discrete observations, or time series.[1] However, noise and uncertainty pose a notorious problem. Due to weather conditions, failing equipment, technological limitations like observation errors of

---

[1] A time series is a sequence of discrete data points, measured typically at successive points in time spaced at uniform time intervals.

sensor equipment, observations may be (and often are) inaccurate, incomplete or even invalid. Another challenge is integrating local observations from multiple observers to develop a coherent and consistent global picture of a complex situation as it unfolds. Additionally, differences between the local view of individual observers and the resulting global view of the system require conflict resolution algorithms to be deployed for maintaining a coherent and consistent global picture.

Concepts and Terminology: We outline here the main concepts and vocabulary of our situation analysis system model.

- Environment refers to the part of the external physical world that interacts with the system. The environment has an impact on the margin of accuracy of sensor values. In that regard, the environment is left implicit; however, we see its effects in the accuracy and reliability of sensory input.
- Observation refers to a data point in a time series, normally associated with a single object in the environment. We generally assume that each observation enters the system with an associated timestamp.
- Objects are all entities in the environment that can be observed by sensors and are typically divided into two basic types: objects of interest (OOI) (e.g., a vessel) and other objects (e.g., a log drifting in the water). Objects that are not an OOI at a given time are considered noise. The type of an object may change from OOI to noise and vice versa as the situation unfolds.
- Sensors are interfaces of the system with the external world. Often, multiple heterogeneous sensors are used in combination to provide a collection of observations in any given state. We distinguish between *logical* and *physical* sensors. A logical sensor is an input source for an observer and belongs to only one observer. More than one logical sensor can be mapped onto one physical sensor. Typical sensor types include radar, satellites, sensor units on aircrafts, drones, the human eye, etc.
- Trajectory refers to a time series that describes how an object moves through space as a function of time, assuming that data points may as well specify additional object characteristics beyond geospatial and kinematic aspects of observable behavior such as position, heading and speed over ground.
- Monitoring means continued reading of sensor values and is subject to uncertainty due to possible malfunctioning of sensors, weather conditions, and other possible factors.
- Observers are computational agents processing sensor output values. Each observer has a number of sensors which generate local observations. They may be mobile or stationary, where stationary observers are considered a special case of mobile ones.
- Reasoners are computational agents with a more global view of the environment and typically have access to additional contextual and background information. Each reasoner has a number of observers which provide input values.

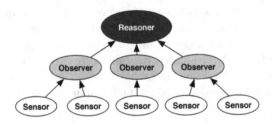

Fig. 2. Logical Network Structure

- Confidence refers to a real number in the closed interval $[0, 1]$ and reflects the likelihood of a sensor to produce reliable output relative to other sensors in a system. Confidence values are assigned by observers and may vary dynamically, depending on the type of data it senses. Observers may have a confidence value too, assigned by their associated reasoners.

There are more attributes of perceived information such as completeness and timeliness of information, which are outside of the scope of this paper. The relevant concepts introduced above are carefully selected in consultation with domain experts, based on their knowledge and experience (see [1], [19]).

Physical Network vs. Logical Network: We generally distinguish between the *logical network*, which defines the interactions between different conceptual components of the system, i.e., connections and interactions between sensors, observers, and reasoners, and the *physical network*, which refers to the network situated in the environment. As Figure 2 illustrates, in the logical network, which constitutes the generic system architecture of the framework, a set of observers is connected to a reasoner. Each observer has a number of logical sensors providing observations as input to the observer. Each logical sensor belongs to only one observer, and each observer reports to only one reasoner. In Figure 2, the reasoners constitute a flat hierarchical structure; however, the framework provides flexibility for structuring the level of hierarchy of reasoner interactions from flat and decentralized to more hierarchical paradigms, resulting in architectures that can have additional layers. A fourth layer, not explicitly shown in Figure 2, is the human decision-maker who relies on the information from reasoners (see [4]).

Figure 3 illustrates the physical network structure. Observers are located geographically close to the area they monitor. Sensors have possibly overlapping observation areas, which may differ in size, and different sensing capabilities. Observers assign appropriate confidence values to sensors based on what they observe and how accurate the measurements are relative to the observation of the same object by other sensors. These confidence values are used for conflict resolutions and inconsistency management of conflicting observations. Observer agents may change these confidence values over time, as they perform internal examination and reconfiguration during the operation of the system.

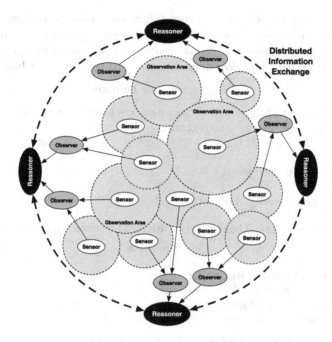

Fig. 3. Physical Network Structure

Global situation awareness of such a distributed collaborative system is more than the sum of the bits and pieces from local awareness associated with each of the collaborating members. For forming global situation awareness, the DSA system incorporates autonomous reasoning of reasoner and observer agents. Formation of the global situation awareness is achieved from local observations and local understandings, through communication, decentralized decision making, and distributed agreement.

5   Formal Semantic Framework

This section introduces and formally describes our generic DSA system model in terms of a distributed ASM with multiple autonomously operating agents, which is intended to serve as a ground model [6] of a semantic framework for defining the meaning of the abstract system concepts identified in the previous section. We assume an asynchronous computation model with mobile and stationary agents, where each agent is identified with either an observer entity or a reasoner entity in the network. Additionally, a single distinguished agent represents the part of the physical environment that interacts with the system. The respective agents are referred to as: *observer agent*, *reasoner agent*, and *environment agent*.

Sections 5.1 and 5.2 present the control units associated with observer and reasoner agents, respectively, while Section 5.3 addresses data and information flow in the system. Because the underlying concepts are fairly abstract and thus require some explanation, we can present here only part of our formalization. Certain parts are left open intentionally for a clear separation of concerns between the generic DSA model and instantiations of this model in a given domain context (see Sect. 6). The below domains and universes represent basic entities.

Universes and Domains

AGENT $\equiv$ OBSERVERAGENT $\cup$ REASONERAGENT $\cup$ ENVIRONMENTAGENT
SENSOR $\equiv$ RADAR $\cup$ SONAR $\cup$ SATELLITE $\cup \cdots \cup$ HUMAN
OBJECT
OBSERVATION
CONFIDENCE $\equiv$ REALNUMBER $\in [0 \ldots 1]$
TRAJECTORY
MODE $\equiv \{Idle, Cleaning, Association, \ldots\}$ // Operating modes of control units

Characteristic attributes of basic entities are ASM functions as illustrated below.

Functions

$reasoner$: OBSERVER $\mapsto$ REASONER   // Network structure
$observers$: REASONER $\mapsto$ SET(OBSERVER)
$observer$: SENSOR $\mapsto$ OBSERVER
$sensors$: OBSERVER $\mapsto$ SET(SENSOR)

$object$: OBSERVATION $\mapsto$ OBJECT   // Observed object
$location$: OBSERVATION $\mapsto$ POSITION   // Object coordinates
$time$: OBSERVATION $\mapsto$ TIME   // Observation timestamp
$observationConfidence$: OBSERVATION $\mapsto$ CONFIDENCE // Confidence values

The *initial state* of the DSA model when the system starts is monitoring, in which dynamic ASM functions for observations, objects, and trajectories are set to *undef*. A distinguished monitored function *now* represents the logical system time, assuming that the values of *now* increase monotonically over system runs. Actual *physical time* constraints depend on operational aspects like speed of moving objects, size of the observation area, observation time window, et cetera.

5.1   Observer Agent Controller

Observer agents run the *observer agent controller* program, which is formulated as a Control State ASM [5]. Figure 4 shows the graphical representation. At the highest level of abstraction, this ASM program consists of the following four components (i.e., ASM rules) for analyzing and processing time series.

- **Detection & Registration:** Each observer agent receives raw sensor data from its sensors in a format that depends on the sensor type. Data is subject to inconsistencies, corrupted and missing data, and noise. At this level, the agent detects observed objects and pre-processes data to generate collections of *cleaned observations* (see Section 5.3) as input for the next step.
- **Object Association:** Processing the *cleaned observations*, an observer agent determines for each observed object whether the object occurs for the first

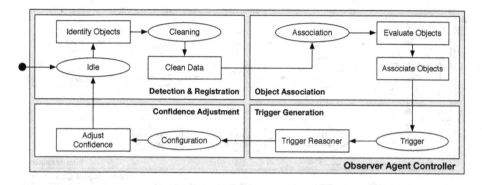

Fig. 4. Control State ASM of Observer Agent

time or has already been spotted before in a previous state. First, multiple observations of the same object by different sensors are fused into a *merged observation*. In this process, any inconsistencies are resolved by taking into account the confidence values of sensors either to determine the most likely observation or to assign a combined confidence value to the resulting observation. Next, the agent updates its set of visible objects and produces *processed observations* by associating with each object a history of related observations over time, which constitutes part of the object trajectory. For details on the **seq**-construct, see [5, Chapter 4.1].

ObjectAssociation$(O : \text{OBSERVERAGENT}, T : \text{TIME}) \equiv$
if $mode(O) = Association$ then

$\quad mergedObs(O,T) \leftarrow$
$\qquad \text{EvaluateObjects}(\{\bigcup_{s \in sensors(O)} cleanedObs(s,T)\}, O, T)$ **seq**
$\quad processedObs(O,T) \leftarrow$
$\qquad \text{AssociateObjects}(\{mergedObs(O,T)\}, \{\bigcup_{T-\delta \leq t < T} processedObs(O,t)\}, O, T)$

$\quad mode(O) := Trigger$

The ObjectAssociation rule above consists of two consecutive steps. EvaluateObjects merges different observations of the same objects into a single observation. AssociateObjects then checks each object being observed in the current state, trying to link the object with matching observations (for the same object) in a previous state, and effectively associates objects with their related history of observations, where δ limits the history being considered. The first step of vertical refinement of AssociateObjects is illustrated below.

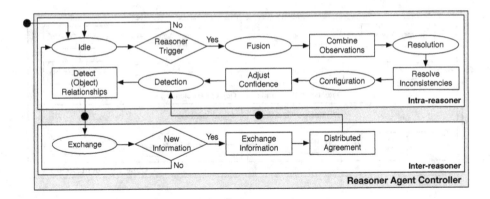

Fig. 5. Control State ASM of Reasoner Agent

AssociateObjects($mergedObs$, $processedObs$: SET(OBSERVATION),
$\qquad$ O : OBSERVERAGENT, T : TIME) $\equiv$
$objectSet(O) := \{object(x) \mid x \in (\bigcup_{t<T} processedObs(O,t))\}$
$\quad$**forall** obs **in** $mergedObs(O,T)$ **do**
$\quad\quad$**if** $\nexists object(obs) \in objectSet(O)$ **then** $\qquad$ // New object detected
$\quad\quad\quad objectStatus(object(obs)) := initiated$
$\quad\quad\quad processedObs(O,T) := processedObs(O,T) \cup \{obs\}$
$\quad\quad$**else**
$\quad\quad\quad objectStatus(object(obs)) := updated$ $\qquad$ // Update known object

$\quad\quad\quad tempAssociatedObs \leftarrow$ AssociateObservations(obs,
$\quad\quad\quad\quad \{x \mid object(obs) = object(x) \wedge x \in (\bigcup_{t<T} processedObs(O,t))\}$) **seq**
$\quad\quad processedObs(O,T) := processedObs(O,T) \cup tempAssociatedObs$
$\quad\quad$// Update history of known objects observed in current state

- **Trigger Generation**: Each observer agent notifies its associated reasoner agent once the *processed observations* are ready. The reasoner agent starts processing its input once it has received notifications from all its observers.

- **Confidence Adjustment**: Observer agents dynamically update confidence values associated with sensors based on accuracy assessment of the sensor observations by comparing the current and previously *processed observations*.

5.2   Reasoner Agent Controller

Reasoner agents run the *reasoner agent controller* program specified as Control State ASM in Figure 5. At the highest level of abstraction, the behavior is defined in terms of a synchronous parallel composition of two main components: *intra-reasoner controller* and *inter-reasoner controller*. Each of these controllers is explained and formally defined below.

Intra-Reasoner Controller

- **Combine Observations**: A reasoner agent fuses information received from its observer agents by combining the *processed observations* for the respective objects being identified so as to eliminate any redundancy. The following ASM specification of the CombineObservations rule describes in more detail the process of combining observations.

  CombineObservations(R : REASONERAGENT, T : TIME) $\equiv$
  if $mode(R) = Fusion$ **then**

  $tempCategorizedObs(R,T)\leftarrow$
  $\quad$ CategorizeObservations($\{ \bigcup\limits_{o\in observer(R)} processedObs(o,T)\}, R, T$) **seq**
  $combinedObs(R,T)\leftarrow$ FuseObservations($\{tempCategorizedObs(R,T)\}$)
  $\quad mode(R) := Resolution$

  First, CategorizeObservations orders in a list the processed observations received from observer agents according to the observed objects ($\oplus$ adds an element at the end of a list). Next, FuseObservations joins and blends multiple observations of the same object into a single, combined observation.

  CategorizeObservations($processedObs$: SET(OBSERVATION),
  $\qquad\qquad\qquad R$: REASONERAGENT, T : TIME) $\equiv$
  $objectSet(R) := \{object(x) \mid x \in (\bigcup\limits_{o\in observer(R)} processedObs(O,T))\}$
  $\quad$ **forall** obj **in** $objectSet(R)$ **do**
  $\qquad$ **forall** obs **in** $\{(\bigcup\limits_{o\in observer(R)} processedObs(o,T)) \mid object(obs) = obj\}$ **do**
  $\qquad tempCategorizedObs(R,T) := tempCategorizedObs(R,T) \oplus obs$

- **Resolve Inconsistencies**: Processing of *combined observations* requires to resolve any inconsistencies that may arise from conflicting observations in overlapping observation areas. Factors include different sets of sensors used by different observers, weather conditions, sensor limitations, transmission problems, etc. The reasoner agent manages inconsistencies by relying on the confidence values of reported observations and other relevant parameters.

- **Adjust Confidence**: Reasoner agents dynamically update the confidence values associated with their observer agents. This is a more complex computational procedure that we intend to address in detail outside of this paper.

- **Detect (Object) Relationships**: The agent uses *consistent observations* to detect relationships between the objects. Further, the agent incorporates contextual information and other additional sources to detect whether an interaction between two or more objects is suspicious and requires special attention or even interjection [20].

Inter-Reasoner Controller

- **Exchange Information**: Each reasoner agent communicates with a set of neighbors (i.e., reasoner agents), which may vary from a single one to all the others. If there is any new relevant information to share with neighbors after performing the *Detect Relationships* rule, the agent prepares the new information and communicates it to its neighbors in the logical network.
- **Distributed Agreement**: In every state, reasoner agents assess the newly received information. If this conflicts with their own local "perception" of a situation, they apply conflict resolution strategies using confidence values and eventually agree on the situation. Upon reaching mutual agreement, the updated information is shared with neighbors, eventually forming coherent *global observations*. Different distributed agreement protocols exist in literature, and the selection of a suitable protocol is context dependent (e.g., [21]).

In the transition from Object Assessment to Situation Assessment (see Fig. 1), the focus shifts from single observations of objects to object trajectories as a fundamental concept in defining a situation analysis process. The meaning of trajectory, as associated with each observable object o in a state S_t of the DSA system model, is uniquely defined by the history of global observations for o shared among all reasoner agents, up to state S_{t-1}. That is, a trajectory for a given object is equivalent to the global view of the history of observations related to this object; this history will be be extended in the next step by fusing the collection of observations reported for this object in the current state into a single observation. We define the trajectory of an object o inductively as follows.

Trajectory

trajectory : OBJECT $\rightarrow$ TRAJECTORY
derived *trajectory*$(o) \equiv$ *trajectory*$^*(o, time(now - 1))$

where the inductive definition of *trajectory** is as follows:
  trajectory* : OBJECT $\times$ TIME $\rightarrow$ LIST(OBSERVATION) // list of *globalObs*
  trajectory$^*(o, t) :=$ *trajectory*$^*(o, t - 1) \oplus \{o'|o' \in globalObs(t) \land object(o') = o\}$
  trajectory$^*(o, 0) := \varnothing$

Reasoner agents use trajectories of objects along with additional contextual and background information to reason about complex situations. The concept of trajectory, as defined above, has the following properties:

Corollary 1. The function *trajectory* is a bijective mapping between the set of objects in any given DSA state and their history of global observations.

Corollary 2. For each object o in a DSA state, *trajectory*(o) is a time series of coherent and, in a probabilistic sense based on confidence values, consistent observations of o over a run of the DSA model.[2]

---

[2] Intuitively, the function *trajectory* yields the best possible situational evidence one can compute under uncertainty from multiple observations of how objects move.

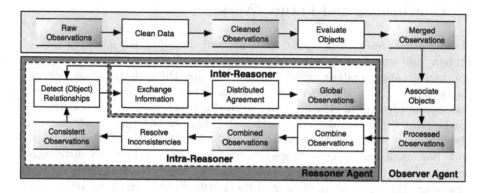

Fig. 6. Data & Information Flow

5.3   Data and Information Flow

A vital aspect of any situation analysis system is fusion of raw data (after filtering out noise) at the sensor level into information that is relevant to decision makers. Figure 6 illustrates the data and information flow in the proposed framework. As explained in Sections 5.1 and 5.2, an observer agent receives raw observations from its sensors and, through its operational components, produces processed observations for each object. The reasoner agent then combines those observations, seeking to maintain consistency in the presence of uncertainty.

6   Lessons Learned

Research collaborations with MDA Systems Ltd. and Defence R&D Canada in several funded projects on maritime safety and security have inspired the work we present here [7, 13, 22]. Maritime situation awareness [23] is critical for Canada to secure its vast coast lines against a variety of threats and illegal activities. The DSA model applies directly to maritime surveillance operations, for instance, to detect suspicious multi-vessel interactions that point to a threat (e.g. piracy, smuggling) or a dangerous situation (e.g. imminent collision). This is a research project to develop probabilistic anomaly-detection algorithms using clustering methods for grouping vessels and Hidden Markov Models for activity and scenario detection [20]. In our experiments, we use US Coast Guard AIS data from marine traffic in coastal waters around North America, which amounts to billions of data points (one every few minutes for each vessel: position, speed, course, type, etc.) forming complex time series [20]. Striving for what works best in real-world situations, we often face modeling decisions that inevitably require "getting one's hands dirty" and combine qualitative with quantitative modeling to evaluate solutions. Large-scale situation analysis calls for design and engineering methods and tools that are distinguished by their mathematical rigor and facilitate evolutionary processes for analyzing the suitability and the validity of abstract models by observation and experimentation with real data.

7   Conclusions and Future Work

For the work presented here, the ASM method simplified the daunting task of bridging the gap between intuitive understanding of abstract requirements and a formal situation analysis model that specifies how the underlying concepts are interpreted in terms of a DSA *ground model*. This model only addresses Object Assessment (detection, identification and tracking) and Situation Assessment (in combination with domain-specific methods) to describe how time series related to observation of objects under uncertainty can be fused into trajectories associated with the monitored objects. Propagation of confidence values and distributed agreement are beyond the scope of this paper as well. We are working on such extensions through horizontal and vertical refinements, which introduce new challenges since trajectories are more complex entities than objects. We also work on integrating distributed agreement and conflict resolution algorithms. After all, our model provides a robust and scalable mathematical framework for turning common knowledge and understanding of situation analysis processes, in the maritime domain and beyond, into computational descriptions with a precise semantic foundation for defining the meaning of the underlying concepts in an explicit and tangible form to analyze and reason about key attributes prior to construction of systems. Finally, we aim at a fully executable DSA model in CoreASM [24] for the maritime security domain, and encourage others to explore other application fields by extending and evolving our model.

Acknowledgements. We would like to sincerely thank four anonymous reviewers for their detailed and valuable feedback contributing substantially to the final version of this paper.

References

1. Bossé, É., Roy, J., Wark, S.: Concepts, Models, and Tools for Information Fusion. Artech House Publishers (2007)
2. Endsley, M.R.: Theoretical Underpinnings of Situation Awareness: A Critical Review. In: Situation Awareness Analysis and Measurement, pp. 3–32 (2000)
3. Nemati, H.R., Steiger, D.M., Iyer, L.S., Herschel, R.T.: Knowledge Warehouse: An Architectural Integration of Knowledge Management, Decision Support, Artificial Intelligence & Data Warehousing. Decision Support Systems 33(2), 143–161 (2002)
4. Klashner, R., Sabet, S.: A DSS Design Model for Complex Problems: Lessons from Mission Critical Infrastructure. Decision Support Systems 43(3), 990–1013 (2007)
5. Börger, E., Stärk, R.: Abstract State Machines: A Method for High-Level System Design and Analysis. Springer (2007)
6. Börger, E.: Construction and Analysis of Ground Models and their Refinements as a Foundation for Validating Computer Based Systems. Formal Aspects of Computing 19(2), 225–241 (2007)
7. Farahbod, R., Glässer, U., Bossé, É., Guitouni, A.: Integrating Abstract State Machines and Interpreted Systems for Situation Analysis Decision Support Design. In: 11th Int'l Conference on Information Fusion, pp. 1–8 (2008)

8. Baader, F., et al.: A Novel Architecture for Situation Awareness Systems. In: Giese, M., Waaler, A. (eds.) TABLEAUX 2009. LNCS, vol. 5607, pp. 77–92. Springer, Heidelberg (2009)

9. Chmielewski, M.: Ontology Applications for Achieving Situation Awareness in Military Decision Support Systems. In: Nguyen, N.T., Kowalczyk, R., Chen, S.-M. (eds.) ICCCI 2009. LNCS, vol. 5796, pp. 528–539. Springer, Heidelberg (2009)

10. Jakobson, G., Lewis, L., Buford, J., Sherman, C.E.: Battlespace Situation Analysis: The Dynamic CBR Approach. In: Military Communications Conference, vol. 2, pp. 941–947 (2004)

11. Lambert, D.A.: A Blueprint for Higher-level Fusion Systems. Information Fusion 10(1), 6–24 (2009)

12. Jousselme, A.L., Maupin, P.: Interpreted Systems for Situation Analysis. In: 10th Int'l Conference on Information Fusion, pp. 1–11 (2007)

13. Farahbod, R., Avram, V., Glässer, U., Guitouni, A.: A Formal Engineering Approach to High-Level Design of Situation Analysis Decision Support Systems. In: Qin, S., Qiu, Z. (eds.) ICFEM 2011. LNCS, vol. 6991, pp. 211–226. Springer, Heidelberg (2011)

14. McDermid, J.: Science of Software Design: Architectures for Evolvable, Dependable Systems. In: NSF Workshop on the Science of Design: Software & Software-Intensive Systems (2003)

15. Smith, R.G., Davis, R.: Frameworks for Cooperation in Distributed Problem Solving. IEEE Transactions on Systems, Man and Cybernetics 11(1), 61–70 (1981)

16. Wooldridge, M.: An Introduction to MultiAgent Systems. Wiley (2009)

17. Wooldridge, M., Jennings, N.: Formalizing the Cooperative Problem Solving Process. In: 13th Int'l Workshop on Distributed Artificial Intelligence, pp. 403–417 (1994)

18. Salmon, P.M., Stanton, N.A., Walker, G.H., Jenkins, D.P.: Distributed Situation Awareness: Theory, Measurement and Application to Teamwork. Ashgate Publishing (2009)

19. Roy, J.: Automated Reasoning for Maritime Anomaly Detection. In: Workshop on Data Fusion and Anomaly Detection for Maritime Situational Awareness (2009)

20. Yaghoubi Shahir, H., Glässer, U., Nalbandyan, N., Wehn, H.: Maritime Situation Analysis. In: 2013 IEEE Int'l Conference on Intelligence and Security Informatics, pp. 230–232 (2013)

21. Aikebaier, A., Enokido, T., Takizawa, M.: Trustworthiness among Peer Processes in Distributed Agreement Protocol. In: 24th Int'l Conference on Advanced Information Networking and Applications, pp. 565–572 (2010)

22. Yaghoubi Shahir, H., Glässer, U., Farahbod, R., Jackson, P., Wehn, H.: Generating Test-Cases for Marine Safety and Security Scenarios: A Composition Framework. Security Informatics 1(1), 1–21 (2012)

23. Glandrup, M.: Improving Situation Awareness in the Maritime Domain. In: van de Laar, P., Tretmans, J., Borth, M. (eds.) Situation Awareness with Systems of Systems, pp. 21–38 (2013)

24. Farahbod, R., Gervasi, V., Glässer, U.: Executable Formal Specifications of Complex Distributed Systems with CoreASM. Science of Computer Programming 79(1), 23–38 (2014)

Introducing Aspect–Oriented Specification for Abstract State Machines

Marcel Dausend and Alexander Raschke

Institute of Software Engineering and Compiler Construction
Ulm University, Ulm, Germany

Abstract. With the paradigm of aspect orientation, a developer is able to separate the code of so-called cross-cutting concerns from the rest of the program logic. This possibility is useful for formal specifications, too. For example, security aspects can be separated from the rest of the specification. Another use case for aspect orientation in specifications is the extension of specifications without touching the original ones. The definition of formal semantics for UML profiles without changing the original UML specification is an example for this application. This paper describes the implementation of the aspect oriented approach in Abstract State Machines. We introduce an aspect language with its syntax and formal semantics. It allows for specifying pointcuts where an original specification is augmented with aspect specification. Besides the general overview of this language extension, some ASM specific features of the realization are depicted in detail.

1 Introduction

Aspect-oriented programming (AOP) [1] facilitates a clear separation and modularization of cross-cutting concerns. A cross-cutting concern is an issue like logging that cannot be encapsulated in one function or class, but its corresponding code has to be spread over the whole program [2]. According to AspectJ [3], the most common representative of AOP, such a concern can be integrated in an aspect. An aspect consists of pointcuts, advices and new local or global definitions. A pointcut declares in a formal language where and under which conditions the advices are inserted into the original program code.

This paper presents an approach of adapting the ideas of AOP for Abstract State Machine (ASM) specifications. Besides the advantage of making it possible to specify cross-cutting concerns like logging or security at one joint place, aspect-orientation in ASMs allows for extending existing specifications without touching the original ASM. An example for this use case is the definition of the semantics for an Unified Modeling Language (UML) profile. The UML offers extensions to its syntax and semantics via so called profiles. Profiles can be used to add new elements to UML models in order to define a domain specific language for a special purpose.

Since there exist formal specifications of the semantics of significant parts of the UML in ASMs ([4,5]), it is reasonable to define the semantics of UML

Y. Ait Ameur and K.-D. Schewe (Eds.): ABZ 2014, LNCS 8477, pp. 174–187, 2014.

profiles with ASMs, too. However, the UML specification "does not allow for modifying existing metamodels" [6] through defining profiles. In this context, aspect-orientation provides a comfortable solution to integrate the semantics of profile elements into the UML. An aspect defines the partial semantics of new elements and where they have to be deployed into the original specification. An application of this approach is described in [7].

Aspect-oriented ASMs (AoASM) comprises the most important concepts of AOP that are part of AspectJ (cf. [3,8]). Obviously, we have to restrict the language features of AoASM to those that can be mapped from AspectJ to ASM. The semantics of AoASM is defined by ASM and makes only use of existing language features so that we can state that AoASM is a conservative extension to ASM. Moreover AoASM introduces ASM specific concepts, e. g. dealing with agents and parallelism.

1.1   Fundamental Terms of Aspect–Oriented Programming

In order to understand the core of AOP, the most important terms and concepts are introduced in the following. As every AOP language, AoASM is based on a join point model consisting of **join points**, "point[s] in the execution of a program together with a view into the execution context when that point[s] occur[s]" [8] and pointcuts "that picks out join points and expose[s] data from the execution context of those join points." [8]. An **advice** declares a specification fragment, given by regular ASM statements, to be inserted.

In other words, a join point depicts specific points during the execution of a program, where the execution of the aspect code will be inserted. Thus, a join point describes a dynamic situation of an execution.

In order to specify a dynamic join point, a **pointcut** expression describes static positions in the code together with constraints that must hold at this position during execution. An appropriate orchestration of the original code ensures that if the constraints hold, the code of the **advice** is executed at the specified situations. The concrete static position in the code or specification is called a **weaving candidate**, because the aspect execution is "weaved" into the original execution. A **pointcut language** defines well-formed pointcut expressions.

An **aspect** is a unit of pointcuts, advices, and new definitions. In AoASM, an aspect is allowed to introduce new universes, (derived) functions, and rules.

1.2   Intuitive Use of Aspect-Oriented ASMs

In the following, we show by an example from [9] how the AoASM pointcut language can be used to specify a join point. The rule PROCESSMONEYREQUEST shown in spec. 1 describes how an automated teller machine (ATM) processes a request for a given amount of money *In* and a current card *CurrCard*. If the request is *allowed* the ATM will grant the money and change the balance of the card account *MoneyLeft*. Otherwise, the card will be rejected and the ATM will output a message that gives a reason for the refusal.

If one would like to extend this specification in the sense of AoASM, one has to identify join points where the extension should take place. In spec. 1 possible extension points can be determined by pointcut expressions (cf. annotations) of different kinds.

within, cflow ····· 1 ⌈ PROCESSMONEYREQUEST(In) = ▪ ················· execution
                     2 │   if $allowed(\overline{In,\ CurrCard})$ then ··············· args
this ············   3 │ $\overline{\text{GRANTMONEY}(In)}$ ····················· call
set ············   4 │ $MoneyLeft := MoneyLeft - In$ ················· get
                     5 ⌊   else $Out := \{NotAllowedMsg,\ CurrCard\}$

ASM-specification 1. The rule PROCESSMONEYREQUEST [9]

The pointcut language defines pointcut expressions by a couple of patterns that describe different constructs of the base language, in our case ASM, including some properties of these constructs like the name or parameters.

Spec. 2 defines an aspect COMMUNICATEMONEYGRANT that extends spec. 1 in order to prevent money laundering. Whenever an amount of money of more than Euro 10.000 shall be granted, the finance authority has to be informed.

```
 1    aspect COMMUNICATEMONEYGRANT =
 2        pointcut ptcGrantMoney(currIn) :
 3            within("ProcessMoneyRequest", ".*")  and
 4            call("GrantMoney", ".*")  and
 5            args(currIn)
 6
 7        advice COMMUNICATEMONEYREQUESTTOFINANCEAUTHORITY(currIn)
 8            after : ptcGrantMoney(currIn) =
 9                if currIn > 10.000 then
10                    INFORMFINANCEAUTHORITY(accountHolder, currIn, date)
```

ASM-specification 2. The aspect COMMUNICATEMONEYGRANT.

The pointcut *ptcGrantMoney* conjuncts three pointcut expression. The keyword **within** restricts the static context for the extension to a rule or derived function of the given name and arguments, here PROCESSMONEYREQUEST and the regular expression pattern ".*" for one arbitrary argument. The name and the arguments are string based regular expressions that are used for matching the textual representation of the original ASM.

Additionally, **call** determines that the extension will be executed at rule calls or derived function calls of the given name and arguments, here GRANTMONEY with one arbitrary argument. Now, the join point is exactly determined because the macro rule call GRANTMONEY(In) matches exactly the defined pointcut.

Last, the pointcut expression **args** exposes the context, here the one and only argument of GRANTMONEY(In), so that it can be accessed within the advice. Therefore, the pointcut definition uses this argument as an argument by its own. A binding between In and *currIn* is created so that the actual value can be considered in the advice's body.

The advice (spec. 2, ll. 6-9) uses the pointcut *ptcGrantMoney(currIn)* and defines the behavior, that is inserted at matching weaving candidates in the original specification. The locator **after** states that the advice's body has to be executed after the matching weaving candidate. In order to make it possible that the advice may be affected by updates of the matching weaving candidate, the macro call and the advice are combined by sequential execution replacing the original macro call.

The pointcut expressions **get** and **set** in spec. 1 expose the context of functions analogously to the **args** pointcut expression. The **execution** pointcut expression describes the entry point of the body of a derived function or a rule.

The **cflow** pointcut expression gets an arbitrary pointcut expression as parameter. It matches any join point that is evaluated in the control flow defined by the given pointcut expression, i.e. **cflow(call**("ProcessMoneyRequest", ".*")) **and call**("GrantMoney", ".*") matches the same join point as the pointcut *ptcGrantMoney*, but does it dynamically. **This** dynamically restricts the execution of an extension to specific agents.

2   Ground Model of the Semantics of AoASM

In this section, we present a ground model according to [9] for the semantics of AoASM. The semantics is described by operations on abstract syntax trees (ASTs) of the specifications. We use a faintly extended AST representation by ASM functions and universes as used by [10] to describe the semantics of the ASM interpreter CoreASM. Differently from that work, we define our semantics not by giving new interpretations but rather by analyzing, creating, and manipulating ASTs.

The **aspect weaving** activity analyses the original specification (α), taking into account the aspect-oriented specification (δ) in order to create the resulting specification (ω). Most of the weaving can be done statically, but some pointcut expressions (e.g. **cflow**) have to be checked during runtime. Therefore, it is necessary to orchestrate ω, so that runtime information can be derived and accessed for dynamic conditions (e.g. the callstack). This complex activity of weaving is broken into the following self-contained tasks:

1. initialization (*INIT*)
2. identification of weaving candidates (*IDENTIFY*)
3. orchestration of the resulting specification (*ORCHESTRATE*)
4. weaving (*WEAVE*)

The ground model consists of an ASM rule WEAVING that controls these four different modes (cf. spec. 3). Each mode is described in detail in the following subsections.

2.1   Initialization

The static part of the AoASM semantics is defined by ASM functions that enable the navigation through the AST, like parent and firstChild, and functions that provide information about a node, like its kind of node in the function grammarRule, its source specification spec (position in the textual representation), and its token (identifier or symbol). The function grammarRule defines the kind of a node. Fig. 1 shows the AST representing the update rule of spec. 1.

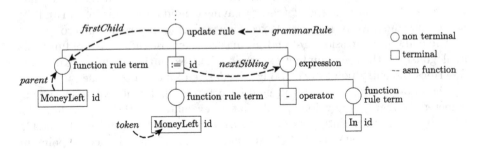

Fig. 1. AST representation of the update rule of PROCESSMONEYREQUEST

Different universes are introduced to mimic the hierarchy of node types. Each node becomes a member of the universe NODE and if it is a non terminal it is added to the universe ASTNODE, too. In particular, each node is a member of the universe that corresponds to its grammar rule, e. g. the node with the token *In* (see Figure 1) will be a member of the universe NODE as well as ID. The representation of the AST for δ is built analogously.

In order to store the information collected during the identification phase, two new universes JOINPOINT and BINDING are defined.

Analogously, we introduce some functions to keep track of dynamic information that is required during the weaving process. If a join point in universe JOINPOINT is created, its corresponding weaving candidate, advice, and binding, are stored by dynamic functions, here jpCandidate, jpAdvice, and jpBinding.

The dynamic part of the initialization prepares the abstract state for the weaving process. All functions representing the ASTs of α and δ are initialized. Next, the initial resulting specification ω is created as a copy of α. The corresponding ASM specification is omitted because it is straight forward.

2.2 Identification of Weaving Candidates

During the identification phase, the **universe** JOINPOINT will be extended with new join points for each pointcut (cf. spec. 3, ll. 11-12) that successfully matches a weaving candidate.

The semantics of finding weaving candidates and creating corresponding join points is defined in spec. 3. Every ASTNODE from the specification ω is matched against each advice's pointcut from δ (see subsection 3.1 for the refinement of rule MATCHING). The result is a tuple of matchingResult and a binding.

A binding is a tuple of the name of the argument respectively the context and the name of this argument inside the pointcut. It is added to the universe BINDING and can be **undef** if the matching was not successful or no context has to be exposed for this particular pointcut.

```
1   rule WEAVING =
2     [..]
3     else if mode(self) = IDENTIFY then
4        forall can ∈ ASTNODE with spec(can) = ω
5          forall ptc ∈ POINTCUT
6            let adv = parent(ptc) in
7              local matchingResult, binding in
8                let
9                  (matchingResult, binding) ← MATCHING(can, ptc)
10               in
11                 if matchingResult then
12                   CREATEJOINPOINT(can, adv, binding)
13       mode(self) := ORCHESTRATE
14    [..]
```

ASM-specification 3. Identification of Pointcuts.

Each successful matching creates a join point based on its candidate, the corresponding advice, and the binding. This join point is added to the universe JOINPOINT by the **rule** CREATEJOINPOINT. Finally, mode(self) is set to *ORCHESTRATE* in order to continue with the orchestration.

2.3   Orchestration

The orchestration prepares ω for requests about the abstract state at runtime. These requests are necessary to implement the pointcut expressions **cflow** and **this**. The refinement in subsection 3.2 illustrates the orchestration for **cflow**. Specification 4 shows the relevant excerpt of the ground model within the rule WEAVING. The specification ω is extended by rule ADDCALLSTACKSUPPORT and rule ADDSTATEQUERYSUPPORT. The second rule ADDSTATEQUERYSUPPORT integrates auxiliary rules into ω that are used to dynamically query the state or call stack. Last, mode(self) is set to *WEAVE* in order to proceed with the weaving phase that is partitioned into the two parts *SORT* and *INSERT*.

```
1   rule WEAVING =
2     [..]
3     else if mode(self) = ORCHESTRATE then
4        seq
5          ADDCALLSTACKSUPPORT
6        seq
7          ADDSTATEQUERYSUPPORT
8        mode(self) := WEAVE
9        weavingPhase(self) := SORT
10    [..]
```

ASM-specification 4. Orchestrating the resulting specification.

2.4   Weaving

In phase *SORT*, all join points are split into three sets for each weaving candidate according to the locator ("before", "around", or "after") of the corresponding advice. The sets for each weaving candidate are used to create subtrees that are woven into ω at their specific location during the second phase INSERT.

The semantics of the defined locators **before**, **around**, and **after** is a crucial issue: What does **before** mean in the context of an ASM statement in AoASM? The execution context of a single ASM statement is the current step of the machine. In order to be able to guarantee that the inserted advice statements will affect the corresponding statement of its weaving candidate or vice versa, both have to be encapsulated into a sequential rule construct (cf. [9]).

The order regarding the sequential execution is determined by the locator, e. g. **before** means that the advice call precedes the call referenced by the corresponding weaving candidate. The locator **around** states that the call of the weaving candidate has to be replaced by the advice's call. An **around** advice can be used in combination with the **proceed** construct to call the replaced rule one or multiple times from within the advice's body, even with different parameters.

We introduce the locator **parallel** which reflects the basic semantics of ASMs best and should be used to avoid over-specification. This locator means that additional behavior will be woven into a parallel block relative to a matching join point. The parallel locator's semantics can be reduced to the semantics of **around** if the corresponding advice's body is a parallel block that contains exactly one proceed call that reuses the original parameters of the weaving candidate beside the additional advice behavior. Therefore and for the sake of clarity, we omit the detailed description of the parallel locator in the following.

```
1   rule WEAVING =
2       [..]
3       else if mode(self) = WEAVE then
4           if weavingPhase(self) = SORT then
5               forall jp ∈ JOINPOINT do
6                   seq
7                       advCall ← CREATEADVICECALL(jpAdvice(jp), jpBinding(jp))
8                   seq
9                       let can = jpCandidate(jp) in
10                          case locator(jpAdvice(jp)) of
11                              "before" :  add advCall to beforeAdvs(can)
12                              "around" :  add advCall to aroundAdvs(can)
13                              "after" :   add advCall to afterAdvs(can)
14                  weavingPhase(self) := INSERT
15      [..]
```

ASM-specification 5. Weaving phase SORT: All advices of a join point are collected.

The insertion of advice calls into ω is prepared when the weaving phase is SORT. We create advice calls taking into account an existing binding for each join point. If the execution of advice's statements depends on the current state, e. g. in case of a **cflow** pointcut, the call is encapsulated into a guard that reflects the dynamic parts of the pointcut's condition (cf. spec. 5, l. 7). Those guarded calls are sorted into sets, one set for each kind of locator (cf. spec. 5, ll. 9-13).

The first task of the weaving phase *INSERT* is to add the additional definitions of aspects to the definitions of ω. After that, the sets of advice calls are used to create a sequential rule that consists of three parallel blocks, one for each kind of locator (cf. spec. 6). All advice calls from beforeAdvs(*can*) build the first block, the ones from aroundAdvs(*can*) the second block, and the ones from afterAdvs(*can*) form the last block. The rule CREATEPARBLOCK creates a parallel block for the given set of advice calls and returns the root node of this block (see fig. 2).

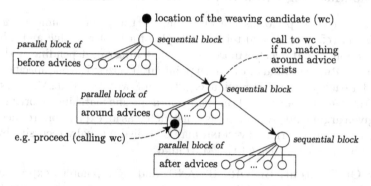

Fig. 2. Insertion template for different locators of advice calls

A sequential block is created for the three parallel blocks (cf. spec. 6, ll. 17). Last, the original call referenced by the join point is replaced by the sequential block of advice calls (cf. spec. 6, ll. 19).

```
1   rule WEAVING =
2      [..]
3      else if mode(self) = WEAVE then
4         [..] //phase SORT
5         else if weavingPhase(self) = INSERT then
6            INSERTASPECTDEFINITIONS
7            forall can ∈ {c | c ∈ ASTNODE, jp ∈ JOINPOINT
8               with c = jpCandidate(jp)}
9            seq
10              beforeBlock  ← CREATEPARBLOCK(beforeAdvs(can))
11              if aroundAdvs(can) = undef then
12                 aroundBlock := can
13              else
14                 aroundBlock ← CREATEPARBLOCK(aroundAdvs(can))
15              afterBlock  ← CREATEPARBLOCK(afterAdvs(can))
16            seq
17              seqBlock ← CREATESEQBLOCK([beforeBlock, aroundBlock,
         afterBlock])
18            seq
19              EXCHANGENODE(can, seqBlock)
20      [..]
```

ASM-specification 6. Insert joinpoints' advice calls into the resulting specification ω.

3   Excerpts of the Refined Semantics of AoASM

In this section, we pick out some important concepts of the semantics of AoASM and provide a more detailed description for the matching and the orchestration by means of an example as well as a refinement of the semantics of those concepts.

3.1   The Matching - How to Find Weaving Candidates

The rule MATCHING performs the matching for the given arguments *candidate* and *pointcut* (cf. spec. 3) and returns a tuple of a boolean value and a binding (cf. subsection 2.2). It selects a more specific matching rule depending on the function grammarRule, the kind of the given pointcut. There exists one specific matching rule for each pointcut expression, e. g. BINORMATCHING, CALLMATCHING, and CFLOWMATCHING. The selected rule performs a specific matching algorithm based on the given arguments. A pointcut is constructed as an expression tree according to spec. 7, so that the rule MATCHING may be called mutually recursive by more specific matching rules, like CFLOWMATCHING.

Binary Or-Pointcut. BINOR-, BINAND- and NOT-pointcut expressions are the top-level nodes for building expression trees. The tail-recursive EBNF-grammar for expression trees of pointcuts is given in lst. 7.

```
1   PointcutExpressionTree ::= BinOr
2   BinOr ::= BinAnd [ or BinOr ]
3   BinAnd ::= PointcutExpression [ and BinAnd ]
4   PointcutExpression ::= PointcutTerm | ( BinOr )
5   PointcutTerm ::= call(SignaturePattern) | args(SignaturePattern) |
        get(SignaturePattern) | set(SignaturePattern) | not BinOr |
        within(SignaturePattern) | cflow(BinOr) | this(IdPattern) | ...
6   SignaturePattern ::= Pattern ( , Pattern )*
7   Pattern ::= (RulePattern | FunctionPattern | IdPattern) [ as Identifier ]
8   FunctionKind ::= ( monitored | controlled | in | out ) | derived
9   FunctionPattern ::= [ FunctionKind ] SignaturePattern [ from UnivPattern ]
10  RulePattern ::= SignaturePattern [ ( with | without) ( return | result ) ]
11  IdPattern ::= Identifier | RegularExpressionString
12  ...
```

Listing 7. Partial grammar of pointcut expressions.

Of course, the expressions defined by AspectJ have been adapted with respect to ASMs. Hence, the FunctionPattern can restrict matching to functions of a certain kind or demand the value of a function to be a member of specific universes. RulePattern specifies whether side effects (as defined by [10, p. 108]) are forbidden, allowed, or gratuitous and whether a return value is expected. The grammar rule Pattern can expose the context join points similar to the **args** pointcut expression by writing **as** followed by one of the pointcut's parameters to bind a value of a join point to a parameter of the pointcut's declaration. For example, the combination of **call** and **args** pointcut expressions in spec. 2 can be rewritten by the expression **call**("GrantMoney", ".*" **as** *currIn*). This extension to the binding concept makes its application more flexible.

Since BinOr is the root of the expression tree, the rule BINORMATCHING(*can*, *ptc*) is always called first. The matching result depends on the number of children of the BinOr-node *ptc*. If *ptc* has one child, the result is the matching result of this child (which is a BinAnd-node) and the given candidate *can*. In case of two children, the matching result is the conjunction of the children's results. The derived function *astChildren(ptc)* returns an order preserving list of all children of *ptc*.

```
1   rule BINORMATCHING(can, ptc) =
2      if | astChildren(ptc) | = 1 then
3         result ← MATCHING(can, firstChild(ptc))
4      else if | astChildren(ptc) | = 2 then
5         result ← MATCHING(can, firstChild(ptc)) or
6            MATCHING(can, nextSibling(firstChild(ptc)))
```

ASM-specification 8. The rule BINORMATCHING

If the pointcut consists of a single call, the rule MATCHING will be called mutually recursive so that BINANDMATCHING will be performed similar to the semantics of BINORMATCHING. We assume that the pointcut expression **call** will be evaluated next.

Call-Pointcut The call pointcut matches calls of derived functions and macro call rules. The matching takes into account the name and the parameters of the call given by *can* and the **call** pointcut expression *ptc* with its signature pattern (cf. spec. 9). Pattern matching is performed between the children of *can* and *ptc*. Thereby, each child of *can* is compared against a regular expression given by a child of *ptc* with a corresponding position.

```
1    rule CALLMATCHING(can, ptc) =
2       //ptc has the addition keyword 'call'
3       if can ∉ DerivedFunction or can ∉ MacroCallRule
4          or | astChildren(can) | ≠ | astChildren(ptc) | + 1
5       then
6          result := (false, undef)
7       else
8          if ∀c ∈ astChildren(can), p ∈ tail(astChildren(ptc)):
9             pos(c) = pos(p) ∧ matches(toString(c), toString(p))
10         then
11            result ← (true, CREATEBINDING(can, ptc) )
12         else
13            result ← (false, undef)
```

ASM-specification 9. The rule call matches a derived function or a macro call rule against a call-pointcut.

3.2 Orchestration - Introducing a Call Stack

The semantics of the **cflow** pointcut requires information about the current call stack during interpretation. Therefore, one step of orchestration is to implement a call stack into ω (cf. spec. 4, l. 4). We use sequential steps to introduce a kind of entry and exit behavior for any macro call rule body. The function callStack is defined to keep track of each agent's call of macro rule calls and derived functions. Whenever a rule or function is called, we have to add its signature to the current agent's callStack and whenever the rule or function has been interpreted, this signature has to be removed from the callStack again. This can be guaranteed if each rule body is encapsulated into a turbo ASM sequential rule block. This block starts with **seq add** «signature of current call» **to** callStack (**self**), continues with **seq** «original rule block», and ends with **next remove** head **from** callStack(**self**).

Introducing the additional statements does not change the semantics of the current specification, because the sequential rule block will never produce an update according to the regular ASM step. The updates resulting from **add .. to** and **remove .. from** cancel each other out.

```
1   function callStack : AGENTS → LIST //of signatures
2
3   derived getSignature(def) =
4       [sig | sig ∈ astChildren(def), grammarRule(sig) = ID]
5
6   rule ADDCALLSTACKSUPPORT =
7       forall def ∈ RULEDECLARATION with spec(def) = ω  do
8           choose ruleblock ∈ astChildren(def) with nextSibling(def) = undef do
9           local add, rem, newRuleBody in
10              seq
11                  add ← CreateSeqElem(ADDTOCALLSTACK(GETSIGNATURE(def)))
12                  rem ← CreateSeqElem(REMOVECALLSTACKHEAD)
13              seq //also creates an update for parent(def)
14                  newRuleBody ← CREATESEQBLOCK([add, def, rem])
15              seq //change specification's ast of ω
16                  EXCHANGENODE(def, newRuleBody))
17
18  derived OnCallStack(call) =
19      return match in
20          seq
21              callSig := getSignature(call)
22          seq
23              choose sig ∈ callStack(self) do
24                  if ∀ nb ∈ [0 .. |sig|] : |sig| = |callSig| ∧
25                      matchingParameter(nth(sig, nb), nth(callSig, nb))
26                  then  match := true
27                  else  match := false
```

ASM-specification 10. Introducing a Call Stack.

Certainly, during the interpretation, the function callStack can be used as intended, e. g. guards for cflow pointcuts can use this information. The rule ADDCALLSTACKSUPPORT changes a specification ω by introducing the described constructs (cf. spec. 10).

Specification 11 shows the result of our orchestration by means of an example applying the orchestration for **cflow** on the rule PROCESSMONEYREQUEST.

```
1   rule PROCESSMONEYREQUEST(In) =
2   seq
3     if callStack(self) = undef then
4       callStack(self) := [ ["ProcessMoneyRequest", "In"] ]
5     else
6       callStack(self) := cons (["ProcessMoneyRequest","In"], callStack(self))
7   seq
8     if allowed(In, CurrCard) then
9       GRANTMONEY(In)
10      MoneyLeft := MoneyLeft − In
11    else Out := {NotAllowedMsg, CurrCard}
12  seq
13    if | callStack(self) | = 1 then
14      callStack(self) = undef
15    else
16    callStack(self) := tail( callStack(self) )
```

ASM-specification 11. The result of orchestration for cflow of the rule PROCESSMONEYREQUEST (cf. spec. 1).

4   Discussion and Validation

There exists some work on the semantics of AOP (cf. [11],[12]), although AspectJ has not yet a formal semantics. From our point of view, a vast number of existing approaches generally tend to clarify specific aspects of AOP or reasoning about AOP by defining their own base- and aspect language (cf. [13]), but are not intended for practical use.

Our approach is aimed at application in a real context starting with the requirements phase in order to enable continuous separation of concerns. The idea of aspect-oriented specification has been conceived by [14], but never has been lead to results in the sense of an applicable aspect-oriented extension to any common specification language. The latest approach by [15] has been a case study using metadata and templates to introduce AOP into the early phases of software development, but did not base on any or offer a well-formed specification language.

The validation of AoASM is twofold: First, AoASM is implemented as a plugin for CoreASM [10]. Second, AoASM is used to specify the semantics of an extension of UML for multi-modal interaction.

The AoASM plugin for CoreASM implements a parser for AoASM, a weaver according to our semantics, and provides a set of tools to make the specification process as easy as possible. This is done by refining the ground model specification down to the implementation. Further, we make sure that the refined semantics of the ground model takes into account compatibility with the Core-ASM implementation. Specifications that are generated by the AoASM plugin can be inspected either as text or as AST. Meaningful warnings, error messages and tooltips are provided to support the user. In the future, weaving candidates will be marked in the original specification so that these markers can be used as links to matching advices or vice versa.

The second part of the validation will be to apply AoASM on the extension of UML for multi-modal interactive systems as described in [7]. This approach is based on the formal semantics of UML by [4] on the one hand and defines a semantics for a UML-profile on the other hand. The combination of those semantics is given in a semi formal manner in [7]. We are going to apply AoASM to precisely define the combination of both semantics. This work will benefit from the opportunity to separate concerns as well as extend the basic semantics of UML without touching it.

5   Conclusion and Future Work

In this paper, we introduce an ASM model of AoASM as a conservative extension to the formal specification language ASM. We define AoASM's operational semantics and describe some interesting refinements. AoASM covers a comprehensive subset of concepts of modern aspect-oriented programming languages like AspectJ [3,8] and introduces some expansions that take into account the characteristics of ASMs [9]. AoASM introduces several ASM-specific extensions to optimize the exploit for its application. The pointcut language is extended to support guards that restrict advice executions to specific agents, functions and its locations inside conditional pointcut expression. Finally, the locator **parallel** exposes new possibilities offered by ASMs.

Our pointcut language uses well known regular expressions as a basis for the matching process based on the text of an ASM specification. These kinds of matchings can be performed statically, or as type based matchings that are performed on programs of strongly typed languages. As ASM-domains can be interpreted in terms of types (like in [5]), one could introduce conditional pointcuts to dynamically check if a certain parameter is a member of a certain universe.

The orchestration could be smoothly integrated into ASM specifications. Especially the call stack, which is needed for control flow based pointcuts, is implemented so that it does not influence the original specification as it is built and decomposed during each step of the interpreter, so that the update set of every step will not contain any update of the call stack.

Currently, the defined semantics for AoASM is validated by implementing an extension for CoreASM. This extension brings some comfort to users applying aspect-oriented specification with ASM. For instance, the tool does not only perform the weaving according to our semantics, but it also allows the inspection of the resulting specification, and offers some support during the specification of aspects, e. g. for pointcut design.

References

1. Kiczales, G., Lamping, J., Mendhekar, A., Maeda, C., Lopes, C.V., Loingtier, J.-M., Irwin, J.: Aspect-Oriented Programming. In: Akşit, M., Matsuoka, S. (eds.) ECOOP 1997. LNCS, vol. 1241, pp. 220–242. Springer, Heidelberg (1997)
2. Dijkstra, E.W.: A Discipline of Programming, 1st edn. Prentice Hall PTR, Upper Saddle River (1997)
3. Laddad, R.: AspectJ in Action - Enterprise AOP with Spring Applications. Manning, G. (2010)
4. Kohlmeyer, J.: Eine formale Semantik für die Verknüpfung von Verhaltensbeschreibungen in der UML 2. PhD thesis, Universität Ulm (2009)
5. Sarstedt, S.: Semantic Foundation and Tool Support for Model-Driven Development with UML 2 Activity Diagrams. PhD thesis, Ulm University (2005)
6. OMG Unified Modeling Language (OMG UML) Superstructure v2.4.1 (2011)
7. Dausend, M.: Towards a UML Profile based on Formal Semantics for Modelling Multimodal Interactive Systems. Technical report, Ulm University, Ulm (2011)
8. The AspectJ Team. The AspectJ Project (2011), http://www.eclipse.org/aspectj/ (last access January 2014)
9. Börger, E., Stärk, R.: Abstract State Machines – A Method for High-Level System Design and Analysis. Springer (2003)
10. Farahbod, R.: CoreASM: An Extensible Modeling Framework & Tool Environment for High-level Design and Analysis of Distributed Systems. PhD thesis, Simon Fraser University, Burnaby, Canada (2009)
11. Wand, M.: A Semantics for Advice and Dynamic Join Points in Aspect-Oriented Programming. In: Taha, W. (ed.) SAIG 2001. LNCS, vol. 2196, pp. 45–46. Springer, Heidelberg (2001)
12. Avgustinov, P., Hajiyev, E., Ongkingco, N., de Moor, O., Sereni, D., Tibble, J., Verbaere, M.: Semantics of Static Pointcuts in Aspect. In: Proceedings of the 34th Annual ACM SIGPLAN-SIGACT Symposium on Principles of Programming Languages, vol. 42(1), pp. 11–23 (2007)
13. Clifton, C., Leavens, G.T.: MiniMAO: An imperative core language for studying aspect-oriented reasoning. Science of Computer Programming (2006)
14. Blair, L., Blair, G.S., Andersen, A.: Separating Functional Behaviour and Performance Constraints: AspectOriented Specification. Technical report, Computing Department, Lancaster University, Bailrigg, Lancaster (1998)
15. Agostinho, S., Moreira, A., Marques, A., Araújo, J., Brito, I., Ferreira, R., Raminhos, R., Kovacevic, J., Ribeiro, R., Chevalley, P.: A Metadata-Driven Approach for Aspect-Oriented Requirements Analysis. In: International Conference on Enterprise Information Systems, pp. 129–136 (2008)

Modular Refinement for Submachines of ASMs

Gidon Ernst, Jörg Pfähler, Gerhard Schellhorn, and Wolfgang Reif

Institute for Software & Systems Engineering
University of Augsburg, Germany
{ernst,joerg.pfaehler,schellhorn,reif}@informatik.uni-augsburg.de

Abstract. We describe and formalize a compositional, contract-based submachine refinement for a variant of Abstract State Machines. We motivate the approach by models of the Flash file system case study, where it is infeasible to refine a complete machine as a whole.

1   Introduction

Abstract State Machines (ASMs, [5]) are a general software development method for state-based systems. By integrating them with refinement [4] and the algebraic specification of data types they provide a rigorous framework for verifying correctness critical applications.

This paper contributes a formally defined instance of ASM refinement theory, which is *compositional* for submachines that respect information hiding. Formally, we prove that a machine $\mathcal{M}$ that calls the operations of a submachine $\mathcal{L}$ satisfies the following substitution law: If $\mathcal{K}$ is a correct refinement of $\mathcal{L}$, then substituting calls to $\mathcal{L}$ with calls to $\mathcal{K}$ in $\mathcal{M}$ is a refinement of $\mathcal{M}$. The theorem allows to refine submachines independently of the context formed by $\mathcal{M}$.

This work is strongly motivated by our current effort to construct a verified file system for flash memory. This challenge has been proposed by NASA [15] in response to problems with the flash file system of the Mars Rover "Spirit" [20].

As a consequence, the syntax and semantics of the variant of ASMs we consider here differs and is somewhat restricted compared to traditional ASMs. In particular, we define both an *atomic semantics* and an *non-atomic semantics* for the rules. The former is intended for the environment of rules, e.g. the caller of a POSIX operation like "create directory". It is also the semantics of calls to submachines. The non-atomic semantics is necessary to study the effects of power failures while an operation (i.e., a rule of a control-state ASM which goes through a potentially infinite sequence of intermediate states) is running: the recovery mechanism that runs when rebooting after a power failure must restore a consistent state from any intermediate state.

On the syntactic level our approach currently considers sequential constructs only, since we have not investigated concurrent execution for the Flash file system. Thus, the atomic semantics is an instance of sequential ASMs, the non-atomic semantics is an instance of control-state ASMs with a single control state. An extension to several control states and interleaved execution is possible, at the price that reasoning with the wp-calculus has to be replaced with the more

Y. Ait Ameur and K.-D. Schewe (Eds.): ABZ 2014, LNCS 8477, pp. 188–203, 2014.

complex reasoning using the temporal logic RGITL we define in [26]. The ASM rules we use here could also be called (a subset of) RGITL programs.

Our earlier work on ASM refinement [22,24] assumed that rules have *guards*, and that an ASM chooses to invoke rules only when the guard is true. This view is not suitable for submachines, where no guarantee can be given that calls to the submachine will respect the guards, nor that a refined submachine (viewed in isolation) has fewer runs than the original submachine with the guard view. Instead we will use *preconditions* for rules, where a call with precondition false is possible, but results in arbitrary behavior. Our approach takes up ideas from *contract refinement* as used in Z [29], and adapts it to our scenario.

We motivate the need for compositional refinement of submachines in Sec. 2, by giving an overview over the structure of our development. Sec. 3 demonstrates why preconditions are necessary instead of guards. For reasons of space, we only sketch the machines needed. For an overview over the project we refer the reader to [25] (this volume). The full details and a list of previous publications is available online [9]. Sec. 4 defines syntax and semantics of the ASM rules we use. We contribute two compatible semantic definitions: a *non-atomic* view, where execution of an ASM rule results in a sequence of steps; and an *atomic* view, on which we base the definition of runs of an ASM, the semantics of submachine calls, and the weakest precondition calculus we use for deduction in our prover KIV [21]. Sec. 5 defines refinement for our setting, gives the proof obligations for forward simulation, and proves that refinement is modular for submachines. Sec. 6 gives related work and Sec. 7 concludes.

2   Submachines in the Flash File System

In this section, we briefly show the topmost refinement of the refinement hierarchy and motivate that the file system challenge is inherently compositional. Figure 2 displays the structure of the topmost part of the project, where boxes represent components, and layers respectively, connected by refinement (dotted lines). These are formally given by Abstract State Machines (ASMs) with algebraic states. The grey boxes are the leaves of the hierarchy from which we will generate the final code.

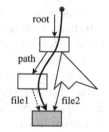

Fig. 1. FS graph

At the toplevel, POSIX [28] specifies the requirements: The file system (FS) is a graph consisting of directories (internal nodes) and files (leaves), an example is shown in Fig. 1. Files can be referred to by multiple directories under different names ("hard-links"), consequently, names are attached to edges of the graph. The directory part is a proper tree. The POSIX interface is based on *paths*. Our formal POSIX model can be found in [11]. As an example, Fig. 4 shows the specification of the unlink operation, which removes one link to the file denoted by *path*, and also deletes the file's content once it is unreferenced.

Real file system implementations consist of two parts. Generic aspects, i.e. traversing paths and checking access rights are realized in Linux by the *Virtual*

```
posix_unlink(path; err)
   tree := tree − path
   // conditionally delete content
```

Fig. 4. POSIX specification of unlink

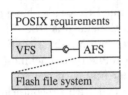

Fig. 2. FFS upper layers

```
vfs_unlink(path; err)
   ino := ROOT_INO;
   while path.length > 1 do {
      let n = path.head
      in afs_lookup(ino, n; ino', err);
      path := path.tail, ino := ino'
   }
   let dent = negdentry(path.head)
   in afs_unlink(ino; dent, err)
   // conditionally evict ino

pre: dirs[ino] ≠ undef ∧ ...
afs_unlink(ino; dent, err)
   let name = dent.name in
   dirs[ino].entries[name] := undef
```

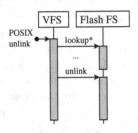

Fig. 3. Call sequence in the final composed file system code

Fig. 5. VFS/AFS rules (no error-handling, submachine calls underlined)

Filesystem Switch (VFS). Windows' "Installable File System" serves the same purpose. Concepts specific to individual file system implementations are realized by the individual file systems (FS). For standard magnetic discs ext4 or ReiserFS would be such file systems, for flash memory we use UBIFS [14] as a design blueprint for our formal models.

VFS communicates with individual file systems through a well-defined interface visualized by the symbol —◦— in Fig. 2. The main data structure used in this interface is called *inodes* in Linux. To specify this interface we define an ASM called *Abstract File System* (AFS). Technically, AFS is a *submachine* of VFS.

A typical ASM rule for the VFS operation unlink is sketched in Fig. 5 (full models can be found in [10]). Several calls of afs_lookup are used to traverse the path, checking that the individual directories exist with suitable access rights. Finally, afs_unlink is called for the actual removal of the link in the target directory. Operation afs_unlink has a *precondition* to characterize valid inputs, which needs to be checked at every call site. The use of an abstract AFS specification makes the proof that VFS is a proper refinement of the POSIX specification [11] *independent* of the actual file system.

Figure 3 shows the corresponding sequence of operations in the final *composed code* we generate (marked grey in Fig. 2). In this code calls to abstract AFS operations have been replaced by calling the concrete FS code. The main theorem we prove in this paper, is that this methodology is sound, i.e., the composed code refines the top-level specification (see Theorem 2 in Sec. 5).

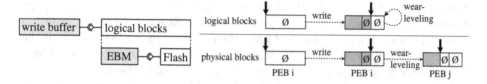

Fig. 6. FFS lower layers   **Fig. 7.** Abstract/concrete run with wear-leveling

As expected, the size of implementation components is much larger than the size of their respective specifications (POSIX: 50 lines, VFS: 500 lines, AFS: 100 lines, abstract specification of UBIFS again calling some submachines: 500 lines). This pattern repeats all the way down to the hardware interface, forming a *deep* hierarchy. An approach that exploits the compositional structure is key to make the verification of the whole file system tractable.

3   Contracts in the Eraseblock Management Layer

This section motivates why we need preconditions, not guards for our refinements. In the following we present a (simplified) example taken from the lower layers of our flash file system and show how this technique is employed to simplify the abstract layer.

Figure 6 shows a section of these layers. We have a model of flash hardware at the bottom. A device is divided into *physical erase blocks* (PEBs). The basic limitation of flash hardware is that only *sequential* writes within a block are allowed and overwriting is not possible. Space can only be reclaimed by *erasing* a full block. Erasing is slow and physically degrades the memory, i.e., after 10^4-10^6 erasures a block is unusable. In order to increase the reliability and lifetime of the device, the layer directly above the hardware, the Erase Block Management (EBM), performs *wear-leveling* in the background. Wear-leveling moves stale data to new blocks, in order to spread the erases evenly amongst the blocks. This is implemented transparently by introducing a mapping from *logical* to *physical* erase blocks. The mapping is managed by the EBM internally, the client can only access logical erase blocks (LEBs). We then abstract this implementation to a layer similar to the flash device ("logical blocks" in the figure). The aim of the following is to show how preconditions can be used to hide wear-leveling from upper layers. For technical details not covered here, the reader is referred to [18].

The basic idea to specify the limitation to sequential writes on physical and their abstraction to logical blocks is to associate an offset called *fillcount* with each block. It stores how far the block is already programmed with data. This offset can not be accessed by the EBM and is only used to express the precondition of a write (to logical as well as physical blocks): the offset of the write is above the *fillcount* of the target block. The vertical arrows ↓ in Fig. 7 denote the *fillcount*. The upper part of the figure shows a logical block and how it is affected by a write operation and a subsequent wear-leveling cycle. The lower half depicts the PEB mapped for this LEB at any point in time. During writing the *fillcount* fields are affected in the same way. For efficiency the implementation of wear-leveling, however, does not copy the entire contents of the block (if the remainder is already empty). This means that in general the target block j may have a lower *fillcount* than the source block i, as shown in the figure.

If the guard-semantics is used, the EBM model has more runs than its abstraction: In Fig. 7 for example after wear-leveling additional writes to physical block j are possible, which the abstract model can not reproduce. These additional runs, however, will never be exploited by a client, since their existence crucially depends on wear-leveling, which is not triggered by the client but performed non-deterministically in the background.

In conclusion, preconditions are more appropriate for submachines than the guards. The reason is that operations of a submachine are explicitly called and are not triggered internally. Therefore, an implementation may have a more liberal precondition than the abstract system. This can be used to simplify the abstract system by strengthening the abstract precondition and thereby hiding inconsequential runs of the concrete system. Refinement can still be expressed as the usual trace inclusion if all possible runs (including divergence) are added to the semantics of an operation outside of its precondition (see Def. 3 and 5).

4   Syntax and Semantics of ASMs with Submachines

Section 4.1 defines the syntax of ASMs without submachines. Roughly, an ASM consists of a number of rules with preconditions (called "operations"). The rules are given a non-atomic semantics in Sec. 4.2 that is similar to the one of control state ASMs, however, we never use an explicit control state. We then abstract to an atomic view, which is used to define runs of ASMs in Sec. 4.3 and calls to submachine operations in Sec. 4.4. Finally, Sec. 4.5 gives the semantics of wp-formulas that we use to define and verify properties.

4.1   Syntax

This subsection defines the syntax of the ASMs we use. We assume the reader is familiar with first-order logic, where based on a signature $SIG = (F, P)$ with functions $f \in F$ and predicates $p \in P$ terms t, formulas φ and boolean expressions ε (= quantifier-free formulas) can be defined. The semantics $[\![t]\!](s)$ of terms t and the semantics $s \models \varphi$ of formulas φ is defined over a state s consisting of an algebra and a valuation for variables x as usual.

We assume the signature is partitioned into four parts: a *static* signature (no updates allowed), an *input* signature (that is only read by ASM rules), an *output* signature (that is only written by ASM rules), and a *controlled* signature that may be read and written by ASM rules.

We use the general convention to underline sequences of elements, i.e., $\underline{a}$ stands for a sequence $(a_1, \ldots, a_n)$ for some $n \geq 0$. We write $s\{x \mapsto a\}$ for the modified state, where variable x now maps to value a, and $s\{f(\underline{t}) \mapsto a\}$ for the state, where function f has been updated to have value a for arguments $[\![t]\!](s)$. A location loc is either a variable x or $f(\underline{t})$, so $s\{loc \mapsto a\}$ denotes a generic update. We introduce the abbreviation $s\{loc \mapsto t\} = s\{loc \mapsto [\![t]\!](s)\}$ for terms t, and the generalization $s\{\underline{loc} \mapsto \underline{t}\}$ to a parallel update, when all locations are different. The leading symbol of a location is x and f, respectively. An input resp. output location is a location $f(\underline{t})$ where the leading symbol f is in the input resp. output signature. We use the following syntax for our rules α, β:

$$\alpha ::= \quad \underline{loc} := \underline{t} \mid \alpha; \beta \mid \textbf{if } \varepsilon \textbf{ then } \alpha \textbf{ else } \beta \mid$$
$$\textbf{while } \varepsilon \textbf{ do } \alpha \mid \textbf{choose } \underline{x} \textbf{ with } \varphi \textbf{ in } \alpha \textbf{ ifnone } \beta$$

For parallel updates we require that the leading symbols of $\underline{loc}$ are all distinct (so no clashes are possible) and writable, i.e., are local variables or part of the controlled or output signature. We write **skip** for an empty parallel update.

The **choose** constructs binds local variables $\underline{x}$ to values such that φ is satisfied and executes α. If there is no possible choice (e.g. if $\varphi \equiv \texttt{false}$) then β is executed instead. Standard local variable declarations are defined as

$$\textbf{let } \underline{x} = \underline{t} \textbf{ in } \alpha \equiv \textbf{choose } y \textbf{ with } \underline{y} = \underline{t} \textbf{ in } \alpha\frac{y}{\underline{x}} \textbf{ ifnone skip}$$

where y are new variables and $\alpha\frac{y}{\underline{x}}$ denotes the substitution of $\underline{x}$ with y in α.[1] Note that **ifnone skip** is never executed here, and we will we drop such irrelevant ifnone-clauses as well as random choice (i.e., **with true**) in the following.

Based on the syntax of rules we define abstract state machines.

Definition 1. *A (data type-like) ASM* $\mathcal{M} = (SIG, Ax, Init, \{\texttt{Op}_j\}_{j \in J})$ *consists of a signature* SIG, *a set* Ax *of predicate logic axioms for the static part of the signature, a predicate* $Init$ *to characterize initial states, and a set of operations for indices* $j \in J$. *Each operation* $\texttt{Op}_j = (pre_j, \underline{in}_j, \alpha_j, \underline{out}_j)$ *consists of an ASM rule* α_j *that describes possible state transitions, provided precondition* pre_j *holds. It reads input from a vector* $\underline{in}_j$ *of input locations, and writes output to a vector* $\underline{out}_j$ *of output locations. It may modify local variables, controlled locations and the locations of* $\underline{out}_j$. *The rules should have no non-local variables.*[2]

In concrete code like the one given in Fig. 4 each operation $\texttt{Op}_j$ has a *name* (instead of using an index j), the precondition is given after keyword **pre**, and the other components are given in the form of $name(\underline{in}_j; \underline{out}_j)\{ \alpha_j \}$.

4.2 Non-atomic Semantics of Rules

This section gives a semantics to rules that assumes that they are executed non-atomically: each update and each test of a condition is executed as a separate step. The semantics of rules is therefore based on sequences $I = (I(0), I(1), \ldots)$ of states $I(k)$, which may be finite or infinite. Such sequences are called *intervals*. Formally, $I \models \alpha$ expresses that the interval I is a possible execution of α.

---

[1] The renaming avoids conflicts when $\underline{x}$ is used in $\underline{t}$.

[2] Thus, states of $\mathcal{M}$ are just SIG-Algebras; the values of variables are irrelevant.

We introduce some auxiliary notation: The length of an interval $\#I$ is in $\mathbb{N} \cup \{\infty\}$. If I is finite it consists of $\#I + 1$ states. In particular, the smallest interval with $\#I = 0$ has one state only. We also lift modification of states to intervals: given a vector of variables $\underline{x}$ and a sequence of value vectors $\underline{a} = (\underline{a}_0, \underline{a}_1, \ldots)$ of the same length as the interval (where each element $\underline{a}_k$ has the length of vector $\underline{x}$), then $I\{\underline{x} \mapsto \underline{a}\}$ is the modified interval, where $I(k)(\underline{x})$ is $\underline{a}_k$.

For sequential execution we need the sequential composition of intervals I_1 and I_2, written $I_1 \,\substack{\circ \\ \circ}\, I_2$, which is defined in two cases. For finite I_1, the last state of I_1 (written $I_1.\text{last}$) must agree with the first of I_2: $I_1.\text{last} = I_2(0)$, and the result is $(I_1(0), \ldots, I_1.\text{last}, I_2(1), I_2(2), \ldots)$, i.e., the duplicate middle state is removed. If I_1 is infinite, then $I_1 \,\substack{\circ \\ \circ}\, I_2 := I_1$.

Definition 2.

$I \models \underline{loc} := \underline{t}$ iff $I = (s, s')$ and $s' = s\{\underline{loc} \mapsto \underline{t}\}$

$I \models \alpha; \beta$ iff there are I_1, I_2 such that $I_1 \models \alpha$, $I_2 \models \beta$ and $I = I_1 \,\substack{\circ \\ \circ}\, I_2$

$I \models \textbf{if } \varepsilon \textbf{ then } \alpha \textbf{ else } \beta$
     iff either $I(0) \models \varepsilon$ and $I \models \textbf{skip}; \alpha$ or $I(0) \not\models \varepsilon$ and $I \models \textbf{skip}; \beta$

$I \models \textbf{choose } \underline{x} \textbf{ with } \varphi \textbf{ in } \alpha \textbf{ ifnone } \beta$
     iff either $I(0)\{\underline{x} \mapsto \underline{a}_0\} \models \varphi$ and $I\{\underline{x} \mapsto \underline{a}\} \models \textbf{skip}; \alpha$

         for some $\underline{a} = (\underline{a}_0, \underline{a}_1, \ldots)$

     or $I \models \textbf{skip}; \beta$ and there are no values $\underline{a}$ with $I(0)\{\underline{x} \mapsto \underline{a}\} \models \varphi$

$I \models \textbf{while } \varepsilon \textbf{ do } \alpha$
     iff $I \in \nu(\lambda \mathcal{I}. \{I_0 \mid$
         either $I_0(0) \not\models \varepsilon$ and $I_0 \models \textbf{skip}$
         or $\#I_0 = \infty$ and $I_0(0) \models \varepsilon, I_0 \models \textbf{skip}; \alpha$
         or $I_0 = I_1 \,\substack{\circ \\ \circ}\, I_2$ with $\#I_1 < \infty, I_1(0) \models \varepsilon, I_1 \models \textbf{skip}; \alpha, I_2 \in \mathcal{I}\})$

Most of the clauses should be intuitive. The **skips** in the clauses for **if, while** and **choose** indicate that evaluating the test is done in a separate step.[3] In the first disjunct of the semantics of choose, the sequence of states $\underline{a}$ captures the values $\underline{x}$ in the entire interval of α, not just in the first state. The set of runs of a while loop is defined as the greatest fixpoint ν[4] of interval sets $\mathcal{I}$ whose elements I_0 denote different possibilities to execute the loop. Informally, an interval I is a run of the while loop, if it can be split into a (finite or infinite) sequence of adjacent pieces. Each piece I_1 must be finite and execute the loop body (last case $I_1(0) \models \varepsilon, I_1 \models \textbf{skip}; \alpha$), the only exception being the last interval, when the sequence is finite. This interval may either be a nonterminating (infinite) execution of the loop body (second case of the definition), or it may be one **skip** step, where the loop test evaluates to false (first case of the definition).

---

[3] It is possible (and sometimes useful; e.g., to define atomic test-and-set instructions), to define the semantics such that evaluations of tests takes no additional step.

[4] The greatest fixpoint $\nu(\lambda \mathcal{I}. \{I \mid \varphi(I, \mathcal{I})\})$ can be understood as the *union* of all sets $\mathcal{I}$ whose elements satisfy the recursive property φ. The more commonly used least fixpoint is inadequate here, since it gives finite executions only.

4.3   Semantics of ASMs

The interval semantics of rules is relevant when we want to study the effect of power cuts, which will interrupt execution and produce prefixes of the intervals. It is also relevant when we add calls to submachines in the next subsection.

To define the runs of a machine $\mathcal{M}$, the fine-grained semantics of rules can be abstracted to an atomic view of the execution of an operation. It is also much easier to reason about the atomic semantics (using wp-calculus, see Sec. 4.5). The atomic view is therefore a relation over the set of states augmented with a bottom element, to indicate nontermination: $S_\perp := S \cup \{\perp\}$.

Definition 3. *The atomic semantics* $[\![\mathrm{Op}]\!] \subseteq S_\perp \times S_\perp$ *of an ASM operation* $\mathrm{Op} = (pre, \underline{in}, \alpha, \underline{out})$ *is defined as:*

$(s, s') \in [\![\mathrm{Op}]\!]$ iff either $s \neq \perp$ and $s \not\models pre$ (and s' is arbitrary from $S_\perp$)

or $s \neq \perp, s \models pre$ and there is I with $I(0) = s, I \models \alpha$

and if I is finite then $s' = I.\text{last}$, otherwise $s' = \perp$

or $s = s' = \perp$

The first line of the definition gives the idea of a precondition: if it is violated in state s, then calls to α may result in any successor state (including nontermination). The second clause collapses terminating runs I of α to their first and last state. Infinite runs yield $\perp$. The last line allows to define the semantics of calling two operations sequentially as relational composition: If the first operation does not terminate (gives $\perp$), then attempting to call another operation is not possible and will also give $\perp$.

Based on the semantics of single operations we can define runs of a machine:

Definition 4. *An ASM program over a machine* $\mathcal{M}$ *is a possibly infinite sequence* $\underline{j} = (j_0, j_1, \ldots)$ *of (indices or names of) operation calls. An execution of the program* $\underline{j}$ *is an interval* $\boldsymbol{I}$ *with states in* $S_\perp$ *and* $\#\boldsymbol{I} = \#\underline{j}$,[5] *where*

$$(s, s') \in [\![\mathrm{Op}_{j_k}]\!]   \text{ for } s = \boldsymbol{I}(k) \text{ and } s' = \boldsymbol{I}(k+1)\{\underline{in}_{j_k} \mapsto s(\underline{in}_{j_k})\}$$

holds for all $0 \leq k < \#\boldsymbol{I}$. *An execution is a run of the program, written* $\boldsymbol{I} \in runs^{\mathcal{M}}(\underline{j})$ *if it starts with an initial state* $\boldsymbol{I}(0) \neq \perp$, $\boldsymbol{I}(0) \models Init$.

The definition of runs of programs mimics the definition of runs of data types, although we consider both finite and infinite runs. Note that state $\boldsymbol{I}(k)$ stores the input $s(\underline{in}_{j_k})$ for calling Op_{j_k}. The operation itself does not change it, so state s' still stores the old input. Instead, the environment of the ASM is assumed to modify the input arbitrarily to the next one stored in $\boldsymbol{I}(k+1)$.

In contrast to the standard definition of guarded rules, runs as defined here may well call an operation with the precondition being false. According to the semantics of one operation (Def. 3) the rest of the run is unpredictable then: either the operation diverges, or execution may continue with an arbitrary state.

---

[5] We write intervals that may contain $\perp$ in bold.

4.4   Submachines

Now we define ASMs $\mathcal{M} = (SIG, Ax, Init, \{\mathtt{Op}_j\}_{j \in J})$ that call operations of a submachine $\mathcal{L} = (SIG^{\mathcal{L}}, Ax^{\mathcal{L}}, Init^{\mathcal{L}}, \{\mathtt{Op}_k^{\mathcal{L}}\}_{k \in K})$. We require that $\mathcal{M}$ uses $\mathcal{L}$ *properly*, indicated by the notation $\mathcal{M}(\mathcal{L})$. The following conditions must be satisfied:

- $\mathcal{M}$ extends $\mathcal{L}$'s signature and axioms: $SIG^{\mathcal{L}} \subseteq SIG$ and $Ax^{\mathcal{L}} \subseteq Ax$,
- initialization of $\mathcal{M}$ includes initialization of $\mathcal{L}$, i.e., $Init \to Init^{\mathcal{L}}$ holds and
- $\mathcal{M}$ respects information hiding: The signature of $\mathcal{L}$ is never accessed directly by operations of $\mathcal{M}$, i.e., $\mathcal{M}$ can only read and update the signature of $\mathcal{L}$ indirectly via calls to operations of $\mathcal{L}$.

The latter means that the local state of $\mathcal{L}$, consisting of the locations in the input, output and controlled signature of $\mathcal{L}$, may not be used in updates or tests of $\mathcal{M}$ operations. We write ls for this local state (and similarly ls' for the local part of s'). The global state gs is the state of $\mathcal{M}$ without ls.

Rules α_j of the ASM may now contain calls to operations of $\mathcal{L}$. We extend the syntax of rules of Sec. 4.1 with

$$\alpha ::= \ldots \mid \mathtt{Op}_k^{\mathcal{L}}(\underline{t}; \underline{loc})$$

The call copies (values of) actual input parameter terms $\underline{t}$ to the input locations $\underline{in}_k^{\mathcal{L}}$ of $\mathtt{Op}_k^{\mathcal{L}}$, executes the rule $\mathtt{Op}_k^{\mathcal{L}}$, and finally copies $\underline{out}_k^{\mathcal{L}}$ back to actual outputs $\underline{loc}$, which must be writable locations of $\mathcal{M}$. Within the run of α_j the call to $\mathtt{Op}_k^{\mathcal{L}}$ is considered as one atomic step. The semantics is therefore defined as

$$I \models \mathtt{Op}_k^{\mathcal{L}}(\underline{t}; \underline{loc})$$
$$\text{iff} \quad I = (s, s'\{\underline{loc} \mapsto \underline{out}_k^{\mathcal{L}}\}) \text{ and } gs' = gs \text{ and } (ls\{\underline{in}_k^{\mathcal{L}} \mapsto \underline{t}\}, ls') \in [\![\mathtt{Op}_k^{\mathcal{L}}]\!]$$
$$\text{or } \#I = \infty, I(0) = s \text{ and } (ls\{\underline{in}_k^{\mathcal{L}} \mapsto \underline{t}\}, \bot) \in [\![\mathtt{Op}_k^{\mathcal{L}}]\!]$$

Note that we avoid adding $\bot$ to the non-atomic semantics here: if the call of $\mathtt{Op}_k^{\mathcal{L}}$ does not terminate, then the resulting interval for $\mathtt{Op}_k^{\mathcal{L}}$ is infinite, implying that the operation calling $\mathtt{Op}_k^{\mathcal{L}}$ does not terminate, too. Intervals $I \in runs^{\mathcal{M}}(j)$ therefore contain the outputs of the submachine (in the formal outputs $\underline{out}_k^{\mathcal{L}}$).

Given an interval $I \models \alpha$, where α calls operations of a submachine $\mathcal{L}$, it is possible to extract the execution $I\big|_{\mathcal{L}}$ of the submachine. It is an interval over the state space of $\mathcal{L}$ including $\bot$ and its length matches the number of submachine calls in I. For every call (s, s') in I, $I\big|_{\mathcal{L}}$ has a state transition (ls, ls'), all other transitions are left out. If the last call to an operation of $\mathcal{L}$ starting in a state s does not terminate, a transition $(ls, \bot)$ is added. Note that merging the intervals of consecutive calls to $\mathcal{L}$ is possible because the local state is not altered in between the calls. We lift this definition to an execution $\boldsymbol{I}$ of a program j on $\mathcal{M}(\mathcal{L})$: $\boldsymbol{I}\big|_{\mathcal{L}}$ is the concatenation of the submachine intervals of each of the operations of $\underline{j}$. It follows:

Lemma 1. *Given an execution $\boldsymbol{I}$ of a program of $\mathcal{M}(\mathcal{L})$, then $\boldsymbol{I}\big|_{\mathcal{L}}$ is an execution of $\mathcal{L}$. It is a run if $\boldsymbol{I}$ is.* $\square$

4.5   Calculus

To define and to verify properties of ASM rules we use the wp-calculus. The calculus defines two program formulas $\langle\!\langle\alpha\rangle\!\rangle\,\varphi$ and $\langle\alpha\rangle\,\varphi$ as follows:

$s \models \langle\!\langle\alpha\rangle\!\rangle\,\varphi$ iff all intervals $I \models \alpha$ with $I(0) = s$ have $\#I < \infty$ and $I.\text{last} \models \varphi$

$s \models \langle\alpha\rangle\,\varphi$ iff there is a finite interval $I \models \alpha$ with $I(0) = s$ and $I.\text{last} \models \varphi$

Formula $\langle\!\langle\alpha\rangle\!\rangle\,\varphi$ expresses the weakest precondition for rule α to be guaranteed to terminate and to establish postcondition φ (which is often written $\mathbf{wp}(\alpha,\varphi)$ in the literature). Formula $\langle\alpha\rangle\,\varphi$ is from Dynamic Logic [12] and expresses that α has a terminating run after which φ holds.[6]

Note that in contrast to standard wp-calculus formula φ is not restricted to a predicate logic formula, but may be another program formula. This will be exploited in the proof obligation for simulations (see Theorem 1). The wp-calculus has simple symbolic execution rules for reasoning about rules α (some of these rules can e.g. be found in [21]; an extension of symbolic execution to temporal logic formulas is described in [26]).

5   Refinement of ASMs and of Submachines

5.1   Contract Refinement for ASMs

ASM refinement between an abstract machine $\mathcal{A} = (SIG^{\mathcal{A}}, Init^{\mathcal{A}}, Ax^{\mathcal{A}}, \{Op_j^{\mathcal{A}}\}_{j \in J})$ and a concrete machine $\mathcal{C} = (SIG^{\mathcal{C}}, Init^{\mathcal{C}}, Ax^{\mathcal{C}}, \{Op_j^{\mathcal{C}}\}_{j \in J})$ with the same operation set J is defined relative to a relation IO ("input/output correspondence") over the input and output part of the two algebras. It specifies what matching inputs and outputs are. Often IO requires identity for input and output locations, but more general cases are possible. $IO(as, cs)$ is given syntactically as a formula over the combined signature $SIG^{\mathcal{A}} \,\dot\cup\, SIG^{\mathcal{C}}$. Correspondence of two executions $\mathbf{I}^{\mathcal{C}}$ and $\mathbf{I}^{\mathcal{A}}$ of $\mathcal{C}$ and $\mathcal{A}$ ("$\mathbf{I}^{\mathcal{C}}$ matches $\mathbf{I}^{\mathcal{A}}$ via IO") is defined as

$\mathbf{I}^{\mathcal{C}} \sqsubseteq_{IO} \mathbf{I}^{\mathcal{A}}$ iff $\#\mathbf{I}^{\mathcal{C}} = \#\mathbf{I}^{\mathcal{A}}$ and for all $k < \#\mathbf{I}^{\mathcal{C}}$:

  either $\mathbf{I}^{\mathcal{A}}(k) = \bot$ (and $\mathbf{I}^{\mathcal{C}}(k)$ is arbitrary)

  or $\mathbf{I}^{\mathcal{C}}(k) \neq \bot, \mathbf{I}^{\mathcal{A}}(k) \neq \bot$ and $IO(\mathbf{I}^{\mathcal{A}}(k), \mathbf{I}^{\mathcal{C}}(k))$ holds.

Refinement relative to IO is then defined as follows.

Definition 5. *Machine $\mathcal{C}$ refines $\mathcal{A}$ relative to IO, written $\mathcal{C} \sqsubseteq_{IO} \mathcal{A}$, if for every program j and every $\mathbf{I}^{\mathcal{C}} \in runs^{\mathcal{C}}(j)$ an abstract run $\mathbf{I}^{\mathcal{A}} \in runs^{\mathcal{A}}(j)$ exists, such that $\mathbf{I}^{\mathcal{C}} \sqsubseteq_{IO} \mathbf{I}^{\mathcal{A}}$ holds.*

The refinement definition allows to refine an abstract run, which calls a diverging operation (i.e., one where the precondition is violated) with a terminating run: the state $\mathbf{I}^{\mathcal{A}}(k)$ after the diverging operation (and all subsequent states) will be $\bot$, and match any concrete state.

Proofs of refinement are done with forward simulation:

---

[6] Dynamic Logic writes $\mathbf{wlp}(\alpha,\varphi)$ as $[\alpha]\,\varphi$; $\langle\alpha\rangle\,\varphi$ is equivalent to $\neg\,[\alpha]\,\neg\,\varphi$.

Theorem 1 (Forward Simulation). $\mathcal{C} \sqsubseteq_{IO} \mathcal{A}$ *follows from a forward simulation* $R \subseteq IO$ *that satisfies*

Initialization: $Init^{\mathcal{C}}(cs) \rightarrow \exists as.\ Init^{\mathcal{A}}(as) \wedge R(as, cs)$

Correctness: $R(as, cs) \wedge pre_j^{\mathcal{A}}(as) \wedge \langle\!\langle \alpha_j^{\mathcal{A}} \rangle\!\rangle\ \mathbf{true}$

$\rightarrow pre_j^{\mathcal{C}}(cs) \wedge \langle\!\langle \alpha_j^{\mathcal{C}} \rangle\!\rangle \langle \alpha_j^{\mathcal{A}} \rangle\ R(as, cs)$ for all $j \in J$

Proof. For finite runs the proof is by induction over its length. Infinite runs require a simple diagonalization argument.   □

5.2   Refinement of Submachines

In this section we show that refinement is modular in the following sense: Given a machine $\mathcal{M}(\mathcal{L})$, and a refinement $\mathcal{K} \sqsubseteq_{LIO} \mathcal{L}$, then replacing calls for $\mathcal{L}$ in operations of $\mathcal{M}$ with calls to $\mathcal{K}$ gives a machine $\mathcal{C} := \mathcal{M}(\mathcal{K})$ that refines $\mathcal{A} := \mathcal{M}(\mathcal{L})$. The result needs one additional restriction compared to general refinement: relation LIO is the identity relation over the input and output parameters of $\mathcal{L}$ and $\mathcal{K}$. Otherwise calls could not just be replaced. The replacement of $\mathcal{L}$ by $\mathcal{K}$ in $\mathcal{M}(\mathcal{L})$ is defined as follows: The signature is $(SIG \setminus SIG^{\mathcal{L}}) \dot{\cup} SIG^{\mathcal{K}}$ and the initialization condition is

$$Init^{\mathcal{M}(\mathcal{K})}(ks, gs) \leftrightarrow Init^{\mathcal{K}} \wedge \exists\ ls.\ R(ls, ks) \wedge Init^{\mathcal{M}(\mathcal{L})}(ls, gs),$$

where ks is the local state of $\mathcal{K}$. The I/O correspondence IO extends LIO to the entire set of input/output parameters of $\mathcal{C}$ and $\mathcal{A}$ by identity.

In order to express the modularity of refinement on the level of intervals, we define the substitution $I' := I\{I|_{\mathcal{K}} \mapsto \boldsymbol{I}^{\mathcal{L}}\}$ of all calls to a submachine $\mathcal{K}$ by corresponding calls to $\mathcal{L}$ taken from $\boldsymbol{I}^{\mathcal{L}}$ assuming $\boldsymbol{I}^{\mathcal{L}}$ satisfies $I|_{\mathcal{K}} \sqsubseteq_{LIO} \boldsymbol{I}^{\mathcal{L}}$. For each transition $\tau = (gs, ks, gs', ks')$ in I after k calls to the submachine, the corresponding transition of the substitution I' is $(gs, \boldsymbol{I}^{\mathcal{L}}(k), gs', ls')$ with

$$ls' = \begin{cases} \boldsymbol{I}^{\mathcal{L}}(k), & \text{if } \tau \text{ is not a call} \\ \boldsymbol{I}^{\mathcal{L}}(k+1), & \text{if } \tau \text{ is a call and } \boldsymbol{I}^{\mathcal{L}}(k+1) \neq \bot \\ \text{arbitrary}, & \text{if } \tau \text{ is a call and } \boldsymbol{I}^{\mathcal{L}}(k+1) = \bot \end{cases}$$

In the last case the $(k+1)$-th call to $\mathcal{L}$ did not terminate and we additionally demand that the interval I' is infinite and may be arbitrary after the call. According to $\sqsubseteq_{LIO}$, $\#\boldsymbol{I}^{\mathcal{L}} = \#I|_{\mathcal{K}}$ and $\boldsymbol{I}^{\mathcal{L}}$ reaches $\bot$ before $I|_{\mathcal{K}}$ (if at all). After lifting this substitution to an execution $\boldsymbol{I}$ of a program $\underline{j}$ it follows:

Lemma 2. *Given an execution* $\boldsymbol{I}$ *of a program* $\underline{j}$ *on* $\mathcal{M}(\mathcal{K})$ *and* $\boldsymbol{I}^{\mathcal{L}}$ *with* $\boldsymbol{I}|_{\mathcal{K}} \sqsubseteq_{LIO}$ $\boldsymbol{I}^{\mathcal{L}}$, *then* $\boldsymbol{I}' := \boldsymbol{I}\{\boldsymbol{I}|_{\mathcal{K}} \mapsto \boldsymbol{I}^{\mathcal{L}}\}$ *is an execution of* $\underline{j}$ *on* $\mathcal{M}(\mathcal{L})$ *with* $\boldsymbol{I} \sqsubseteq_{IO} \boldsymbol{I}'$. $\boldsymbol{I}'$ *is a run if* $\boldsymbol{I}$ *is.*

Proof. The proof is by inspecting the (non-atomic) runs of each operation. For a single operation induction over rule complexity gives the desired result.   □

Given these prerequisites we prove the compositionality theorem:

Theorem 2 (Compositionality). $\mathcal{K} \sqsubseteq_{LIO} \mathcal{L}$ *implies* $\mathcal{M}(\mathcal{K}) \sqsubseteq_{IO} \mathcal{M}(\mathcal{L})$

Proof. Let $I \in \mathit{runs}^{\mathcal{C}}(j)$ be arbitrary. According to Lemma 1 $I\big|_{\mathcal{K}}$ is a run of the submachine $\mathcal{K}$. By assumption there is a run $I^{\mathcal{L}}$ of $\mathcal{L}$ with $I\big|_{\mathcal{K}} \sqsubseteq_{LIO} I^{\mathcal{L}}$. The matching abstract run then is $I\{I\big|_{\mathcal{K}} \mapsto I^{\mathcal{L}}\}$ by Lemma 2. □

6 Related Work

In general our approach is based on (iterated) refinement, following the general idea of ASM refinement [4]. We prefer this over an approach that just annotates code with pre- and postconditions. With such an approach all the abstract layers would become ghost code and extra ghost state, cluttering the implementation with annotations. This would be particularly problematic for the Flash file system, since its refinement hierarchy is *deep*, at least a dozen layers and various submachines are necessary to conceptually isolate the relevant building blocks. Verifying that the whole implementation is a refinement of the POSIX specification in one step is practically infeasible.

The specific instance of refinement defined here is based on data refinement [13], in particular the contract-based approach of Z [29] (see [8] for other approaches, and [23] for a comparison to ASM refinement). It can be viewed as an adaption of this approach to the setting of ASMs. We prefer the operational style of ASM rules over the relational style of Z operations, since ASMs can be executed (and we think that they are easier to understand).

Nevertheless, our atomic semantics (Def. 3) of ASM operations parallels the contract embedding of Z relations into states with bottom, except that we do not add $\{\perp\} \times S_{\perp}$, but just $\{\perp\} \times \{\perp\}$ to preserve the meaning of $\perp$ as "nontermination" (not "unspecified"). [23] argues that for both embeddings the same refinements are correct. As a result the proof obligations for forward simulations are similar to those of Z refinement. As a minor difference our theory allows an operation to have diverging runs, even when its precondition is satisfied, though we have not exploited this in the Flash project (we always prove termination). The generalization results in the extra precondition $(\!|\alpha_j^{\mathcal{A}}|\!)$ **true** in the correctness proof obligation.

It is a folklore theorem of data refinement that proof obligations for individual operations are sufficient to allow substitution of abstract with concrete operations in any reasonable context, i.e., one that does not access the local state of operations. Our formal proof of Theorem 2 shows that ASM rules are one suitable context. In [7] an analogous result is proved on a semantic level using relations of μ-calculus as context.

It should however be noted that the contract approach [29] itself is not sufficient for such a result, since it considers finite sequences of operation calls only, while our context (the main rule of an ASM) may be a loop calling operations of the submachine an infinite number of times. Considering finite runs only has the advantage that forward and backward simulation together give a complete proof technique. Here, as in most refinement definitions that consider infinite runs, backward simulation is not sound: it may result in implementing a terminating run of a rule of $\mathcal{M}(\mathcal{L})$ with a non-terminating run of $\mathcal{M}(\mathcal{K})$, when the abstract machine has infinite nondeterminism (i.e., has a **choose** from some infinite domain). Most of our ASMs have infinite nondeterminism.

The refinement concept discussed here differs from our earlier formalisation [22,24], and from Event-B [3] in that it uses *preconditions*, not *guards* (the earlier B formalism [2] had both preconditions and guards). Whether one needs one or the other concept is application dependent: when rules are "called" by the environment (as here), the precondition approach is appropriate, while applications, where the machine itself chooses a rule (e.g., an interpreter for a programming language, where the next rule is chosen according to the next statement to interpret), then the guard interpretation is appropriate.

The definition given here is on the one hand more liberal than the one in [24], as it allows one to implement a diverging operation on the abstract level with any run on the concrete level (since for $\boldsymbol{I}^A(k) = \bot$ any concrete state is allowed). On the other hand it is more strict, as it forbids general $m : n$ diagrams where m abstract operations are implemented with n concrete ones. The case $m > 1$ is disallowed here, since any sequence of submachine calls must be verified. The case $m = 0$ can be simulated by adding an abstract skip operation that does nothing. Diagrams with $n \neq 1$ are still implicitly possible, by using a concrete rule that takes n atomic steps (in the fine-grained semantics) to complete.

With respect to ASMs, our syntax only uses a fragment of the syntax available in [5]. In particular we use parallel updates only in the atomic updates, while control state ASMs allow arbitrary ASM rules. It would be possible to generalize the atomic steps to general ASM rules, however this would have two drawbacks. Code generation would become more difficult, and simple symbolic execution rules would be precluded since parallel rules may have clashes. These require a complex axiomatization of update and consistency predicates, even when nondeterministic choice (that we often use for specification purposes) is omitted (see [27] and Chapter 8 of [5]).

For the atomic semantics given in Def. 3 it is not difficult to show that it agrees with standard rule semantics of ASM rules, when $\alpha; \beta$ is interpreted as α **seq** β in the following sense: $(s, s') \in [\![\mathtt{Op}]\!]$ corresponds to a successful computation of a consistent set of updates of a Turbo ASM rule in [5], Chapter 4. $(s, \bot) \in [\![\mathtt{Op}]\!]$ corresponds to either a diverging computation of updates, or to the computation of an inconsistent set. Our largest fixpoint for the non-atomic semantics of **while** reduces to the least fixpoint definition 4.1.2 of **iterate** that is used to define the semantics of **while** with deterministic body in [5]. In general, using a largest fixpoint is unavoidable to characterize guaranteed termination for rules with infinite nondeterminism.

The non-atomic semantics we give in Def. 2 is based on Interval Temporal Logic (ITL [16,17]). We prefer this alternative over a structural operational semantics (SOS, [19]), since SOS must model an explicit stack of local variables which is unnecessary for a direct interval semantics.

The non-atomic semantics in this paper is a simplified version of the one we give in [26], which additionally handles interleaved concurrency and temporal operators.

Our definition of submachines is different from the one in [5]. A submachine there is a subrule that may be called within a rule, with the purpose to support mutual recursion in Turbo ASMs. These are similar to the calls of submachine operations, however, submachines as defined here are full ASMs (with initialization, signature etc.). Additionally, information hiding constraints have to be satisfied for modular refinement. To be able to check these constraints syntactically we use input and output parameters passed by value, whereas subrules in [5] use call by name. This extension does not give additional expressivity: a declaration $Op(\underline{x}, \underline{y})\{\alpha\}$ could be replaced by $Op(; \underline{x}, \underline{y})\{\alpha\}$ using reference parameters only. Calls $Op(\underline{t}; \underline{z})$ for a submachine operation with declaration would have to be replaced with let $\underline{in} = \underline{t}$ in $Op(; \underline{in}, \underline{z})$.

The ASM formalism is also strong enough such that preconditions are definable. A rule $RULE$ working on dynamic functions $f_1, \ldots f_n$ with precondition pre is equivalent to the extended rule **if** pre **then** $RULE$ **else** $CHAOS$ where

$$CHAOS = \textbf{choose } diverge? \textbf{ in if } diverge? \textbf{ then abort}$$
$$\textbf{else } RANDOM(f_1)$$
$$\cdots$$
$$RANDOM(f_n)$$

and
$$RANDOM(f_i) = \textbf{forall } args_i \textbf{ choose } val \textbf{ in } f_i(args_i) := val$$

Rule $CHAOS$ either diverges (when $diverge?$ is true) or chooses a random next state by overwriting each f_i with a new function in $RANDOM(f_i)$.

Event-B has two decomposition concepts for machines that roughly correspond to interleaved [1] and synchronous parallel execution of rules [6]. It is not immediately clear how our submachine concept could be encoded by such a decomposition, since events in Event-B have no internal control structure (although a construction with program counters and explicit call/return events for subrules may be possible).

7   Conclusion

We have defined a refinement theory for ASMs with submachines, which respect information hiding. The theory has been key to enable modular, incremental development of the Flash case study.

So far we have used forward simulations for our proof only. As noted in related work, backward simulation is not a sound proof technique in the presence of infinite runs. A completeness proof will therefore be possible only along the lines of [24], by replacing **choose** with choice functions, but such a proof is still future work.

Although we have no need for guards in the Flash case study (the toplevel POSIX specification has total operations, all intermediate layers have preconditions), it would be interesting to analyze, whether our refinement definition for submachines is compatible with the main machine having guards.

Finally, it should be noted that our definition of refinement does not solve all problems in the Flash case study. One important extension is necessary to deal with power failures and recovery. A paper on this issue based on the same semantic setting (with the idea that runs of a rule may be aborted in any intermediate state) is currently in preparation. Another important issue, which we will have to consider, is that the actual implementation uses concurrency to do work in the background: As an example, actually erasing blocks is done in a concurrent thread that calls back to the main thread, when it has finished.

References

1. Abrial, J.-R., Hallerstede, S.: Refinement, Decomposition, and Instantiation of Discrete Models: Application to Event-B. Fundamenta Informaticae 77 (2007)
2. Abrial, J.-R.: The B Book - Assigning Programs to Meanings. Cambridge University Press (1996)
3. Abrial, J.-R.: Modeling in Event-B. Cambridge University Press (2010)
4. Börger, E.: The ASM Refinement Method. Formal Aspects of Computing 15(1-2), 237–257 (2003)
5. Börger, E., Stärk, R.F.: Abstract State Machines — A Method for High-Level System Design and Analysis. Springer (2003)
6. Butler, M.: Decomposition Structures for Event-B. In: Leuschel, M., Wehrheim, H. (eds.) IFM 2009. LNCS, vol. 5423, pp. 20–38. Springer, Heidelberg (2009)
7. de Roever, W., Engelhardt, K.: Data Refinement: Model-Oriented Proof Methods and their Comparison. Cambridge Tracts in Theoretical Computer Science, vol. 47. Cambridge University Press (1998)
8. Derrick, J., Boiten, E.: Refinement in Z and in Object-Z: Foundations and Advanced Applications. FACIT, 2nd revised edn. Springer (2014)
9. Ernst, G., Pfähler, J., Schellhorn, G.: Web presentation of the Flash Filesystem (2014), https://swt.informatik.uni-augsburg.de/swt/projects/flash.html
10. Ernst, G., Schellhorn, G., Haneberg, D., Pfähler, J., Reif, W.: A Formal Model of a Virtual Filesystem Switch. In: Proc. of Software and Systems Modeling (SSV), pp. 33–45 (2012)
11. Ernst, G., Schellhorn, G., Haneberg, D., Pfähler, J., Reif, W.: Verification of a Virtual Filesystem Switch. In: Cohen, E., Rybalchenko, A. (eds.) VSTTE 2013. LNCS, vol. 8164, pp. 242–261. Springer, Heidelberg (2014)
12. Harel, D., Kozen, D., Tiuryn, J.: Dynamic Logic. MIT Press (2000)
13. He, J., Hoare, C.A.R., Sanders, J.W.: Data refinement refined. In: Robinet, B., Wilhelm, R. (eds.) ESOP 1986. LNCS, vol. 213, pp. 187–196. Springer, Heidelberg (1986)
14. Hunter, A.: A brief introduction to the design of UBIFS (2008), http://www.linux-mtd.infradead.org/doc/ubifs_whitepaper.pdf
15. Joshi, R., Holzmann, G.J.: A mini challenge: build a verifiable filesystem. Formal Aspects of Computing 19(2) (June 2007)
16. Moszkowski, B.: Executing Temporal Logic Programs. Cambr. Univ. Press (1986)
17. Moszkowski, B.C.: An automata-theoretic completeness proof for Interval Temporal Logic. In: Welzl, E., Montanari, U., Rolim, J.D.P. (eds.) ICALP 2000. LNCS, vol. 1853, pp. 223–234. Springer, Heidelberg (2000)

18. Pfähler, J., Ernst, G., Schellhorn, G., Haneberg, D., Reif, W.: Formal Specification of an Erase Block Management Layer for Flash Memory. In: Bertacco, V., Legay, A. (eds.) HVC 2013. LNCS, vol. 8244, pp. 214–229. Springer, Heidelberg (2013)
19. Plotkin, G.D.: A structural approach to operational semantics. Technical Report DAIMI FN-19, Aarhus University (1981)
20. Reeves, G., Neilson, T.: The Mars Rover Spirit FLASH anomaly. In: Aerospace Conference, pp. 4186–4199. IEEE Computer Society (2005)
21. Reif, W., Schellhorn, G., Stenzel, K., Balser, M.: Structured specifications and interactive proofs with KIV. In: Bibel, W., Schmitt, P. (eds.) Automated Deduction—A Basis for Applications, vol. II, pp. 13–39. Kluwer, Dordrecht (1998)
22. Schellhorn, G.: Verification of ASM Refinements Using Generalized Forward Simulation. Journal of Universal Computer Science (J.UCS) 7(11), 952–979 (2001), http://www.jucs.org
23. Schellhorn, G.: ASM Refinement and Generalizations of Forward Simulation in Data Refinement: A Comparison. Journal of Theoretical Computer Science 336(2-3), 403–435 (2005)
24. Schellhorn, G.: Completeness of Fair ASM Refinement. Science of Computer Programming, 76(9) (2009)
25. Schellhorn, G., Ernst, G., Pfähler, J., Haneberg, D., Reif, W.: Development of a Verified Flash File System. In: Ait Ameur, Y., Schewe, K.-D. (eds.) ABZ 2014. LNCS, vol. 8477, pp. 9–24. Springer, Heidelberg (2014)
26. Schellhorn, G., Tofan, B., Ernst, G., Pfähler, J., Reif, W.: RGITL: A Temporal Logic Framework for Compositional Reasoning about Interleaved Programs. In: AMAI (2014), appeared online first, draft available at https://swt. informatik.uni-augsburg.de/swt/projects/RGITL.html
27. Stärk, R.F., Nanchen, S.: A Complete Logic for Abstract State Machines. Journal of Universal Computer Science (J.UCS) 7(11), 981–1006 (2001)
28. The Open Group. The Open Group Base Specifications Issue 7, IEEE Std 1003.1, 2008 Edition (2008), http://www.unix.org/version3/online.html (login required)
29. Woodcock, J.C.P., Davies, J.: Using Z: Specification, Proof and Refinement. Prentice Hall International Series in Computer Science (1996)

Towards ASM-Based Formal Specification of Self-Adaptive Systems

Elvinia Riccobene[1] and Patrizia Scandurra[2]

[1] Computer Science Department, Università degli Studi di Milano, Italy
[2] Engineering Department, Università degli Studi di Bergamo, Italy

Abstract. This paper shows how to use multi-agent Abstract State Machines to specify self-adaptive behavior in a decentralized adaptation control system. A traffic monitoring system is taken as case study.

1 Introduction

Modern software systems typically operate in dynamic environments and are required to deal with changing operational conditions: components can appear and disappear, may become temporarily or permanently unavailable, may change their behavior, etc. Self-adaptation (SA) has been widely recognized [4,6] as an effective approach to deal with the increasing complexity, uncertainty and dynamics of these advanced systems. A well recognized engineering approach to realize self-adaptation is by means of a feedback control loop conceived as a sequence of four computations: Monitor-Analyze-Plan-Execute [6].

One major challenge in self-adaptive systems is to assure the required quality properties (e.g., flexibility, robustness, etc.). Formal methods are an attractive option for solving this problem as they provide a means to precisely model and reason about the behaviors of self-adaptive systems. The survey in [7] shows that the attention for self-adaptive software systems is gradually increasing, but the number of studies that employ formal methods remains low, and is mainly related to runtime verification. However, formally founded *design* models that cover structural and behavioral aspects of self-adaptation, and of approaches to validate behavioral properties are of extreme importance in order to provide guarantees about qualities at the early stages of the system design.

By exploiting the theoretical framework of the Multi-Agent Abstract State Machines (ASM) [2], we here show how to model the behavior of self-adaptive distributed systems with decentralized adaptation control, where the MAPE control loop is naturally formalized in terms of agents' actions (transition rules). A traffic monitoring application, inspired from [5], is taken as case study.

This is a first work of our ongoing research activity on answering the request of precise models that help reasoning about adaptation at design time. In the conclusion we report some lessons learned from our experience that reveal the high potentiality of the ASMs in the context of self-adaptive systems.

Y. Ait Ameur and K.-D. Schewe (Eds.): ABZ 2014, LNCS 8477, pp. 204–209, 2014.

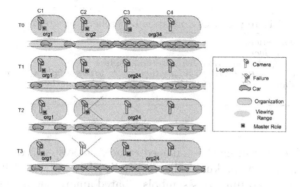

Fig. 1. Adaptation scenarios (adapted from [5])

2   The Traffic Monitoring Case Study

We present a traffic monitoring application inspired by the case study in [5].

A number of intelligent cameras are distributed along a road, each with a limited viewing range (see Fig. 1). Cameras are equipped with a data processing unit capable of processing the monitored data, and a communication unit to communicate with other cameras. Traffic jams can span the viewing range of multiple cameras and can dynamically grow and dissolve. Each camera monitors the traffic state within its viewing range. Because there is no central point of control, cameras have to aggregate the monitored data to determine the position of the traffic jam on the basis of the head and tail of it. Cameras enter or leave the collaboration whenever the traffic jam enters or leaves their viewing range.

There are two main adaptation concerns. The first is system *flexibility* for the dynamic adaptation of an organization. See, e.g., the scenario in Fig. 1 from configuration T0 to T1, where camera 2 joins the organization of cameras 3 and 4 after it monitors a traffic jam. The second is related to *robustness* due to camera failures, i.e., when a failing camera becomes unresponsive. This scenario is shown in Fig. 1 from T2 to T3, where camera 2 fails.

3   Multi-Agent ASM Specification

Because of the distributed nature of adaptive systems, we use the notion of *multi-agent ASMs* where multiple agents interact in parallel in a synchronous/asynchronous way. Each agent executes its own (possibly the same but differently instantiated) ASM-based program that specifies the agent's behavior. Some agents form the *managing ASM* part encapsulating the logic of self-adaptation. Some other agents form the *managed ASM* part encapsulating the functional logic.

For the traffic monitoring application, we introduce four ASM agents: the agents *OrganizationController* and *SelfHealingController* representing the managing components, an agent *Camera* and an agent *TrafficMonitor* (the sensor that in case of "congestion" or "no longer congestion" notifies the organization

Listing 1.1. Organization controller's program

```
macro rule r_organizationControl =
  seq //MAPE control loops
    r_selfFailureAdapt[] //Adaptation due to internal failure
    r_failureAdapt[] //ROBUSTNESS: Adaptation due to external failure (silent nodes)
    r_congestionAdapt[] //FLEXIBILITY: Adaptation due to congestion
  endseq
```

controller) both representing the managed camera subsystem. An ASM module, called *knowledge*, is used as knowledge for the MAPE loops to define the ASM signature (domains and functions symbols) shared among the managing agents. We specify the self-adaptive behavior of the managing components of a camera using two main MAPE loops: the first loop deals with flexibility concerns to restructure organizations in case of congestion, the second loop deals with robustness concerns to restructure organizations in case of failing cameras. Both the two loops start with the program associated to the agent *OrganizationController*, however due to the decentralized nature of MAPE computations, part of the monitoring functionality of the second loop is on the *SelfHealingController* behavior. A further MAPE loop to deal with internal failures of the camera is also executed by the two managing agents.

For the lack of space, we describe only part of the behavior of the *OrganizationController*. The complete specification is available online[1].

Organization middleware for Flexibility. An organization controller runs on each camera and is responsible for managing organizations depending on the data it gets from the traffic monitor and from the self-healing controller of the camera. A master/slave control model is adopted to structure organizations in case of congestion. Each camera has a unique ID (a static integer-valued function *id*). To keep the master election policy simple, we assume the camera ID is monotonically increasing on the traffic direction and the camera with the lowest ID becomes master. Traditional election algorithms (like the Bully algorithm and the Ring algorithm) or new ones are out of the scope of this paper.

Each camera has four basic states (the function *state*). In normal operation, the camera can be master with no slaves (i.e., master of a single organization), master of an organization with slaves, or it can be slave. Additionally, the camera can be in the failed state, representing the status of the camera after a silent node failure. Initially, all cameras are master. A camera state is changed by the organization controller as part of the adaptation logic. The organization controller has the same four basic states of the camera it manages. The organization controller's program (see the rule r_organizationControl in Listing 1.1) executes sequentially the three MAPE control loops.

We here focus on the third MAPE loop for adapting organizations in case of traffic congestion notified by the traffic monitor. Such a behavior is represented by the rule r_congestionAdapt defined in Listing 1.2.

---

[1] See the examples directory in the ASMETA repository http://asmeta.sf.net/

Listing 1.2. ASM rule `r_congestionAdapt`

```
//@M_c context—aware monitoring
macro rule r_congestionAdapt =
par
 if state(self) = MASTER
 then
     if ( cong(camera(self)) and not congested(self) )//Congestion detected!
     then //@P Planning
          par
          congested(self) := true
          cong(camera(self)) := false
          if isDef(next(camera(self))) then s_offer(next(camera(self))) := true endif
          endpar
     else if congested(self) then r_analyzeCongestion[] endif endif endif
 if state(self) = SLAVE
 then
     if no_cong(camera(self)) //No longer congested!
     then //@P Planning
          par
          no_cong(camera(self)):= false
          congested(self) := false
          slaveGone(getMaster(camera(self)),camera(self)) := true
          r_turnMaster[]
          endpar
     else
          r_receiveOrgSignals[] endif endif
 if state(self) = MASTERWITHSLAVES
 then if no_cong(camera(self)) //No longer congested!
     then //@P Planning
          par r_removeSlavesTurningMaster[]
          no_cong(camera(self)):= false
          congested(self):= false endpar
     else r_analyzeOrganization[] endif endif
endpar
```

In the role of master of a single member organization, when a congestion is detected (the signal *cong*) the organization controller sends a request (the predicate *s_offer*) to the next alive camera (if any) in the direction of the traffic flow to join the organization as slave. Depending on the traffic condition of the next camera and its role, the organizations may be restructured according to the rule *r_analyzeCongestion* reported in Listing 1.3. If traffic is not jammed (the controlled predicate *congested* is false) for the next camera, organizations are not changed, otherwise organizations are joined. The next camera becomes slave of the requester camera by executing the rule `r_turnSlave` that changes the camera state, sets the requester camera as new master and informs back it by setting the shared function *newSlave* and (indirectly) the derived predicate *m_offer* to true. When the *m_offer* signal is set the requester camera becomes master of the joined organization executing the rule `r_turnMasterWithSlaves` (see Listing 1.3) to concretely add the new slave to its list and change state.

In the role of slave, if the traffic in the viewing range of the camera is no longer jammed (the signal *no_cong*), the organization controller leaves the organization it belongs to (by setting the function *slaveGone*) and becomes master of a single member organization (by executing the rule `r_turnMaster` in Listing 1.2). Otherwise (still congested), the organization controller waits (by

Listing 1.3. ASM rules for analysis computations

```
//@A Analyzing
macro rule r_analyzeCongestion =
 if m_offer(camera(self)) then r_turnMasterWithSlaves[]
 else if s_offer(camera(self)) then r_turnSlave[prev(camera(self))] endif endif
//@A Analyzing
macro rule r_receiveOrgSignals =
   par
     if change_master(camera(self)) then //Master changed!
           //@P Planning
           par
              r_setMaster[prev(getMaster(camera(self)))]
              newSlave(prev(getMaster(camera(self))),camera(self)) := true
              change_master(camera(self)) := false
           endpar endif
     if masterGone(camera(self)) then r_turnMaster[] endif
     if m_offer(camera(self)) then r_notifyPendingSlavesMasterChanged[] endif
   endpar
//@A Analyzing
macro rule r_analyzeOrganization =
if m_offer(camera(self)) then r_addNewSlave[]
else if isEmpty(slaves(camera(self))) //Simply turn master
     then r_turnMaster[]
     else if (s_offer(camera(self)) and congested(self))
          then r_turnSlave[prev(camera(self))] endif endif endif
//@P Planning
macro rule r_addNewSlave =
   forall $s in Camera with newSlave(camera(self),$s) do
     par
        r_addSlave[$s]
        newSlave(camera(self),$s):= false
        s_offer($s):=false
     endpar
```

the rule r_receiveOrgSignals) for a trigger from its master. The rule r_-receiveOrgSignals is reported in Listing 1.3. If (in the slave role) the controller receives a signal *change_master* as effect of a restructuring of the organization, it is responsible for planning adaptations to change its master to the new master. If it receives that the master is gone (by the shared predicate *masterGone*), it restarts the camera as master of a single member organization (by invoking the rule r_turnMaster[] already shown in Listing 1.2). Finally, if it receives an *m_offer* signal, it means there are slaves not effectively engaged when in the role of master it asked them to join the organization as slave. In this last case it is responsible for notifying them that the master changed by executing the rule r_notifyPendingSlavesMasterChanged in Listing 1.3.

Finally, in the role of master with slaves, when the traffic is no longer jammed (the signal *no_cong*), the organization controller notifies all its depending slaves that the master is gone (setting the predicate *masterGone* to true) and leaves the organization becoming master of a single member organization (see rule r_-removeSlavesTurningMaster in Listing 1.2). Otherwise (still congested), the organization controller analyzes the organization by the rule r_analyzeOrganization (see Listing 1.3) to add and remove slaves dynamically. When no slaves remain, the master with slaves becomes master of a single member organization again. During analysis of its organization, it has also to wait for a trigger *s_offer*, and if notified it has to plan to become slave of the requester camera by executing `r_turnSlave` in Listing 1.2.

For the lack of space, macro rules (such as `r_turnSlave`, `r_turnMaster`,etc.) and those annotated with @E for atomic adaptation actions in a master/slave organization (such as `r_clearSlaves`, `r_addSlave`, etc.) are not reported.

4   Conclusion and Future Directions

Besides modeling, we were also able to validate the Traffic Monitoring case study by exploiting model simulation and scenario construction. We focused on two qualities: flexibility (i.e., the ability of the system to adapt dynamically with changing conditions in the environment), and robustness (i.e., the ability of the system to cope autonomously with errors during execution). By means of the ASM tools[1,3], we simulated different scenarios with increasing number of cameras. In particular, we reproduced the adaptations scenarios shown in Fig. 3 from T0 to T2 for flexibility, and from T2 to T3 for robustness.

From modeling and validation, we learned some lessons briefly reported. We were able to achieve a clear separation of concerns: (i) separation between adaptation logic and function logic, (ii) separation between behavior of managing and managed components, (ii) separation between the specification of the MAPE functions. This helps the designer to focus on one adaptation concern at a time, and, for each concern, separate the adapting parts from the adapted ones.

In the future, we plan to define a formal framework providing high level constructs for expressing context-awareness, self-awareness, adaptation actions, distributed communication patterns. We plan to investigate on the verification of self adaptive systems by using the ASMETA tool set.

References

1. Arcaini, P., Gargantini, A., Riccobene, E., Scandurra, P.: A model-driven process for engineering a toolset for a formal method. SPE J. 41(2), 155–166 (2011)
2. Börger, E., Stärk, R.: Abstract State Machines: A Method for High-Level System Design and Analysis. Springer (2003)
3. Carioni, A., Gargantini, A., Riccobene, E., Scandurra, P.: Scenario-Based Validation Language for ASMs. In: Börger, E., Butler, M., Bowen, J.P., Boca, P. (eds.) ABZ 2008. LNCS, vol. 5238, pp. 71–84. Springer, Heidelberg (2008)
4. de Lemos, R., et al.: Software engineering for self-adaptive systems: A second research roadmap. In: de Lemos, R., Giese, H., Müller, H.A., Shaw, M. (eds.) Software Engineering for Self-Adaptive Systems. LNCS, vol. 7475, pp. 1–32. Springer, Heidelberg (2013)
5. Iftikhar, M.U., Weyns, D.: A case study on formal verification of self-adaptive behaviors in a decentralized system. In: Kokash, N., Ravara, A. (eds.) FOCLASA. EPTCS, vol. 91, pp. 45–62 (2012)
6. Kephart, J.O., Chess, D.M.: The vision of autonomic computing. IEEE Computer 36(1), 41–50 (2003)
7. Weyns, D., Iftikhar, M.U., de la Iglesia, D.G., Ahmad, T.: A survey of formal methods in self-adaptive systems. In: C3S2E, pp. 67–79. ACM (2012)

Distributed ASM - Pitfalls and Solutions

Andreas Prinz[1] and Edel Sherratt[2]

[1] Department of ICT, University of Agder
andreas.prinz@uia.no
[2] Department of Computer Science, Aberystwyth University
eds@aber.ac.uk

Abstract. While sequential Abstract State Machines (ASM) capture the essence of sequential computation, it is not clear that this is true of distributed ASM. This paper looks at two kinds of distributed process, one based on a global state and one based on variable access. Their commonalities are extracted and conclusions for the general understanding of distributed computation are drawn, providing integration between global state and variable access.

1 Introduction

For many years, models and languages of astonishing variety and depth have been developed to describe distributed computation, and still its essence is far from understood.

Distributed Abstract State Machines (ASM) [8] are a key part of a drive to establish a distributed ASM thesis analogous to the successful sequential ASM thesis [7]. This work has not yet led to a final result, although Glausch and Reisig in [5] have established that distributed algorithms that fulfil certain criteria are captured by DASM.

This paper looks at characteristics of distributed computations and scenarios that are not fully captured by distributed ASM. Based on the work of Lamport [9], a new ASM model is proposed that captures more of these scenarios.

The paper is structured as follows. Section 2 introduces distributed ASM. Section 3 argues that they do not fully capture distributed algorithms. The Lamport model is presented in section 4. Section 5 extracts essential properties of distributed computations, and section 6 proposes how the global view can be combined with the local variables view. Section 7 concludes the paper.

2 Asynchronous Multi-agent (Distributed) ASMs

A distributed ASM (DASM) is a family of pairs $(a; Module(a))$ with pairwise different agents, elements of a possibly dynamic finite set *Agent*, each equipped with a sequential ASM $Module(a)$. Each sequential ASM provides a set of states (first order structures over the same vocabulary), a set of initial states and a state transition function which can only take into account a bounded number of elements.

Y. Ait Ameur and K.-D. Schewe (Eds.): ABZ 2014, LNCS 8477, pp. 210–215, 2014.
© Springer-Verlag Berlin Heidelberg 2014

Definition 1 ((global) DASM run). *A partially ordered run of a DASM is a partially ordered set $(M; \prec)$ of moves m (rule applications) of its agents agent(m) with a state function s satisfying the following conditions [8]:*

- *finite history: each move has only finitely many predecessors, i.e. $\{m' \in M | m' \prec m\}$ is finite for each $m \in M$.*
- *sequentiality of agents: for each agent a the set of its moves is linearly ordered, i.e. $agent(m) = agent(m')$ implies $m \prec m'$ or $m' \prec m$.*
- *coherence: each finite initial segment (downward closed subset) I of $(M; \prec)$ has an associated state $s(I)$ – think of it as the result of all moves in I – which for every maximal element $m_{max} \in I$ is the result of applying the state transition function of $agent(m_{max})$ in state $s(I - \{m_{max}\})$.*

This definition implies a global state accessible to all agents, where each agent has its own local view given by the variables read by the agent. The definition does not say how moves are to be scheduled in a run; moves can be performed in parallel, or by interleaving the moves of different agents. However, every run leads to the same end state.

Proposition 1. *All linearizations of the same finite initial segment of a DASM run have the same final state [8].*

This means that each ASM run is essentially sequential and we conclude.

Proposition 2. *If DASM runs are the most general way to look at distributed computation, then distributed computation is essentially sequential.*

3   Distributed ASM Do Not Capture Distributed Algorithms

The distributed ASM thesis is still open, because there are many distributed scenarios that are not properly captured by distributed ASM.

1. *Context switching* between threads can occur between a read and a write. In ASM, an update is performed instantaneously, which means that the state is read, the answer is computed and the result is written as a single atomic action.
2. In larger distributed systems, *inconsistent system states* are possible. With ASM, the system state is always consistent.
3. In parallel computation, two processors can *simultaneously* write the same memory location. Similarly, a write could be at the same time as a read. The ASM consistency condition [4] excludes such conflict, and a more elaborate treatement by [1] treats memory locations as proclets (active processors) in their own right, that do some computing to resolve write conflicts.
4. The meaning of distributed computations varies a lot according to the level of *atomicity* used. DASM have a fixed level of atomicity.

This brings us to the following conclusion.

Proposition 3 (Failed Distributed ASM Thesis). *DASM as defined in section 2 do not capture distributed computation; at least they do not capture the scenarios given above.*

If certain restrictions are accepted, then DASM do capture some kinds of distributed computation[5]. However, those restrictions conflict with our scenarios. A consistent *global state* cannot be assumed for a highly distributed computation. Context switching between reads and writes conflicts with the assumption of *instantaneous actions*. For *autonomicity*, the update sets of [5] introduce constraints by claiming that the input (read) and output (write) locations should be the same. This leads to the impossibility of parallel read, which is not in line with our understanding.[1] Finally it has to be noted, that the concept of DASM run as introduced in [5] is not the same as the traditional DASM run. In particular, the consistency condition is not introduced, and no proof is given that both concepts coincide. As our examples in section 4 show, these two ideas of DASM run do not coincide.

4   Sequentially Consistent Runs

A different way of looking at distributed computation was introduced by Lamport [9]. Here, a distributed execution is a set of sequential executions (one per agent), each being a sequence of reads and writes of locations. Not all Lamport executions are valid. Lamport defines sequential consistency as follows[9].

Definition 2 (sequential consistency). *Consider a computation (execution) composed of several sequential processors accessing a common memory. The computation is sequentially consistent iff the result of any execution is the same as if the operations of all the processors were executed in some sequential order, and the operations of each individual processor appear in this sequence in the order specified by its program.*

A sequentially consistent execution has at least one *witness*, which is a legal interleaving of the reads and writes. Different witnesses may yield different results.

The level of granularity is lower for Lamport reads and writes as opposed to moves for DASM. Reading and writing is implicit in the DASM model. As a contrast, Lamport does not look into the global system state.

4.1   Examples to Compare DASM and Lamport Runs

It might not be obvious how Lamport and DASM runs differ, so we give some small examples with agents A and B, and variables x and y.

---

[1] Please note that [5] is not altogether consistent at this place. In requirement D4 (autonomicity), the parameter values of the locations could be locations themselves. However, this is not used in the examples shown later. But if there are no locations used as parameters, D4 is trivially true. On the other hand, using locations as parameters, the notion of "same location" suddenly becomes quite advanced.

Table 1. ASM and Lamport witnesses - all examples

	No	witness ASM	witness Lamport	result
Ex1	W1	$m_A; m_B$	$\text{write}(x,1); \text{write}(x,2)$	x=2
	W2	$m_B; m_A$	$\text{write}(x,2); \text{write}(x,1)$	x=1
Ex2	W1	$m_A; m_B$	$\text{read}(x,0); \text{write}(x,1); \text{read}(x,1); \text{write}(x,2)$	x=2
	W2	$m_B; m_A$	$\text{read}(x,0); \text{write}(x,1); \text{read}(x,1); \text{write}(x,2)$	x=2
	W3	–	$\text{read}(x,0); \text{read}(x,1); \text{write}(x,1); \text{write}(x,1)$	x=1
Ex3	W1	$m_A; m_B$	$\text{read}(y,0); \text{write}(x,1); \text{read}(x,1); \text{write}(y,2)$	x=1, y=2
	W2	$m_B; m_A$	$\text{read}(x,0); \text{write}(y,2); \text{read}(y,2); \text{write}(x,1)$	x=1, y=2
	W3	–	$\text{read}(x,0); \text{read}(y,0); \text{write}(x,1); \text{write}(y,2)$	x=1, y=2
	W4	–	$\text{read}(x,0); \text{read}(y,0); \text{write}(y,2); \text{write}(x,1)$	x=1, y=2

Example Ex1 : $A : x := 1$; $B : x := 2$; initially $x = 0$.
  ASM: two possible runs: {W1}, {W2}
  Lamport: one possible run: {W1, W2}
Example Ex2 : $A : x := x + 1$; $B : x := x + 1$; initially $x = 0$.
  ASM: two possible runs: {W1}, {W2}
  Lamport: two possible runs: {W1, W2}, {W3}[2]
Example Ex3 : $A : x := y * 0 + 1$; $B : y := x * 0 + 2$; initially $x = y = 0$
  ASM: one possible run: {W1, W2}
  Lamport: three possible runs: {W1}, {W2}, {W3, W4}[3]

4.2   Distributed ASM Runs Are Sequentially Consistent

Since each move of a DASM run writes the same values, regardless of the linearization, it is possible to translate DASM runs into sequentially consistent Lamport runs. Please note that it is not true that each move reads the same values independent of the linearization, see the last example in the previous section.

Thus, one DASM run can produce several Lamport runs and each Lamport run of a DASM run is sequentially consistent.

5   General Properties of Distributed Computation

Distributed computation generally comprises sequential agents that work together. They may use synchronization of memory locations to coordinate their work. However, it is essential that their work has to respect causality (proper synchronization of writes with reads). When conflicts arise, then there is an underlying mechanism to handle inconsistencies between reads and writes of different agents.

---

[2] Observe that W3 is not possible in ASM, although it is not conflicting.
[3] The last two runs have one more witness each where the reads are swapped. As opposed to DASM, [5] would not consider W1 and W2 independent but view them as two different runs.

Property 1 (Sequentiality). *The actions of each agent are sequential.* [4]

Property 2 (Synchronization). *There are (global) memory locations where access is sequential.*

Property 3 (Causality). *It is impossible to read values before they have been written.* [5]

Property 4 (Consistency). *When two agents try to write to the same position, then one of them wins as opposed to having arbitrary outcome. In the same way, also possible conflicts between read and write are solved.*

With these requirements in mind, we will describe a local state model that captures our idea of distributed computation.

6   Localized State

We introduce a localized DASM model where a memory location can be updated by one agent, and its value can take some time before it is available to other agents. This is addressed with reference to persistent queries in [2,3], where a query is accompanied by the location where its result is to be deposited.

Definition 3 ((localized) DASM run). *A localized partially ordered run of a DASM is a partially ordered set $(M; \prec)$ of moves m (rule applications) of its agents agent(m) with a state function s satisfying the following conditions:*

1. *finite history: see definition 2*
2. *sequentiality of agents: see definition 2*
3. *The (local) states of an agent before and after a move m are related using the state transition function of agent(m).*
4. *The (local) state of an agent before a move is a combination of all the (local) states after the directly preceding moves. If there are no preceding moves, an initial state is used.*
5. *A combination of two (local) states is done with the following rules.*
   - *When the value of a location is the same in both states, then this value is taken.*
   - *When the value of a location is different in the two states, then the one resulting from the later move with respect to the partial order is taken.*
   - *When the value of a location is different in the two states and both values are coming from moves that are not ordered by the partial order, then an arbitrary value of the two is chosen.*

The new definition brings the following advantages, in particular related to the problems given earlier.

---

[4] Although this property looks innocent enough, it is rejected by the Java memory model for distributed computation [6].

[5] This property is also called no-out-of-thin-air in the context of Java.

1. *Context switching* is implicit in the new model, since the write of a move is taken into account first when the new read is done.
2. Since agents work independent of each other and each agent has its own local state, a global *inconsistent system states* is not only possible but normal.
3. Concurrent *write at the same time* onto the same memory location is possible and would result in one of the written values.
4. Reads and writes are the level of *atomicity*.
5. The new definition does not guarantee sequential consistency. Please note that all examples from section 4 will be captured in one run using the localized model. In all three cases, the moves of agents A and B can be unordered.
6. The new definition provides a higher level of abstraction than Lamport and at the same time brings less restrictions to the runs. It aligns better with the moves of ASM.

7   Summary and Conclusions

In this paper, we have shown that distributed computation is not always easy to understand and that DASM do not capture the essence of distributed computation. We have compared DASM with the Lamport model and have extracted a new model that is not sequential in the bottom. This local state model captures at least the problems indicated with the DASM model.

References

1. Blass, A., Gurevich, Y.: Abstract state machines capture parallel algorithms. ACM Transactions on Computational Logic (TOCL) 4(4), 578–651 (2003)
2. Blass, A., Gurevich, Y.: Persistent queries (2008)
3. Blass, A., Gurevich, Y.: Persistent queries in the behavioral theory of algorithms. ACM Transactions on Computational Logic (TOCL) 12(2), 1–43 (2011)
4. Börger, E., Stärk, R.: Abstract State Machines – a Method for High-Level System Design and Analysis. Springer, Heidelberg (2003)
5. Glausch, A., Reisig, W.: An ASM-characterization of a class of distributed algorithms. In: Abrial, J.-R., Glässer, U. (eds.) Rigorous Methods for Software Construction and Analysis. LNCS, vol. 5115, pp. 50–64. Springer, Heidelberg (2009)
6. Gosling, J., Joy, B., Steele, G., Bracha, G., Buckley, A.: The Java language specification Java SE 7 edition (2013), http://docs.oracle.com/javase/specs/jls/se7/jls7.pdf
7. Gurevich, Y.: The sequential ASM thesis.The Logic in Computer Science Column. Bulletin of European Association for Theoretical Computer Science (1999)
8. Gurevich, Y.: Evolving algebras 1993: Lipari guide. In: Börger (ed.) Specification and Validation Methods. Oxford University Press (1995)
9. Lamport, L.: How to make a multiprocessor computer that correctly executes multiprocess programs. IEEE Transactions on Computers 28(9) (September 1979)

WebASM: An Abstract State Machine Execution Environment for the Web

Simone Zenzaro, Vincenzo Gervasi, and Jacopo Soldani

Dipartimento di Informatica, University of Pisa, Italy

Abstract. We describe WebASM, a web-based environment that embeds the CoreASM execution engine in a web page. WebASM provides several advantages to specification writers: (1) complex behaviour expressed via ASM can be made visible by using the full power of the web-based presentation layer; (2) ASM specifications can be edited and run interactively via any web browser; (3) the full CoreASM environment is made available via zero-install deployment, thus eliminating a major barrier to the adoption of the language.

In this paper, we briefly outline the technicalities of the approach, present an example, and survey possible applications of WebASM.

1 Introduction

Abstract State Machines (ASM) [2] have been demonstrated to be a powerful yet intuitive formalism for describing specifications. A vast number of case studies, including language specifications, microprocessor design, sequential and distributed algorithms, and industrial plant control machines (see [1] for a full survey) have established the practical applicability of ASMs to real-world systems.

In the 30 years history of the ASM method, a number of execution environments have been developed; among the major efforts, we cite [9,8,4,6]. In varying degrees, all these approaches required setting up a moderately complicated programming environment (e.g., using a Gofer interpreter in [9], or a .NET development environment for [8], or using the Eclipse IDE for [4]). Moreover, none of the existing environments are endowed with convenient graphics facilities (although some of them, e.g. AsmL and CoreASM, can make recourse to native calls to platform-specific graphic APIs).

With WebASM, we set to improve on those two aspects by providing a web-based, fully self-contained embodiment of the CoreASM execution environment [4] which can be run in any modern web browser, and that can be controlled via JavaScript so that arbitrarily complex user interfaces and graphical displays can be rendered as a (dynamic) HTML page.

In the following, we first describe the technical approach taken by WebASM; we then present an example, describing the graphical animation of a distributed leader election protocol specified in ASM. A discussion about possible applications of WebASM and some reflections on future work conclude the paper.

Y. Ait Ameur and K.-D. Schewe (Eds.): ABZ 2014, LNCS 8477, pp. 216–221, 2014.

2 Technical Outline

One of the premises of the CoreASM project was that the resulting execution engine should be easy to embed in other applications. In WebASM, we made good on that promise by embedding the whole CoreASM engine (including all the plugins packaged in the official distribution) in a Java applet, which is then connected to the hosting web page through JavaScript bindings.

Security policies restrict what applets can do: in particular, by default no access to the local filesystem is possible (and special configurations are undesirable in a zero-install perspective). As a consequence, dynamic addition of user-developed plugins is not allowed in WebASM. Saving and loading specifications, instead, is managed on the JavaScript side by treating the specification as a string and passing it to the engine for interpretation. In a typical application, the specification text could be obtained from a text editor hosted on the same page, thus allowing the user to write and run ASM specifications in the same environment.

The JavaScript bindings exposed by WebASM include methods to create and initialise ASM machines, to load specifications, to perform an ASM step, and to access the whole abstract state of the machine or a single location.

In particular, access to the abstract state is limited by the concrete representation of values. Indeed, in its full generality the ASM model allows for arbitrary sorts, including those whose values do not have a literal notation. The JavaScript bindings for WebASM allow reading any value in the ASM state (technically, any CoreASM Element instance) as a string; the converse is not always possible (e.g., an element of *Agent* can be printed as a name, but not re-created from its name alone). However, all basic types which are commonly used (e.g., strings and numbers) are fully mapped between ASM and JavaScript.

The final element in our implementation is a map between locations of the ASM state and attributes of DOM elements in the page, optionally transformed by a custom JavaScript function (to account for syntactic differences between ASM and HTML/CSS notations).

What is left to the user is to design an HTML page with suitable graphics to visualise the salient elements of the ASM state, and define a map, as described above, in order to visualise state evolution during the ASM computation. After each ASM step, locations of the ASM state mentioned in the map are read (through the JavaScript bindings), their values are mapped or transformed, and finally applied to attributes of the various corresponding DOM elements, thus updating the DOM state. After each update cycle, the browser re-renders the modified portions of the web page (namely: the part hosting the graphical depiction of the ASM state), and the engine is then ready to execute the next step.

Designing suitable HTML graphics for the desired representation of the ASM state can be a tricky at times, depending on how sophisticated the depiction is. However, HTML design and JavaScript programming skills are much more readily available than what would be required to produce a custom, full-blown application to the same end. Moreover, it is reasonable to assume that in a context where the goal is to teach formal modeling skills, programming skills are already available, so we do not expect this part of the approach to be a significant burden.

In extreme cases, our technique can be extended to manipulate elements of arbitrary SVG vector graphics embedded in a web page, instead of DOM elements, again by using simple JavaScript mappings. We had no need of such an extension in our experiences with the tool.

3 Examples

To exemplify the WebASM approach, we use a specification for a classical distributed algorithm, namely the Extrema Finding by Franklin [5]. In this problem, a number of processes (modelled in ASM as separate agents) are arranged in a ring topology with bidirectional communications, with each process holding a value; their task is to identify the process holding the maximal value.

The corresponding ASM specification, which is a straightforward translation of the algorithm provided in [5], is shown in Figure 1.

$EXTREMAFINDING =$
  if $mode(self) = ACTIVE$ **then**
    if not $isLargest(self)$ **then**
      $rightMsg(l(self)) := id(self)$
      $leftMsg(r(self)) := id(self)$
    if $largerMsgReceived$ **then**
      $mode(self) := INACTIVE$
    if $myMsgReceived$ **then**
      $isLargest(self) := true$
      $notified(r(self)) := true$

  if $mode(self) = INACTIVE$ **then**
    if $notified(self)$ **then**
      $notified(r(self)) := true$
    else
      $rightMsg(l(self)) := rightMsg(self)$
      $leftMsg(r(self)) := leftMsg(self)$
  if $isLargest(self)$
    and $notified(self)$ **then**
    $EXTREMAFOUND$

Fig. 1. The main rule in ExtremaFinding (signature can be seen in Figure 2)

As for the visualisation, we have chosen to show each process (hence, each ASM agent) as a box displaying three figures: at the top, the process' value ($id(self)$); on the bottom left and the right the values received from its left and right neighbour according to the ring topology ($leftMsg(self)$ and $rightMsg(self)$). Moreover, a box's border colour indicates the process state ($mode(self)$, with $ACTIVE \rightarrow$ green and $INACTIVE \rightarrow$ red), and a box's background colour indicates whether the process has been notified ($notified(self)$, $false \rightarrow$ white and $true \rightarrow$ grey). Finally, the border style indicates the value of $isLargest(self)$ ($true \rightarrow$ dashed, $false \rightarrow$ solid).

The user can then experiment running (and modifying) the specification, in a continuous way or step-by-step, while observing its progress through the animation happening on the page. Figure 2 shows the browser-hosted animation environment.

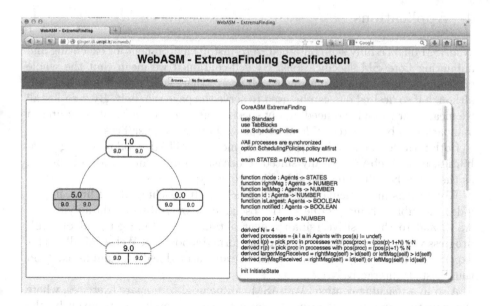

Fig. 2. A screenshot of WebASM animating the Extrema Finding specification

As an additional example, Figure 3 shows three subsequent stages of animation for another classical specification, based on the Distributed Termination Protocol from [3] (the corresponding ASM specification has been published in [7]). Here, each box represents an agent (simulating a different machine), with each machine spontaneously exchanging messages with others. At each step, exactly one machine holds a coloured token (which can be black or white), that is passed around as the protocol progresses. The goal is to determine whether the entire distributed computation has finished, which is detected when a "white" token is returned to the master machine (depicted with a grey background).

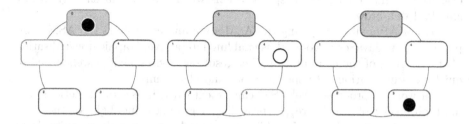

Fig. 3. Three steps of the Distributed Termination Protocol specification

4   Applications

WebASM offers a fully self-contained, zero-install environment for executing and animating CoreASM specifications. By self-contained we mean that the whole environment is contained in a single web page – there is no need of a web server, although if so desired one can be used to serve the page remotely. By zero-install we mean that there is no need of any special software on the client computer: a standard web browser (capable of executing Java applets) suffices.

Both these features, united to the convenience of HTML/CSS rendering capabilities, make WebASM ideally suited to occasional users and ASM newcomers, as they greatly reduce technical barriers to entry.

WebASM is in particular suited for teaching. In teaching algorithms, it provides a double advantage due to the pseudocode-over-abstract data syntax of ASM, and to the positive reinforcement obtained by showing the algorithm's progress via graphical means. In teaching formal modelling, WebASM allows for experimenting with specifications, which can be modified and animated (and thus, tested) interactively.

Also in a teaching context, WebASM can be used to prepare exercises, where the signature and graphic visualisation for a certain problem are given by the instructor, and the task set on students is to write ASM rules to accomplish the desired behaviour. One could envision an entire course based on a number of web pages, each allowing students to experiment with different specifications.

Finally, WebASM provides an alternative to traditional scripting languages for the web. Instead of programming some desired behaviour in languages such as JavaScript or VBScript, a developer could provide a ground model as a CoreASM specification, and have it executed behind the scenes by WebASM. In such a setup, conformance of the implementation to its ASM specification would be guaranteed by construction.

5   Conclusions and Future Work

We have presented WebASM, a self-contained, zero-install, interactive, graphical execution engine for CoreASM specifications which can be run entirely in any standard web browser.

A set of APIs allow accessing the ASM state and controlling the ASM computation from JavaScript code, thus enabling interactive, graphical visualisation of the progress of the computation. The resulting environment is well suited to quick experimentation with specifications and algorithms.

As a work in progress, WebASM can be extended in several directions; the most promising of which are (1) providing a graphical highlight of the rules being executed at each step, in addition to visualising the state resulting from their execution, (2) providing a better editor, with support for syntax highlighting and code completion, and (3) empowering users to visually build graphical representations of the state, by providing a palette of tools to draw graphical elements and link their appearance to elements of the ASM state.

Some of these improvements, and especially (3), we consider as crucial to the full realization of the promises of WebASM. Currently, a modicum of web programming prowess is required to define the graphical representation of the state and the mapping between locations of the ASM state and the DOM elements depicting them. Ideally, the mapping could be defined – at least in most standard cases – by a simple "property sheet"-style editing interface.

In a teaching setting, (3) is geared towards the instructor preparing exercises (on a given problem) for students. In contrast, (1) and (2) are geared towards the students, in that improvements in these areas would directly lead to more effective feedback and ease of experimentation with changing the ASM specification for the given problem. We are currently in the process of implementing (1) and (2) for the first public release of the tool.

References

1. Börger, E.: The origins and the development of the ASM method for high level system design and analysis. Journal of Universal Computer Science 8(1), 2–74 (2002)
2. Börger, E., Stärk, R.: Abstract State Machines – A Method for High-Level System Design and Analysis. Springer (2003)
3. Dijkstra, E., Feijen, W., van Gasteren, A.: Derivation of a termination detection algorithm for distributed computations. Information Processing Letters 16, 217–219 (1983)
4. Farahbod, R., Gervasi, V., Glässer, U.: CoreASM: An extensible ASM execution engine. Fundamenta Informaticae 77, 71–103 (2007)
5. Franklin, R.: On an improved algorithm for decentralized extrema finding in circular configurations of processors. Commun. ACM 25(5), 336–337 (1982)
6. Gargantini, A., Riccobene, E., Scandurra, P.: A metamodel-based language and a simulation engine for abstract state machines. Journal of Universal Computer Science 14(12), 1949–1983 (2008)
7. Gervasi, V., Riccobene, E.: From English to ASM: On the process of deriving a formal specification from a natural language one. Dagstuhl Reports 3(9), 85–90 (2014)
8. Gurevich, Y., Rossman, B., Schulte, W.: Semantic essence of AsmL. Theor. Comput. Sci. 343(3), 370–412 (2005)
9. Schmid, J.: Refinement and Implementation Techniques for Abstract State Machines. PhD thesis, University of Ulm, Germany (2002)

Formal System Modelling
Using Abstract Data Types in Event-B

Andreas Fürst[1], Thai Son Hoang[1], David Basin[1],
Naoto Sato[2], and Kunihiko Miyazaki[2]

[1] Institute of Information Security, ETH-Zurich, Switzerland
{fuersta,htson,basin}@inf.ethz.ch
[2] Yokohama Research Lab, Hitachi Ltd., Japan
{naoto.sato.je,kunihiko.miyazaki.zt}@hitachi.com

Abstract. We present a formal modelling approach using *Abstract Data Types* (ADTs) for developing large-scale systems in Event-B. The novelty of our approach is the combination of refinement and instantiation techniques to manage the complexity of systems under development. With ADTs, we model system components on an abstract level, specifying only the necessary properties of the components. At the same time, we postpone the introduction of their concrete definitions to later development steps. We evaluate our approach using a large-scale case study in train control systems. The results show that our approach helps reduce system details during early development stages and leads to simpler and more automated proofs.

Keywords: Event-B, refinement, abstract data types.

1   Introduction

Event-B [3] is a formalism for developing systems whose components can be modelled as discrete transition systems. An Event-B model contains two parts: a dynamic part (called *machines*) modelled by a transition system and a static part (called *contexts*) capturing the model's parameters and assumptions about them. Event-B's main technique to cope with system complexity is stepwise *refinement*, where design details are gradually introduced into the formal models. Refinement enables abstraction of machines, and since abstract machines contain fewer details than concrete ones, they are usually easier to verify.

However, when developing large, complex systems, refinement alone is often insufficient. Machines containing sufficient details to state and prove relevant safety properties may lead to proofs of unmanageable complexity. We observed this limitation while developing a large-scale train control system by refinement in Event-B. To specify and reason about collision-freeness properties, we needed to model the trains in detail, for example formalising their layout and movement. As a consequence, we had to state numerous complex invariants which resulted in many complicated manual proofs. This motivated an alternative approach to abstract away additional details from the system's model to reduce the complexity and increase the automation of the resulting proofs.

Y. Ait Ameur and K.-D. Schewe (Eds.): ABZ 2014, LNCS 8477, pp. 222–237, 2014.
© Springer-Verlag Berlin Heidelberg 2014

Approach. To model a system at a more abstract level, we introduce the notion of *Abstract Data Types* (ADTs) [16] in Event-B. An ADT is a mathematical model of a class of data structures. It is typically defined in terms of a set of operations that can be performed on the ADT, along with a specification of their effect. By using Event-B contexts to formalise ADTs and their operations, we can subsequently utilise the ADTs to model the system's dynamic behaviour in the machines. We use generic instantiation [5] as a means to further concretise and thereby implement the ADTs. As the ADTs evolve, the machines are also refined accordingly.

We evaluate our approach by developing a substantial industrial case study in the railway domain. Given an informal specification of a train control system, we incrementally develop a formal model of the overall system. This includes modelling the trains, the interlocking system, and the train controller. The complexity of the case study is comparable with that of real train control systems such as CBTC [15] or ETCS Level 3 [6]. We develop the controller all the way to a concrete implementation that runs on specialised hardware. To our knowledge, this is the first published development of a train control system on the system level, *i.e.*, modelling the train controller together with its environment, that is correct-by-construction.

Contribution. Our contribution is the introduction of ADTs in Event-B. We show that reasoning using ADTs can be done purely based on the properties of the ADTs' operations, regardless of how the ADTs will be implemented. As a result, systems specified with ADTs are more abstract and hence easier to verify than systems developed directly without them. In fact, ADTs encapsulate part of the system's dynamic behaviour in the static context of Event-B. This is novel as traditionally Event-B contexts are only used to specify static parameters of a system's model and all dynamic behaviour is modelled as a transition system in the Event-B machines. Furthermore, our use of generic instantiation in Event-B is novel as this technique has until now only been applied to reuse developments, for example in [19]. In contrast, we use generic instantiation as a mechanism to gradually introduce details into the formal models similar to refinement.

The way we introduce ADTs in Event-B allows ADTs to be used alongside Event-B refinement. Hence, one can combine these two different abstraction techniques during development and apply whichever fits better at a particular development stage and results in simpler proofs. In contrast to development strategies that use refinement or ADTs exclusively, our approach is better suited for developing large-scale industrial systems.

Structure. The rest of our paper is structured as follows. In Section 2, we briefly review Event-B, including refinement and instantiation techniques. We motivate and present our approach in Section 3. We evaluate our approach on an industrial case study in Section 4. Finally, we discuss related work in Section 5 and conclude in Section 6.

2   The Event-B Modelling Method

Event-B [3] represents a further evolution of the classical B-method [1], which has been simplified and focused around the general notion of *events*. Event-B has a semantics based on transition systems and simulation between such systems. We will not describe in detail Event-B's semantics here; full details are provided in [3]. Instead, we will describe some Event-B modelling concepts that are important for the later presentation.

Event-B models are organized in terms of the two basic constructs: *contexts* and *machines*.

Contexts. Contexts specify the static part of a model and may contain *carrier sets*, *constants*, *axioms*, and *theorems*. Carrier sets are similar to types. Axioms constrain carrier sets and constants, whereas theorems express properties derivable from axioms. The role of a context is to isolate the parameters of a formal model (carrier sets and constants) and their properties, which are intended to hold for all instances.

Machines. *Machines* specify behavioral properties of Event-B models. Machines may contain *variables*, *invariants*, *theorems*, and *events*. Variables v define the state of a machine. They are constrained by invariants $I(v)$. Theorems are properties derivable from the invariants. Possible state changes are described by events. An event e can be represented by the term

$$e \;\widehat{=}\; \textbf{any } t \textbf{ where } G(t, v) \textbf{ then } S(t, v) \textbf{ end } ,$$

where t is the event's *parameters*, $G(t, v)$ is the event's *guard* (the conjunction of one or more predicates), and $S(t, v)$ is the event's *action*. The guard states the condition under which an event may occur, and the action describes how the state variables evolve when the event occurs. The action of an event is composed of one or more *assignments* of the form $x := E(t, v)$, where x is a variable in v. Assignments in Event-B may also be nondeterministic, but we omit this additional complexity here as it is not used in this paper. All assignments of an action $S(t, v)$ occur simultaneously. A dedicated event without any parameters or guard is used for *initialisation*.

Refinement. *Refinement* provides a means to gradually introduce details about the system's dynamic behaviour into formal models [3]. A machine **CM** can refine another machine **AM**. We call **AM** the *abstract* machine and **CM** the *concrete* machine. The states of the abstract machine are related to the states of the concrete machine by *gluing invariants* $J(v, w)$, where v are the variables of the abstract machine and w are the variables of the concrete machine. A special case of refinement (called superposition refinement) is when v is kept in the refinement, i.e. $v \subseteq w$. Intuitively, any behaviour of **CM** can be simulated by a behaviour of **AM** with respect to the gluing invariant $J(v, w)$.

Refinement can be reasoned about on a per-event basis. Each event e of the abstract machine is *refined* by one or more concrete events f. Simplifying somewhat, we can say that f refines e if f's guard is stronger than e's guard (*guard strengthening*), and the gluing invariants $J(v, w)$ establish a simulation of f by e (*simulation*).

Instantiation. *Instantiation* is a common technique for reusing models by providing concrete values for abstract model parameters. Since an Event-B model is parameterised by the carrier sets and constants, instantiation in Event-B [5,19] amounts to instantiating the contexts.

Suppose we have a generic development with machines $M_1, \ldots, M_n$ building a chain of refinements with carrier sets s and constants c, constrained by axioms $A(s, c)$. Suppose too that we want to reuse the development within another context, specified by (concrete) carrier sets t and constants d, constrained by axioms $B(t, d)$. Let $T(t)$, which must be an Event-B type expression, and $E(t, d)$ be the instantiated values for s and c respectively. Given that the instantiation is correct, *i.e.*, $B(t, d) \Rightarrow A(T(t), E(t, d))$, the instantiated development where s and c are replaced by their corresponding instantiated values is correct-by-construction.

For more details on instantiation in Event-B and its tool support see [5] and [19]. All instantiation steps described in this paper were performed using the generic instantiation plug-in developed by Hitachi and ETH Zurich [13].

3   Abstract Data Types in Event-B

In this section, we describe how to specify and implement ADTs in Event-B. Our approach is based on refinement and generic instantiation. An ADT is typically defined in terms of a set of operations that can be performed on the ADT, along with a specification of their effect. Let us start with the standard example: the *stack* ADT is a last in first out (LIFO) data type that contains a collection of elements.

A stack is characterised by three operations:

- *push*: takes a stack S and an item e, and returns a new stack where e is added to the top of S.
- *pop*: takes a (non-empty) stack S and returns a new stack where S's top element is removed.
- *top*: takes a (non-empty) stack S and returns S's top element.

A special stack is the *empty* stack that contains no elements. Some important constraints for the operations of the stack ADT are as follows. Given a stack S and an element e, $push(S, e) \neq empty$, $pop(push(S, e)) = S$, and $top(push(S, e)) = e$.

Specifying ADTs in Event-B. ADTs and their operations can be modelled using carrier sets, constants and axioms in Event-B. Instantiation can then be used to "implement" the ADTs. The instantiation proofs ensure that the ADTs' implementations satisfy their specifications.

Each ADT **A** is modelled as follows:

- A carrier set A_TYPE defining the type of the **A** objects along with an associated
- set constant $A \subseteq A\_TYPE$ representing all valid **A** objects. [3]
- Each operation is modelled using a constant.
- The constraints on the operations are specified using axioms.

Consider the stack ADT for elements of type $ELEM$. It can be modelled in Event-B as follows.

---

[3] Note that we do not currently support the definition of parameterised ADTs, which would allow one to specify a generic *stack* ADT independent of its elements' type.

sets : $STACK\_TYPE$ constants : $STACK, empty, push, pop, top$

axioms :

axm0_1 : $STACK \subseteq STACK\_TYPE$

axm0_2 : $empty \in STACK$

axm0_3 : $push \in STACK \times ELEM \rightarrow STACK$

axm0_4 : $pop \in STACK \setminus \{empty\} \rightarrow STACK$

axm0_5 : $top \in STACK \setminus \{empty\} \rightarrow ELEM$

axm0_6 : $\forall S, e \cdot S \in STACK \Rightarrow push(S \mapsto e) \neq empty$

axm0_7 : $\forall S, e \cdot S \in STACK \Rightarrow pop(push(S \mapsto e)) = S$

axm0_8 : $\forall S, e \cdot S \in STACK \Rightarrow top(push(S \mapsto e)) = e$

Axioms axm0_7 and axm0_8 specify the relationship between the *pop, top*, and *push* operations. Notice that there is no need to fully specify an ADT. In subsequent examples, we will only define as many axioms as needed to prove the stated properties.

Instantiating ADTs. A possible implementation of the stack ADT is one where a stack is represented as an array. More formally, a stack is represented by a pair (f, n), where n is the stack's size and f is an array of size n representing its content. In other words, we intend to implement the stack ADT by the array datatype. Operations of the array datatype are as follows:

- *append*: takes an array and an element, and returns a new array where the element is appended to the end of the input array.
- *front*: takes an array and returns a new array where the last element of the input array is removed.
- *last*: takes an array and returns the last element of the input array.

The array datatype is specified in Event-B as follows.

constants : $ARRAY, append, front, last$

axioms :

axm1_1 : $ARRAY = \{f \mapsto n \mid n \in \mathbb{N} \wedge f \in 0 .. n - 1 \rightarrow ELEM\}$

axm1_2 : $append = (\lambda (f \mapsto n) \mapsto e \cdot f \mapsto n \in ARRAY \wedge e \in ELEM$
$\mid (f \mathbin{\mkern-5mu\lhd\mkern-5mu} \{n \mapsto e\}) \mapsto n + 1)$

axm1_3 : $front = (\lambda f \mapsto n \cdot f \mapsto n \in ARRAY \wedge n \neq 0$
$\mid ((\{n - 1\} \mathbin{\lhd\mkern-10mu-} f) \mapsto n - 1))$

axm1_4 : $last = (\lambda f \mapsto n \cdot f \mapsto n \in ARRAY \wedge n \neq 0 \mid f(n - 1))$

Notice that at this point all the constants are concretely defined by lambda expressions.

To prove that the array datatype implements the stack ADT, we instantiate $STACK\_TYPE$ with $\mathbb{P}(\mathbb{Z} \times ELEM) \times \mathbb{Z}$, $STACK$ with $ARRAY$, and the operations *push, pop*, and *top* with *append, front*, and *last*, respectively. Constant *empty* is instantiated with $\varnothing \mapsto 0$. We must prove that the instantiated abstract axioms, *i.e.*, axm0_1–axm0_8, are derivable from the concrete axioms, *i.e.*, axm1_1–axm1_4. The proofs can be constructed by expanding the definitions of the concrete constants accordingly.

For more information on how to implement ADTs in Event-B using generic instantiation, we refer the reader to [7].

4   Developing a Train Control System Using ADTs

In this section, we illustrate our approach on an industrial case study. We first briefly describe the system and explain the difficulties when developing such a complex system without ADTs. We then present part of the development where we applied our approach using ADTs. Finally, we evaluate our approach by giving an overview of the entire development of this case study together with some statistics to justify our approach's effectiveness.

4.1   System Description

The scope of our case study is the development of a modern train control system. The main goal of the system is to keep all trains in the railway network a safe distance apart to prevent collisions. The network consists of tracks (divided into sections) and points connecting these tracks. An interlocking system switches the points to connect different tracks with each other, and results in a track layout that dynamically changes. Instead of light signals, the train control system uses radio communication to send the trains the permission to move or stop.

While classic train control systems use trackside hardware to detect whether a section is occupied by a train, our system determines this information from the trains' position and length. The trains themselves determine their positions and send them to the train control system by radio. Based on information on what part of the network is occupied, the controller calculates for every train the area in which it can safely move without collisions. This area is called the *Movement Authority* (MA) and represents the permission for a train to move as long as it does not leave this area. The calculated MAs are then directly sent to the trains where an onboard unit interprets them to calculate the location where the permission to drive ends (*Limit of Authority*, LoA). To prevent driving over the LoA, the onboard unit continuously determines a speed limit and applies the emergency brakes if necessary. An overview of the interacting system components is given in Figure 1.

Collision-freeness between trains is guaranteed by the overall system and relies on two conditions: (1) The trains are always within their assigned movement authorities, and (2) the controller ensures that the MAs issued to the trains do not overlap. In fact, (1) is implementable only if the MAs issued by the controller are never reduced at the front of the trains.

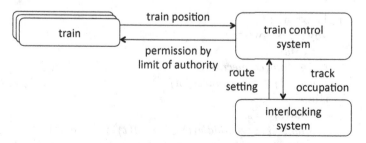

Fig. 1. Train control system with the interlocking system as its environment

4.2   The Need for Abstraction

Our first challenge in developing the train control system is formalising the trains in the network. Figure 2 depicts a train occupying some part of the network. It illustrates a sequence of sections with fully occupied ones in the middle and partially occupied ones at each end of the train.

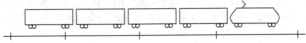

Fig. 2. A train occupying a sequence of sections

In our first attempt at modelling the train control system, we used different variables to denote how the trains occupy the network. Let ids be the set of active trains in the network. We modelled the different aspects of the trains, such as their head, rear, middle, connections, etc., by total functions as follows. For clarity, we omit from the presentation other aspects of the trains, such as the head- and rear-position within a section.

> **variables** : $ids, head, rear, middle, connection, \ldots$
> **invariants** :
> inv0_1 : $head \in ids \rightarrow SECTION$
> inv0_2 : $rear \in ids \rightarrow SECTION$
> inv0_3 : $middle \in ids \rightarrow \mathbb{P}(SECTION)$
> inv0_4 : $connection \in ids \rightarrow (SECTION \rightarrowtail SECTION)$
> inv0_5 : $\forall t \cdot t \in ids \Rightarrow head(t) \notin middle(t)$
> inv0_6 : $\forall t \cdot t \in ids \Rightarrow rear(t) \notin middle(t)$
> inv0_7 : $\forall t \cdot t \in ids \land connection(t) = \varnothing \Rightarrow head(t) = rear(t)$
> $\ldots$

Invariants inv0_5–inv0_7 specify several important properties of trains. For example, inv0_7 specifies that if a train occupies only one single section, its head and rear are in the same section. Note that due to the lack of space, we omit other invariants that ensure that trains are connected and do not contain loops.

To motivate the need for additional abstraction in Event-B, we focus on the event train_extend. Its purpose is to extend the train, denoted by t, to a section, denoted by s. Namely, train_extend prepends s to the head of the train and s becomes the new head. This event is used whenever the train reaches the end of the current head section and moves to the beginning of the next section in front of it.

> train_extend :
>   **any** t, s **where**
> $t \in ids$
> $s \notin \mathrm{dom}(connection(t))$
> $head(t) \notin \mathrm{ran}(connection)$
>   **then**
> $head(t) := s$
> $middle(t) := (middle(t) \cup \{head(t)\}) \setminus \{rear(t)\}$
> $connection(t) := connection(t) \cup \{s \mapsto head(t)\}$
>   **end**

The event's guard ensures that the connection of t remains a partial injective function (inv0_4). When updating $middle(t)$, we remove $rear(t)$ to guarantee that in the case where the train occupies only one section (*i.e.*, $connection(t) = \varnothing$ and hence $head(t) = rear(t)$, according to inv0_7), the train's middle is still empty afterwards.

Proving that train_extend maintains the invariants, in particular inv0_5, requires more invariants, which we omit for clarity. All additional invariants are universally quantified, *i.e.*, of the form "$\forall t \cdot t \in ids \Rightarrow \ldots$" and they express the relationship between different aspects of a train.

Encapsulation. The invariants above describe the trains' layouts that change independently of each other. As a result, the preservation of the invariants should be proven on a per train basis and by hiding the rest of the model. In Event-B, however, invariants are global and all other parts of the system are taken into account during the proof, which increases their complexity. This indicates that some encapsulation for the models of trains will be useful for our proofs.

High-level Properties of Low-level Details. An attempt to specify and prove properties such as collision-freeness at a concrete level like that described above leads to complicated models and difficult proofs. In particular, expressing relationships between sequences, such as "containment" (*e.g.*, a train is always within its movement authority) and "being disjoint" (*e.g.*, the movement authorities of two different trains do not overlap) using information about the sequences' head, rear, middle and connections, is far from trivial. This indicates that we should start modelling the system at an even more abstract level by omitting the detailed aspects of the sequences.

Reuse. In addition to the above mentioned difficulties, another motivation for using ADTs in our development is that modelling the trains' movement authorities is similar to modelling the trains. In fact, both trains and their MAs should be modelled using the same ADT.

4.3   Development Using Abstract Data Types

The Region ADT. Abstracting away the details of sequences, such as head, rear, middle, and connections, we start our modelling with an ADT corresponding to *regions* on a network, focusing on relationships between regions such as "contained" and "disjoint". The region ADT includes the following operations:

- *extend*: takes a region R and a section s, and returns a new region where s is added to R.
- *contained*: binary relation associating a region R_1 with every region R_2 that contains R_1.
- *disjoint*: binary relation associating two regions R_1 and R_2 with each other if they do not overlap.

Note that there are other operations of the region ADT that we omit for clarity.

In Event-B, this ADT is modelled as follows. Constants *contained*, *disjoint*, and *extend* correspond to the operations mentioned above.

sets : $REGION\_TYPE$
constants : $REGION$, $contained$, $disjoint$, $extend$
axioms :
axm0_1 : $REGION \subseteq REGION\_TYPE$
axm0_2 : $contained \in REGION \leftrightarrow REGION$
axm0_3 : $disjoint \in REGION \leftrightarrow REGION$
axm0_4 : $extend \in REGION \times SECTION \nrightarrow REGION$

Constraints on the operations of the region ADT are modelled as axioms. For example, $contained$ is transitive, $disjoint$ is symmetric, $extend$ is strengthening with respect to $contained$. Note that in the following, we use $R_1 \Subset R_2$ to denote $R_1 \mapsto R_2 \in contained$, and $R_1 \cancel{\Subset} R_2$ to denote $R_1 \mapsto R_2 \in disjoint$.

axioms :
axm0_5 : $\forall t_1, t_2, t_3 \cdot t_1 \Subset t_2 \wedge t_2 \Subset t_3 \Rightarrow t_1 \Subset t_3$
axm0_6 : $\forall t_1, t_2 \cdot t_1 \cancel{\Subset} t_2 \Rightarrow t_2 \cancel{\Subset} t_1$
axm0_7 : $\forall t, s \cdot t \mapsto s \in \mathrm{dom}(extend) \Rightarrow t \Subset extend(t \mapsto s)$
axm0_8 : $\forall t_1, t_2, t_3 \cdot t_1 \Subset t_2 \wedge t_2 \cancel{\Subset} t_3 \Rightarrow t_1 \cancel{\Subset} t_3$

The current states of the active trains and their associated movement authorities are represented by a mapping from trains to the set of all possible regions ($train$) and a mapping from movement authorities to the set of all possible regions (ma). Invariant inv0_3 states that the trains always stay within their movement authorities. Invariant inv0_4 states that the movement authorities of any two trains are disjoint.

variables : ids, $train$, ma
invariants :
inv0_1 : $train \in ids \rightarrow REGION$
inv0_2 : $ma \in ids \rightarrow REGION$
inv0_3 : $\forall t \cdot t \in ids \Rightarrow train(t) \Subset ma(t)$
inv0_4 : $\forall t_1, t_2 \cdot t_1 \in ids \wedge t_2 \in ids \wedge t_1 \neq t_2 \Rightarrow ma(t_1) \cancel{\Subset} ma(t_2)$

Importantly, the collision-freeness property, i.e.,

$$\forall t_1, t_2 \cdot t_1 \in ids \wedge t_2 \in ids \wedge t_1 \neq t_2 \Rightarrow train(t_1) \cancel{\Subset} train(t_2) \,,$$

is derivable (as a theorem) from the invariants inv0_3, inv0_4 and the property relating $contained$ and $disjoint$, i.e., axm0_8.

The event train_extend can be specified abstractly as follows. Its last guard ensures that the extended train cannot exceed its assigned movement authority.

train_extend :
    any t, s **where**
        $t \in \mathrm{dom}(train)$
        $train(t) \mapsto s \in \mathrm{dom}(extend)$
        $extend(train(t) \mapsto s) \Subset ma(t)$
    then
        $train(t) := extend(train(t) \mapsto s)$
    end

The Sequence ADT. The model at this stage is abstract in two ways: (1) its dynamic behaviour is not fully described by the machine and (2) it uses the region ADT which is not fully "implemented". For (2), we utilise generic instantiation to introduce more details on how the region ADT and its operations are realised. Similar to refinement, this realisation of ADTs can be split into multiple instantiation steps.

In our development, we first replace the region ADT by the sequence ADT. The sequence ADT includes the following operations:

- *prepend*: takes a sequence S and a section s, and returns a new sequence where s is added to the head of S.
- *head*: takes a sequence S and returns the head section of S.
- *rear*: takes a sequence S and returns the rear section of S.
- *middle*: takes a sequence S and returns the middle sections of S.
- *connection*: takes a sequence S and returns the connection between sections of S.

> **sets :** $SEQUENCE\_TYPE$
> **constants :** $SEQUENCE, prepend, head, rear, middle, connection$
> **axioms :**
> axm1_1 : $SEQUENCE \subseteq SEQUENCE\_TYPE$
> axm1_2 : $prepend \in SEQUENCE \times SECTION \nrightarrow SEQUENCE$
> axm1_3 : $head \in SEQUENCE \rightarrow SECTION$
> axm1_4 : $rear \in SEQUENCE \rightarrow SECTION$
> axm1_5 : $middle \in SEQUENCE \rightarrow \mathbb{P}(SECTION)$
> axm1_6 : $connection \in SEQUENCE \rightarrow (SECTION \nrightarrow SECTION)$
> axm1_7 : $\forall S \cdot S \in SEQUENCE \Rightarrow head(S) \notin middle(S)$
> axm1_8 : $\forall S \cdot S \in SEQUENCE \Rightarrow rear(S) \notin middle(S)$

We prove that the sequence ADT is a valid representation of the region ADT with the instantiation of the set $REGION\_TYPE$ by $SEQUENCE\_TYPE$, the constants $REGION$ by $SEQUENCE$, $extend$ by $prepend$, etc. We replace (instantiate) the operations $contained$ and $disjoint$ using $head$, $rear$, $middle$, and $connection$. For example, $contained$ is instantiated as follows.

$$contained = \left\{ S_1 \mapsto S_2 \; \middle| \; \begin{array}{l} S_1 \in SEQUENCE \wedge S_2 \in SEQUENCE \wedge \\ connection(S_1) \subseteq connection(S_2) \wedge \\ middle(S_1) \subseteq middle(S_2) \wedge \\ head(S_1) \in \{head(S_2)\} \cup middle(S_2) \cup \{rear(S_2)\} \\ rear(S_1) \in \{head(S_2)\} \cup middle(S_2) \cup \{rear(S_2)\} \\ \ldots \end{array} \right\}$$

Note that we omit from our presentation additional conditions related to the exact position of the head and rear within the section.

At this point the sequence ADT is still abstract. In particular, we do not give the exact definition for sequences and we still rely on the operators such as $head$, $rear$, $middle$, and $connection$ and the relationships between them.

Given the instantiation, we subsequently refine the dynamic behaviour of the system (*i.e.*, the machines). For event train_extend, the refinement removes the reference to $contained$ in the guard.

```
train_extend :
    any   t, s   where
        t ∈ dom(train)
        head(train(t)) ≠ head(ma(t))
        ... // other guards related to head/rear positions
    then
        train(t) := prepend(train(t) ↦ s)
    end
```

The Arbitrarily-based Array Data Type. The model based on the sequence ADT is abstract. To ensure that the model is implementable, we must give a representation for the sequence ADT. In our development, we use an arbitrarily-based array data type as the implementation for the sequence ADT. An arbitrarily-based array is an array that starts from an arbitrary index, in contrast to the common zero-based array that always starts from 0. More formally, each arbitrarily-based array can be represented by a tuple (a, b, f), where a and b are the starting and ending indices and f represents the array's content. The operations of the arbitrarily-based array such as *head*, *rear*, *middle*, and *connection* are defined accordingly. For example, the *head* operation is defined as follows.

$$head = (\lambda\, a \mapsto b \mapsto f \cdot a \mapsto b \mapsto f \in ARRAY \mid f(a))$$

The advantage of using arbitrarily-based arrays compared to normal (zero-based) arrays is that there is no need to shift indices when extending or reducing the arrays. For example, the *prepend* operation is defined as follows.

$$prepend = (\lambda\, (a \mapsto b \mapsto f) \mapsto s \cdot a \mapsto b \mapsto f \in ARRAY \,\wedge\, s \in SECTION \,\wedge \dots$$
$$\mid (a-1) \mapsto b \mapsto (f \lhd\!\!\!- \{a - 1 \mapsto s\}))$$

This simplifies the proof that the sequence ADT is correctly implemented by the array data type.

4.4   Development Summary

In our development of the train control system, the transformation of the region ADT into the sequence ADT is carried out in several instantiation steps. The benefit of having steps with small changes in the ADTs is that the machines that are specified using ADTs can also be gradually transformed in small steps. This also serves to decompose the proof of correctness of the systems into small instantiation and refinement steps.

Our development contains five different stages (numbered 0–4), connected by instantiation relationships, where a subsequent stage starts as an instantiation of the previous stage. Each stage contains several refinement steps for developing the system's main functionality.

Stage 0: We formalise the system at the most abstract, generic level, using the region ADT and the network ADT. In the refinement steps, we gradually introduce the active network, the active trains, the trains' movement authorities, the movement authorities calculated by the controller, and the relationships between them.

Stage 1–3: We carry out the transformation from the region ADT to the sequence ADT in three different instantiations. First, we instantiate the *contained* operation. Second, we instantiate the operation "part-of" between the region ADT and the network ADT (stating whether or not a region is *part of* a network). Finally, we instantiate the *disjoint* operation. The refinement steps in these stages have two purposes: (1) they transform the events to use the new data types, and (2) they introduce the design details of the system, including notions like *train ahead*, *train behind*, and *last train within a section*.

Stage 4: We instantiate the sequence ADT by the arbitrarily-based array data type. We also incrementally introduce details on the calculation of the trains' MAs.

Statistics and Comparison. We present statistics for our development in Table 1 and compare the development of the train control system with and without ADTs. Table 1a shows the proof statistics for our first attempt where we did not use ADTs. After 14 refinement steps and 45 difficult manual proofs, we stopped our development with numerous remaining undischarged proof obligations, due to missing invariants. We would have needed additional invariants that are complex to express and lead to even more complex proofs. Considering the proof effort needed up to this point, and the additional effort anticipated to complete the development, we were forced to adapt our development strategy and find additional abstraction techniques to simplify the proofs.

Table 1b shows the proof statistics of the development using ADTs. We distinguish between proofs related to instantiation and proofs related to refinement. Overall, 14% of the proofs are related to instantiation, and the other 86% are related to refinement. As expected, the machines at the more abstract and generic levels are more automated. Most of the manual proofs originating from instantiation (in particular of Stage 4) have a similar structure that includes manually expanding the instantiation definitions. These proof steps could be automated with a dedicated proof strategy, which would increase the amount of proof automation. Overall, the instantiation proofs have a better automation rate (82%) compared to the refinement proofs (58%).

The number of refinement steps as well as the total number of discharged proof obligations indicate that the size and complexity of our case study is significantly higher than typical academic examples. Moreover, given the level of detail in our model, stemming from realistic requirements, this supports our claims about the relevance of our approach for large and complex systems.

5   Related Work

5.1   Instantiation and Data Types

In our approach to introducing ADTs into formal development, generic instantiation [5] is the key technique for realising the ADTs. This differs from [19] where instantiation provides a means to reuse formal models in combination with a composition technique. In particular, to guarantee the correctness of the instantiated model, carrier sets (which are assumed to be non-empty and maximal) must be instantiated by type expressions. This has been overlooked in [5] and [19].

Table 1. Statistics

(a) Development without ADTs

	Obligations	Auto.	Manual	Undischarged
14 Refinements	666	497 (75%)	45 (7%)	124 (18%)

(b) Development using ADTs

		Obligations	Auto.	Manual
Stage 0	8 Refinements	267	267	0
Stage 1	Instantiation	34	24	10
	14 Refinements	632	477	155
Stage 2	Instantiation	165	161	4
	1 Refinement	52	44	8
Stage 3	Instantiation	175	172	3
	16 Refinements	765	314	451
Stage 4	Instantiation	174	90	84
	18 Refinements	1748	891	857
Total		4012	2440 (61%)	1572 (39%)
	Instantiation	548 (14%)	447 (82%)	101 (18%)
	Refinement	3464 (86%)	1993 (58%)	1471 (42%)

Part of our approach was previously published in [7]. There, our main motivation for using ADTs was to encapsulate data and to split the development process into two parts that can be handled by a domain expert and a formal methods expert, respectively. In this paper, we focus more on the need for alternative forms of abstraction when developing large and complex systems in Event-B. We not only use ADTs to abstract away implementation details for the domain expert, but we use them as an integral part from the beginning of our development to simplify the proofs. We describe relations between different ADTs to abstractly specify the system's properties.

The development of the *Theory Plug-in* [11] for Rodin allows users to extend the mathematical language of Event-B, for example, by including new data types. Theorems about new data types can be stated and later used by a dedicated tactic associated with the Theory Plug-in. There is also a clear distinction between the theory modules (capturing data structures and their properties) and the Event-B models using the newly defined data structures. The main difference between the Theory Plug-in and our approach is that the data types in the Theory Plug-in are "concrete". One must give the definitions for the data types and prove theorems about them before using these data types for modelling. This bottom-up approach is in contrast with our top-down approach where the choice of implementations for ADTs can be delayed. More specifically, we can have different implementations for the ADTs. For example, instead of implementing the sequence ADT using arbitrarily-based arrays, we can use standard, zero-based arrays for the same purpose. In fact, we did experiment with both implementations and decided to use arbitrarily-based arrays due to the simpler proofs for the systems.

Our approach of using ADTs in Event-B is similar to work on algebraic specifications [18]. In this domain, a specification contains a collection of *sorts*, *operations*, and *axioms* constraining the operations. Specifications can be *enriched* by additional sorts, operations, or axioms. Furthermore, to develop programs from specifications, the specifications are transformed via a sequence of small "refinement" steps. During these steps, the operations are "coded" until the specification becomes a concrete description of a program. For each such refinement step, one must prove that the code of the operations satisfies the axioms constraining them. An algebraic specification therefore corresponds to an Event-B context, while refinement in algebraic specifications is similar to generic instantiation in Event-B. In contrast to algebraic specifications [12], where the entire functionality of a system is modelled as ADTs (in the form of many-sorted algebras) [18], we use ADTs to abstract only part of our system's functionality. Modelling every aspect of a complex system like our example as an algebraic specification would be very challenging. In addition to the data types, the transition systems must also be encoded as ADTs in the specification. This would require a large number of axioms to describe the transitions.

5.2 Formal Development of Railway Systems

Bjørner gives in [9] a comprehensive overview of formal techniques and tools used for developing software for transportation systems. Beside techniques like model checking and model-based test case generation, he mentions approaches using refinement. The following approaches are of special interest for us.

The development of Metro line 14 in Paris [8,2] is one of the better known industrial application of formal methods. In particular, the safety critical part of the software was developed using the (classical) B Method [1]. The formal reasoning there was only at the software-level, *i.e.*, reasoning about the correctness of the software in isolation. In contrast, in our work we not only model the train control system, but also its environment such as the trains and their movement behaviour. Hence, we can reason on the *system-level* covering the overall structure of the system, its components, and their relationship [4].

In [14], Haxthausen and Peleska present the formal development and verification of a distributed railway control system using the RAISE formal method. Their approach is similar to our work as they also use stepwise refinement and ADTs to cope with the complexity of their system. However, their system is overly simplified at some points which reduced the development challenges that we found to be the most difficult in our work. First, they only consider simple network topologies without loops. Second, they develop a system where sections are either fully occupied or free. Third, their trains can occupy at most two sections. Although they claim that the system can be easily adapted for trains occupying more than two sections, from our experience, this generalisation is a challenging task. Moreover, in their proof they require that if any two events are enabled in a valid state, executing one of the events and therefore changing the state cannot disable the other event's guard. This is a strong property that is cumbersome to verify as one must prove it for all pairs of events. Our model does not require this property in order to guarantee the system's safety.

Platzer and Quesel verify parts of a similar train control system in [17] using their own verification tool KeYmaera. While we developed the functionality of the controller, their work focuses on developing the onboard unit. In their development, the controller belongs to the environment of the onboard unit and they assume that the controller does not issue MAs that are physically impossible for the trains. Our development fulfils this assumption by guaranteeing that the MAs are never reduced.

6 Conclusion

In this paper we presented an approach to building formal models in Event-B using ADTs. ADTs allow us to hide irrelevant details that are unimportant for proving abstract properties. On an abstract level, one can therefore focus on modelling the system's core functionality.

The way we introduce ADTs in our approach allows us to utilise generic instantiation. This handles both the instantiation of an ADT by the chosen data structure as well as the generation of the required proof obligations to guarantee that the chosen structure is a valid instance of the ADT. As a large scale case study we have successfully applied our approach to the development of a realistic train control system. We identified the limitations of only using refinement for this system and showed how we overcome these limitations using ADTs.

As future work we would like to overcome some of the current limitations of our work. As previously mentioned, we cannot presently specify parameterised ADTs. To overcome this limitation, we need to extend the semantics of Event-B contexts and adapt the generic instantiation technique accordingly. The Theory Plug-in might be useful to specify parameterised ADTs.

References

1. Abrial, J.-R.: The B-book: Assigning Programs to Meanings. Cambridge University Press (1996)
2. Abrial, J.-R.: Formal Methods in Industry: Achievements, Problems, Future. In: Osterweil, L.J., Rombach, H.D., Soffa, M.L. (eds.) ICSE, pp. 761–768. ACM (2006)
3. Abrial, J.-R.: Modeling in Event-B: System and Software Engineering. Cambridge University Press (2010)
4. Abrial, J.-R.: From Z to B and then Event-B: Assigning Proofs to Meaningful Programs. In: Johnsen, E.B., Petre, L. (eds.) IFM 2013. LNCS, vol. 7940, pp. 1–15. Springer, Heidelberg (2013)
5. Abrial, J.-R., Hallerstede, S.: Refinement, Decomposition, and Instantiation of Discrete Models: Application to Event-B. Fundam. Inform. 77(1-2), 1–28 (2007)
6. European Railway Agency. ERTMS/ETCS Functional Requirements Specification. European Railway Agency, Valencinnes, France (2007)
7. Basin, D., Fürst, A., Hoang, T.S., Miyazaki, K., Sato, N.: Abstract Data Types in Event-B - An Application of Generic Instantiation. CoRR (2012)
8. Behm, P., Benoit, P., Faivre, A., Meynadier, J.-M.: Météor: A Successful Application of B in a Large Project. In: Wing, J.M., Woodcock, J. (eds.) FM 1999. LNCS, vol. 1708, pp. 369–387. Springer, Heidelberg (1999)
9. Bjørner, D.: New Results and Trends in Formal Techniques & Tools for the Development of Software for Transportation Systems. In: FORMS (2003)
10. Breitman, K., Cavalcanti, A. (eds.): ICFEM 2009. LNCS, vol. 5885. Springer, Heidelberg (2009)

11. Butler, M., Maamria, I.: Practical theory extension in Event-B. In: Liu, Z., Woodcock, J., Zhu, H. (eds.) Theories of Programming and Formal Methods. LNCS, vol. 8051, pp. 67–81. Springer, Heidelberg (2013)

12. Ehrig, H., Mahr, B.: Fundamentals of Algebraic Specification 1: Equations und Initial Semantics. EATCS Monographs on Theoretical Computer Science, vol. 6. Springer (1985)

13. Fürst, A., Desai, K., Hoang, T.S., Sato, N.: Generic Instantiation Plug-in, http://sourceforge.net/projects/gen-inst/

14. Haxthausen, A.E., Peleska, J.: Formal Development and Verification of a Distributed Railway Control System. In: Wing, J.M., Woodcock, J., Davies, J. (eds.) FM 1999. LNCS, vol. 1709, pp. 1546–1563. Springer, Heidelberg (1999)

15. IEEE Std 1474.1-2004. IEEE Standard for Communications-Based Train Control (CBTC) Performance and Functional Requirements. IEEE, New York, USA (2005)

16. Liskov, B., Zilles, S.: Programming with Abstract Data Types. In: Proceedings of the ACM SIGPLAN Symposium on Very High Level Languages, pp. 50–59. ACM, New York (1974)

17. Platzer, A., Quesel, J.-D.: European Train Control System: A Case Study in Formal Verification. In: Breitman, Cavalcanti [10], pp. 246–265

18. Sannella, D., Tarlecki, A.: Essential Concepts of Algebraic Specification and Program Development. Formal Asp. Comput. 9(3), 229–269 (1997)

19. Silva, R., Butler, M.: Supporting Reuse of Event-B Developments through Generic Instantiation. In: Breitman, Cavalcanti [10], pp. 466–484

Formal Derivation of Distributed MapReduce

Inna Pereverzeva[1,2], Michael Butler[3], Asieh Salehi Fathabadi[3],
Linas Laibinis[1], and Elena Troubitsyna[1]

[1] Åbo Akademi University, Turku, Finland
[2] Turku Centre for Computer Science, Turku, Finland
[3] University of Southampton, UK
{inna.pereverzeva,elena.troubitsyna,linas.laibinis}@abo.fi,
{mjb,asf08r}@ecs.soton.ac.uk

Abstract. MapReduce is a powerful distributed data processing model that is currently adopted in a wide range of domains to efficiently handle large volumes of data, i.e., cope with the big data surge. In this paper, we propose an approach to formal derivation of the MapReduce framework. Our approach relies on stepwise refinement in Event-B and, in particular, the *event refinement structure* approach – a diagrammatic notation facilitating formal development. Our approach allows us to derive the system architecture in a systematic and well-structured way. The main principle of MapReduce is to parallelise processing of data by first mapping them to multiple processing nodes and then merging the results. To facilitate this, we formally define interdependencies between the map and reduce stages of MapReduce. This formalisation allows us to propose an alternative architectural solution that weakens blocking between the stages and, as a result, achieves a higher degree of parallelisation of MapReduce computations.

Keywords: formal modelling, Event-B, refinement, event refinement structure, MapReduce.

1 Introduction

MapReduce is a widely used framework for handling large volumes of data [5]. It allows the users to automatically parallelise computations and execute them on large clusters of computers. Essentially, the computation is performed in two stages – map and reduce. The first stage maps the input data to multiple processing nodes, while the second stage performs parallel computations to merge the obtained results. Typically, execution of the map stage is blocking, i.e., execution of the reduce stage does not start until the map stage is completed. Though MapReduce is already a highly performant framework, to keep pace with the drastically increasing volume of data, it would be desirable to loosen the coupling between the stages and hence exploit the potential for parallelisation to the fullest.

In this paper, we undertake a formal study of the MapReduce framework. We formally model the control flow and data interdependencies between the map and

Y. Ait Ameur and K.-D. Schewe (Eds.): ABZ 2014, LNCS 8477, pp. 238–254, 2014.

reduce tasks, as well as derive the conditions under which the execution of the reduce stage can overlap with the execution of the map stage. Our formalisation of the (generic) MapReduce framework relies on the Event-B method and the associated Rodin platform. Event-B [1] is a formal approach that is particularly suitable for the development of distributed systems. The system development in Event-B starts from an abstract specification that is transformed into a detailed specification in a number of correctness-preserving refinement steps. In this paper, the Event Refinement Structure approach [3,6] is used to facilitate the refinement process. The technique provides us with an explicit graphical representation of the relationships between the events at different levels of abstraction and helps to gradually derive the complex MapReduce architecture.

Event-B relies on proof-based verification that is integrated into the development process. The Rodin platform [10] automates development in Event-B by generating the required proof obligations and automatically discharging a part of them. Via abstraction, proof and decomposition, Event-B enables reasoning about system-level properties of complex distributed systems. In particular, it allows us to explicitly define interdependencies between the processed data and derive the conditions under which an execution of the reduce stage can start before completion of the map stage. We believe that the proposed approach provides the designers with a formally grounded insight on the properties of MapReduce and enables fine-tuning of the framework to achieve a higher degree of parallelisation.

The rest of the paper is organised as follows. In Section 2 we describe the generic MapReduce framework and our formalisation of it. In Section 3 we give an overview of the Event-B formalism and the Event Refinement Structure (ERS) approach. In Section 4 we present our formal derivation of the MapReduce framework in Event-B using the ERS approach. As a result, we derive two alternative architectures of the MapReduce framework – blocking and partially blocking. In Section 5 we overview the related work and present some concluding remarks.

2   MapReduce

2.1   Overview of MapReduce

MapReduce is a programming model for processing large data sets. It has been originally proposed by Google [5]. The framework is designed to orchestrate the work on distributed nodes, run various computational tasks in parallel, providing at the same time for redundancy and fault tolerance. Distributed and parallelised computations are the key mechanisms that make the MapReduce framework very attractive to use in a wide range of application areas: data mining, bioinformatics, business intelligence, etc. Nowadays it is becoming increasingly popular in cloud computing. There exist different implementations of MapReduce, among them open-source Hadoop [2], Hive [11], and others.

The MapReduce computational model was inspired by the *map* and *reduce* functions widely used in functional programming. A MapReduce computation is composed of two main steps: the *map stage* and the *reduce stage*. During the map stage, the system inputs are divided into smaller computational tasks, which are then performed in parallel (provided there are enough processors in the cluster). The obtained collective results then become the inputs for the reduce stage, which combines them in some way to produce the overall output. Once again, the reduce inputs are split into smaller computational tasks that can be executed in parallel.

The MapReduce framework can be tuned to perform different data transformations by the user-supplied map and reduce functions. These functions encode basic mapping and reduction tasks to be performed in single nodes. The MapReduce framework then incorporates the provided functions and orchestrates the overall distributed computations based on them.

A typical example illustrating MapReduce computations is counting the word occurrences in a large set of documents. The input data set is split into smaller portions and the user-provided map function is applied to each such data block. The *map* function simply assigns to each word it encounters the value equal to 1. Overall, the map stage produces a collection of (word,1) pairs as intermediate results. Then, during the reduce stage, the user-supplied reduce function takes a portion of these intermediate data related to a particular word and sums all the occurrences of that word. Such a computation is done for each encountered word. The overall result is a set of (word,number) pairs.

2.2 Towards Formal Reasoning about MapReduce

In this section, we present a formalisation of the MapReduce framework. Specifically, we mathematically represent all MapReduce execution stages, i.e., the required data and control flow, and identify the computational (map and reduce) tasks that can be executed in parallel. Moreover, we formally define possible data interdependencies between the map and reduce tasks. The latter allows us to propose an alternative architectural solution, which weakens blocking between the MapReduce phases and, as a result, achieves a higher degree of parallelisation of MapReduce computations. In Section 4, we will propose two alternative formal developments of the MapReduce framework in Event-B, both of which rely on the formalisation presented below.

Let *IData* be an abstract type defining the input data to be processed within the MapReduce framework and *OData* be an abstract type defining the resulting output data. In a nutshell, a MapReduce computation processes the given input data and generates some result. Thus, it can be formally represented as a function:

$$MapReduce \in IData \rightarrow OData.$$

More specifically, a MapReduce computation can be defined as a functional composition of the following phases: *MSplit*, *Map*, *RSplit*, *Reduce*, and *Combine*:

$$MapReduce = MSplit; \ Map; \ RSplit; \ Reduce; \ Combine.$$

Let us note that the phases *MSplit* and *Map* together correspond to the *map stage* mentioned in Section 2.1, while the phases *RSplit* and *Reduce* belong to the *reduce stage*.

The MapReduce process starts with the *MSplit* phase. During this phase, the input data are split into a number of blocks (portions of the input data), which can be handled independently of each other. In the following *Map* phase, the user-provided *map* function is applied to each such input block. Next, in the *RSplit* phase, the MapReduce framework groups together all the intermediate results obtained after the *Map* phase to prepare for the reduce computations. Similarly to the *MSplit* phase, the data are divided into blocks that can be handled separately. After that, the *Reduce* phase is executed, during which the user-supplied *reduce* function is repeatedly applied (once per each block). Finally, in the *Combine* phase, all the obtained results are combined into the final output.

Formalisation of the MapReduce Execution Phases. Next we define all the MapReduce execution phases in more detail. In the *MSplit* phase, the input data are split into a number of blocks that are later supplied to the *map* function. To emphasise the independent nature of map computations, we associate the notion of a *map task* with such a portion of the input data to be processed separately.

Let *MTask* be a set of all possible map tasks and *MData* be an abstract type defining the data obtained after the splitting. Then the *MSplit* phase can be mathematically represented as follows:

$$MSplit \in IData \rightarrow (MTask \nrightarrow MData).$$

Essentially, *MSplit* produces a partitioning of the input data to be used in the *Map* phase among different map tasks. Note that the result of *MSplit* is a partial function since only a subset of *MTask* may be needed for particular input data.

We assume that the input data fully determines the number and the subset of involved map tasks.[1] To extract this information, we use the following functions

$$mtasks \in IData \rightarrow \mathbb{P}_1(MTask), \quad mnum \in IData \rightarrow \mathbb{N}_1$$

defined as

$$\forall idata \in IData \cdot mtasks(idata) = \mathsf{dom}(MSplit(idata)),$$
$$\forall idata \in IData \cdot mnum(idata) = \mathsf{card}(MSplit(idata)),$$

where dom and card are the function domain and set cardinality operators.

The *Map* phase involves transformation of all the data obtained by the *MSplit* phase into the intermediate form to be used in the later phases. Let *RData* be an abstract type defining the intermediate data obtained after the *Map* phase. Then *Map* phase can be mathematically represented as the following function:

$$Map \in (MTask \nrightarrow MData) \rightarrow \mathbb{P}_1(MTask \times RData).$$

Therefore, *Map* takes the map data partitioning produced by *MSplit* and returns the transformed data associated with the map tasks that produced them. These results then become the input data for the following reduce computations.

In our formalisation the *Map* results consist of a set of (*mtask, rdata*) pairs, without assuming any further structure among them. This is done intentionally,

---

[1] This applies only to the involved computational tasks. Actual software components that will be employed to carry out the necessary computations can be dynamically assigned and re-assigned for a specific map or reduce task.

since grouping and partitioning of these data will be performed in the *RSplit* phase.

All the involved map tasks should be performed within the *Map* phase. Formally, this requirement can be formulated as follows:

$$\forall f \in MTask \nrightarrow MData \cdot f \neq \varnothing \Rightarrow \mathsf{dom}(f) = \mathsf{dom}(Map(f)).$$

Next the results obtained by the *Map* phase are grouped together to prepare for reduce computations. Similarly to the *MSplit* phase, they should be first partitioned among the individual *reduce tasks*.

Let *RTask* be a set of all possible reduce tasks. Then the *RSplit* phase can be formally defined as the following function:

$$RSplit \in \mathbb{P}_1(MTask \times RData) \rightarrow (RTask \nrightarrow \mathbb{P}_1(RData)).$$

Essentially, the function takes the intermediate results produced by the *Map* phase and produces data partitioning among the involved reduce tasks.

We can reason about the actual number and the subset of the involved reduce tasks. Once again, this is determined by the original input data. Formally, we introduce the functions

$$rtasks \in IData \rightarrow \mathbb{P}_1(RTask), \quad rnum \in IData \rightarrow \mathbb{N}_1$$

defined as

$$\forall idata \in IData \cdot rtasks(idata) = \mathsf{dom}(RSplit(Map(MSplit(idata)))),$$
$$\forall idata \in IData \cdot rnum(idata) = \mathsf{card}(RSplit(Map(MSplit(idata)))).$$

The *RSplit* phase only rearranges the intermediate data, producing their partitioning among the reduce tasks. Therefore, neither new data should appear nor any of the existing data can disappear during this transformation. Mathematically, this can be formulated as the following property:

$$\forall f \in \mathbb{P}_1(MTask \times RData) \cdot \mathsf{ran}(f) = \left(\bigcup rt \in \mathsf{dom}(RSplit(f)) \mid RSplit(f)(rt)\right),$$

where ran is the function range operator.

The *Reduce* phase is similar to the *Map* phase – it takes as input a data partitioning produced by *RSplit* and returns transformed data:

$$Reduce \in (RTask \nrightarrow \mathbb{P}_1(RData)) \rightarrow \mathbb{P}_1(OData),$$

where *OData* is an abstract type defining the resulting output data.

Finally, the last *Combine* phase can be simply defined as follows:

$$Combine \in \mathbb{P}_1(OData) \rightarrow OData.$$

Formalisation of the map and reduce functions. The *Map* phase is based on repeated invocations of the user-supplied function map. The map function can be formally represented in the following way:

$$\mathsf{map} \in MData \rightarrow \mathbb{P}_1(RData).$$

Thus, it takes an input data from *MData* and produces some intermediate data to be used in reduce computations. The map function and the *Map* phase are tightly linked. To be precise, the union of all the results obtained from all the map function applications should be equal to the overall result of the *Map* phase:

$$Map = \{f \cdot f \in MTask \nrightarrow MData \mid f \mapsto (\bigcup mt \cdot mt \in dom(f) \mid \{mt\} \times map(f(mt)))\}.$$

The user-supplied **reduce** function can be specified as follows:

$$\texttt{reduce} \in \mathbb{P}_1(RData) \rightarrow \mathbb{P}_1(OData).$$

It takes as an input a subset of the reduce data $RData$ and produces some subset of output data from $OData$.

Finally, the overall result of the $Reduce$ phase should be equal to the combined results obtained by repeated application of the **reduce** function:

$$Reduce = \{f \cdot f \in RTask \nrightarrow \mathbb{P}_1(RData) \mid f \mapsto (\bigcup rt \cdot rt \in dom(f) \mid \texttt{reduce}(f(rt)))\}.$$

Essentially, the $Reduce$ definition is directly based on the user-supplied **reduce** function.

Formalisation of Interdependencies between the Map and Reduce Tasks. The main principle of MapReduce is that all the map and reduce computations are distributed to multiple independent processing nodes. The reduce inputs are based on the previously produced map outputs. However, in some cases, the reduce inputs might depend on only particular map outputs. Therefore, the reduce stage can be initiated before all the map computations are finished. To relax the limitation of the original MapReduce computation flow, requiring that the reduce stage starts only after completing the map stage, we formally define the *dependence relation* between the map and reduce tasks as the following function *dep*:

$$dep \in IData \rightarrow \mathbb{P}(RTask \times MTask),$$

with the following property:

$$\forall \, idata \in IData, \, rt \in RTask, \, mt \in MTask \cdot \; rt \mapsto mt \in dep(idata) \quad \Leftrightarrow$$
$$mt \in \mathsf{dom}(MSplit(idata)) \wedge$$
$$(\exists rd \in RData \cdot rt \in \mathsf{dom}(RSplit(Map(MSplit(idata))))) \wedge$$
$$rd \in RSplit(Map(MSplit(idata)))(rt) \wedge mt \mapsto rd \in Map(MSplit(idata))).$$

The property states that for any input data *input*, a map task mt and a reduce task rt are in *dependence relation* (i.e., a reduce task depends on a map task), if and only if some intermediate data rd has been generated for this reduce task rt by the computations of the map task mt during the Map phase. Essentially, the relation *dep* defines the data interdependencies between the map and reduce stages. This formalisation allows us to propose (in Section 4) an alternative architectural solution that weakens blocking between the stages.

Finally, to make it possible for a particular reduce task to start immediately after all the necessary data have been produced by the map tasks related by *dep*, we need a version of *RSplit*, defining a partial split related with a specific reduce task. For a given reduce task, it produces the grouped together results obtained within the Map phase:

$$\texttt{rsplit} \in RTask \nrightarrow (\mathbb{P}_1(MTask \times RData) \rightarrow \mathbb{P}_1(RData)).$$

Again, the union of the results obtained from all the `rsplit` function applications should be the result of the $RSplit$ phase:

$$\forall f \cdot f \in \mathbb{P}_1(MTask \times RData) \Rightarrow$$
$$RSplit(f) = (\bigcup rt \cdot rt \in dom(rsplit) | \{rt \mapsto \mathtt{rsplit}(rt)(f)\}).$$

In Section 4 we will demonstrate that, by relying on the proposed formalisation, we can derive a formal model of the MapReduce framework. There we will propose two models of MapReduce – *blocking* and *partially blocking* models.

3   Formal Development by Refinement: Background

3.1   Event-B

Event-B is a state-based formal approach that promotes the correct-by-construction development paradigm and formal verification by theorem proving [1]. In Event-B, a system model is specified using the notion of an *abstract state machine*. An abstract state machine encapsulates the model state, represented as a collection of variables, and defines operations on the state, i.e., it describes the dynamic behaviour of a modelled system. The variables are strongly typed by the constraining predicates that, together with other important system properties, are defined as model *invariants*. Usually, a machine has an accompanying component, called a *context*, which includes user-defined sets, constants and their properties given as a list of model axioms.

The dynamic behaviour of the system is defined by a collection of atomic *events*. Generally, an event has the following form:

$$e \mathrel{\widehat{=}} \mathbf{any}\ a\ \mathbf{where}\ G_e\ \mathbf{then}\ R_e\ \mathbf{end},$$

where e is the event's name, a is the list of local variables, and (the event *guard*) G_e is a predicate over the model state. The body of an event is defined by a *multiple* (possibly nondeterministic) assignment to the system variables. In Event-B, this assignment is semantically defined as the next-state relation R_e. The event guard defines the conditions under which the event is *enabled*, i.e., its body can be executed. If several events are enabled at the same time, any of them can be chosen for execution nondeterministically.

Event-B employs a top-down refinement-based approach to system development. A development starts from an abstract specification that nondeterministically models the most essential functional requirements. In a sequence of refinement steps, we gradually reduce nondeterminism and introduce detailed design decisions. The consistency of Event-B models, i.e., verification of model well-formedness, invariant preservation as well as correctness of refinement steps, is demonstrated by discharging the relevant proof obligations. The Rodin platform [10] provides an automated support for modelling and verification. In particular, it automatically generates the required proof obligations and attempts to discharge them.

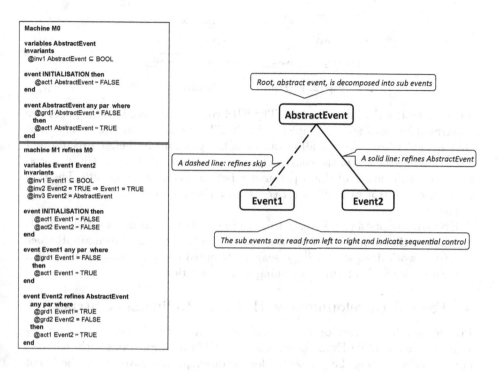

Fig. 1. Event Refinement Structure (ERS) Diagram

3.2 Event Refinement Structure

The Event Refinement Structure (ERS) [3,6] approach augments Event-B refinement with a graphical notation that allows us to explicitly represent the relationships between the events at different abstraction levels as well as define the required event sequence in a model. ERS is illustrated by example in Figure 1. The diagram explicitly shows that *AbstractEvent* is refined by *Event2*, while *Event1* is a new event that refines *skip*. Moreover, the diagram shows that the effect achieved by *AbstractEvent* in the abstract machine is realised in the refining machine by the occurrence of *Event1* followed by *Event2*.

In ERS, the sequential execution of the leaf events is depicted from left to right. The event sequencing is managed by additional control variables introduced into the underlying Event-B model. For instance, for each leaf event (node) represented in Fig. 1, there is one boolean control variable with the same name as the event. When the event Event1 occurs, the corresponding control variable is set to *TRUE*. The following event, *Event2*, can occur only after *Event1*. This is achieved by checking the value of the *Event1* control variable in the guard of *Event2*.

Boolean variables only allow controlling single execution of events. When multiple executions of an event are needed, the event is parameterised and set control variables are used instead of boolean ones. This allows the event to occur many times with different values of its parameter. A parameter can be introduced in

Fig. 2. ERS *all* / *some* Constructors

an event by the ERS constructors. The ERS constructors used in this paper are illustrated by two simple examples in Fig. 2. The use of *all* constructor indicates that *Event1* is executed for all instances of the *p* parameter before execution of *Event2*, while the use of the constructor *some* indicates that *Event1* is executed for some of instances of the *p* parameter before execution of *Event2*. The corresponding control variables for *Event1* and *Event2* are defined as sets in the model.

Event-B adopts an event-based modelling style that facilitates the correct-by-construction development of complex distributed systems. Since MapReduce is a framework designed for large-scale distributed computations, Event-B is a natural choice for its formal modelling and verification.

4   Formal Development with Event Refinement Structure

In this section, we rely on our formalisation presented in Section 2.2 to develop two alternative Event-B models of the MapReduce framework: *blocking* and *partially blocking*. The presented formal developments make use of the Event Refinement Structure (ERS) approach, presented in Section 3.2. Our development strategy is based on gradually unfolding all the MapReduce computational phases by refinement. Such small model transformation steps allow us to efficiently handle the complexity of the MapReduce framework.

Let us note that our development of the MapReduce framework is generic. It relies on the use of abstract functions to represent essential data transformations of MapReduce. These abstract functions can be treated as generic system parameters that can be later instantiated with their concrete instances for the specific MapReduce implementations.

4.1   Blocking Model of MapReduce

The mathematical data structures and their properties from our MapReduce formalisation constitute the basis for defining the Event-B *context* component that is used throughout the whole formal development. Essentially, the whole presented formalisation is incorporated as the context, e.g.

> **axm8:** $MSplit \in IData \rightarrow (MTask \nrightarrow MData)$
>
> **axm9:** $Map \in (MTask \nrightarrow MData) \rightarrow \mathbb{P}_1(MTask \times RData)$
>
> **axm10:** $RSplit \in \mathbb{P}_1(MTask \times RData) \rightarrow (RTask \nrightarrow \mathbb{P}_1(RData)), \dots$

We will constantly rely on these definitions to ensure the correctness of the overall data transformation process within our MapReduce models. Since the formalised definitions are still abstract (generic), our presented development essentially formally describes a family of possible MapReduce implementations.

Due to space limit, we do not present the complete development but rather give its graphical representation using the ERS graphical notation. The full Event-B models of this development can be found in [8].

Abstract model of MapReduce. We start with an abstract model in which the whole MapReduce computation is done in one atomic step. This behaviour is modelled by the event OutputMapReduce:

```
OutputMapReduce ≙
  any    t1, t2, t3, t4
  when t1 = MSplit(idata) ∧ t2 = Map(t1) ∧ t3 = RSplit(t2) ∧ t4 = Reduce(t3)
  then  output := Combine(t4) ∥ done := TRUE   end
```

With help of the ERS approach, we decompose the atomicity of OutputMapReduce into smaller steps. Verification of the refinement proof obligations ensures that the decomposition preserves correctness. Specifically, in the next several consecutive refinement steps, we break the atomicity of the OutputMapReduce event by introducing explicit events for the following MapReduce phases: *MSplit*, *Map*, *RSplit*, and *Reduce*. Fig. 3 presents the ERS diagram of the model.

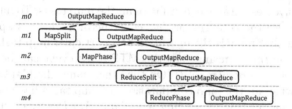

Fig. 3. Blocking model: ERS diagram (for OutputMapReduce)

The new model events MapSplit, MapPhase, ReduceSplit and ReducePhase specify the sequential execution of the MapReduce phases. The sequence between the events is enforced by following the rules given in Section 3.2. It is also specified by the invariant properties on the control variables:

$$OutputMapReduce = TRUE \Rightarrow ReducePhase = TRUE,$$
$$ReducePhase = TRUE \Rightarrow ReduceSplit = TRUE,$$
$$ReduceSplit = TRUE \Rightarrow MapPhase = TRUE,$$
$$MapPhase = TRUE \Rightarrow MapSplit = TRUE.$$

Moreover, to store the intermediate results of separate phases, we introduce a number of variables (*msplit*, *map_result*, *rsplit* and *reduce_result*) that are updated during execution of the corresponding events. The variable updates are also performed according to the formalisation given in Section 2.2. For instance, the variable *msplit* is introduced to store the result of the *MSplit* phase. After the execution of the MapSplit event, *msplit* gets the value equal to *MSplit(idata)*.

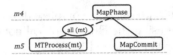

Fig. 4. Blocking model: ERS diagram (for MapPhase)

Machine MapReduce1_m1 **refines** MapReduce1_m0
Variables $idata, output, msplit, MapSplit, ...$
Invariants $OutputMapReduce = TRUE \Rightarrow MapSplit = TRUE \wedge$
$\qquad\qquad MapSplit = TRUE \Rightarrow msplit = MSplit(idata) \wedge ...$
MapSplit $\;\widehat{=}$
  when $MapSplit = FALSE$
  then $MapSplit := TRUE$
      $msplit := MSplit(idata)$
  end
OutputMapReduce **refines** OutputMapReduce $\;\widehat{=}$
  any $\;\; t2, t3, t4$
  where $MapSplit = TRUE \wedge OutputMapReduce = FALSE \wedge$
      $t2 = Map(msplit) \wedge t3 = RSplit(t2) \wedge t4 = Reduce(t3)$
  with $\;t1 = msplit$
  then $\;output := Combine(t4)$
      $OutputMapReduce := TRUE$
  end

Breaking Atomicity of the Map Phase. In the second refinement step, we introduce the event MapPhase that abstractly models the *Map* phase. Essentially, the *Map* phase involves parallel execution of all the map tasks. To introduce such a behaviour, we use the constructor "*all* constructor", which is applied to the MTProcess event that models the execution of a particular map task (see Fig.4). The expression "*all(mt)*" means that the MTProcess event can be enabled for multiple values of $mt \in$ dom($msplit$). On the other hand, the MapCommit event can only occur when all the map computations of map tasks have been finished. In Event-B, we model this by adding a variable $MTProcess$, which is a set containing all possible map tasks that should be processed. The order between the events is ensured by the invariants on the control variables, e.g.,

$$\mathbf{inv4:} \;\; MapCommit = TRUE \Rightarrow MTProcess = \text{dom}(msplit),$$

where dom($msplit$) defines the set of all current map tasks. The invariant states that if the MapCommit event has been executed, then all the map tasks have been completed before it. While specifying the MTProcess event, we rely on the definition of the map function, given in Section 2.2.

Machine MapReduce1_m5 **refines** MapReduce1_m4
MTProcess $\;\widehat{=}$
  any $\;\; mt$
  where $MapSplit = TRUE \wedge mt \in dom(msplit) \wedge mt \notin MTProcess$
  then $MTProcess := MTProcess \cup \{mt\}$
      $MTProcess\_result(mt) := map(msplit(mt))$
  end
MapCommit **refines** MapPhase $\;\widehat{=}$
  when $MapSplit = TRUE \wedge MapCommit = FALSE \wedge MTProcess = dom(msplit)$
  then $MapCommit = TRUE$
      $map\_result := (\bigcup mt \cdot mt \in dom(msplit)|\{mt\} \times MTProcess\_result(mt))$
  end

Further Refinements of the Map Phase. During the MapReduce execution, all the map and reduce tasks are parallelised and distributed to multiple

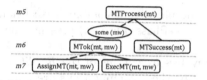

Fig. 5. Blocking model: ERS diagram (for MTProcess)

processing nodes – the actual software components that carry out the computations. We name these components as *map* and *reduce workers*. Moreover, there is a special component – *master* – that controls all the computations and assigns the map and reduce tasks to the workers. The master periodically pings every worker. In case of a worker failure, the master re-assigns tasks from the failed worker to a healthy one. This procedure can be repeated until the master gets the result for a particular map or reduce task from some worker. To introduce such functionality, we carry out several further refinements focusing on the *Map* phase. These refinements elaborate on modelling of map task execution.

Fig.5 illustrates the event MTProcess and its several consecutive levels of atomicity decomposition. First, the abstract event MTProcess is broken into two concrete events, MTok and MTSuccess correspondingly. The MTok event models the execution of the map task *mt* by a particular map worker *mw*. The result of this computation should be approved by the master side, which is modelled by execution of the MTSuccess event. The "*some*" constructor indicates that the event MTok may be executed only for some instances of the *mw* parameter before the MTSuccess event becomes enabled. The *MTSuccess* and *MTok* control variables are defined as sets, which allows for multiple executions of the MTSuccess and MTok events. Later on, in the next refinement step, the atomicity of the MTok event is broken into two events AssignMT and ExecMT. The event AssignMT models an assignment of a map task *mt* to a particular map worker *mw*, while ExecMT models the successful execution of the task by this worker.

Similarly to the *Map* phase, we refine the *Reduce* phase by gradually unfolding its computations. The overall refinement structure is presented on Fig.6.

Let us note that the proposed architecture is *blocking* in the sense that the reduce computations can be only started after all the map computations have been finished. The formal derivation of the blocking model and its dynamics is performed under this condition. Next we propose an alternative architectural solution of the MapReduce framework that weakens blocking between the map and reduce stages and, as a result, achieves a higher degree of parallelisation of the MapReduce computations. For this purpose, we will make use of the *dependence relation* between map and reduce tasks introduced in the Section 2.2. We call this model *partially blocking model*.

4.2   Partially Blocking Model of MapReduce

We start from the same initial specification as for the blocking model, in which the whole MapReduce computation is done in one atomic step, and then refine it in order to introduce the *MSplit* phase. Next, in contrast to the previous derivation, we separate the phase that combines executions of the *RSplit* and *Reduce* phases – RSplitReducePhase. Fig.7 (a) presents the ERS diagram of the refined model.

RSplitReducePhase involves executions of the *RSplit* and *Reduce* phases for all reduce tasks. Essentially, these computations are parallelised. To introduce such behaviour, we use the "*all*" constructor applied to the RTSplitReduceProcess event that, for a particular reduce task *rt*, performs split and then reduce computations (see. Fig.7 (b)). Next, we separate these split and reduce executions of the particular reduce task *rt*. Namely, the event RTSplitReduceProcess is split into two concrete events, RTSplitProcess and RTProcess. Here we again rely on the *rsplit* and *reduce* functions formalised in Section 2.2.

Up to now we did not introduce the Map phase explicitly. However, the results of *MapPhase* are simulated internally, by storing the intermediate results in the local variables of the RTSplitProcess event. To explicitly model the Map phase, the event RTSplitProcess is now split into two events MTProcess and RSplit (see Fig.8). The constructor "all" is parameterised by $(mt \in dep[\{rt\}])$". It means that the event MTProcess is executed for all those map tasks, *mt*, that are in data dependency with the reduce task *rt*. Therefore, to start the *RSplit* phase, we do not need to wait until all the map tasks are completed. Here we are relying on the definition of data interdependency *dep* between the map and reduce stages, formalised in Section 2.2. Finally, the MTProcess and RTProcess events are refined in the same manner as in the blocking model presented in the Section 4.1.

Let us note that the proposed partially blocking model allows us to achieve a higher degree of parallelisation of MapReduce computations. Indeed, for a particular reduce task, when the dependent map tasks have already been executed,

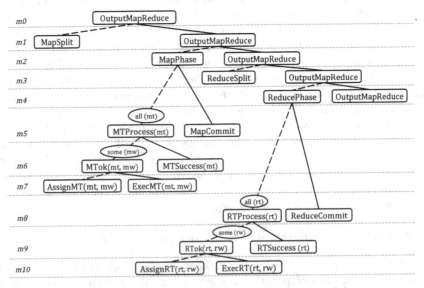

Fig. 6. MapReduce ERS Diagram: blocking model

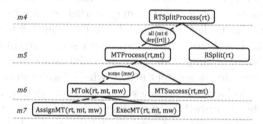

Fig. 7. Partially blocking model: ERS diagrams

the *RSplit* phase for this reduce task can be performed, and then reduce computations can be started. In other words, the computations from three different phases – *Map*, *RSplit*, and *Reduce* – can be performed in parallel, provided the involved data are independent. Therefore, the proposed architectural solution weakens blocking between the stages and, as a result, achieves a higher degree of parallelisation. The overall refinement structure of the partially blocking model is presented on Fig.9.

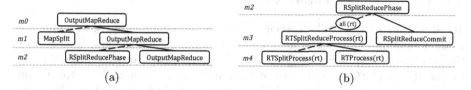

Fig. 8. Partially blocking model: ERS diagram (for RTSplitProcess)

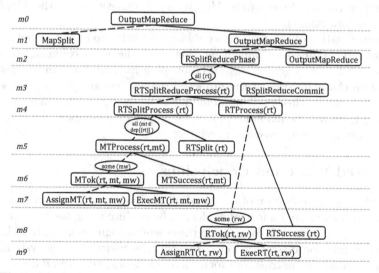

Fig. 9. MapReduce Event Refinement Structure: partially blocking model

4.3 Discussion and Future Work

To verify correctness of the presented models, we have discharged around 270 proof obligations for the first formal development, as well as more than 300 for the second one. Approximately 93% of them have been proved automatically by the Rodin platform and the rest have been proved manually in the Rodin interactive proving environment. With help of the ERS approach, we have decomposed the atomicity of the MapReduce framework and hereby achieved a higher degree of automation in proving. Moreover, the ERS diagrammatic notation has provided us with additional support to represent the model control flow at different abstraction levels and also simplified reasoning about possible refinement strategies. The whole development and proving effort has taken about one person-month.

As a result of the presented refinement chains, we have arrived at two different centralised Event-B models of the distributed MapReduce framework. As a part of the future work, we are planing to derive distributed models by employing the existing decomposition mechanisms of Event-B. This would result in creating separate formal specifications of the involved software components of the MapReduce framework (such as master, map worker, reduce worker, etc.).

The static part of the modelled system is formally defined in the corresponding context component. The definitions of static data structures in the context are mostly very abstract, i.e. they state only essential properties to be satisfied. This makes them generic parameters of the whole formal development. In its turn, such formal development becomes generic, representing a family of the systems that can be described by providing suitable concrete values for the generic parameters. The proposed formal model can be used then as a starting point for future development of a specific MapReduce application. The actual concrete values can be supplied by either the end-user (e.g., the `map` and `reduce` functions) or the developer of the MapReduce framework (e.g., the *MSplit* or *RSplit* transformations).

As a continuation of this work, it would be interesting to create formal models for a concrete MapReduce implementation, e.g., the word counting example, by using the Event-B generic instantiation plug-in. Moreover, to analyse the quantitative characteristics of the proposed models, we are planing to use the Uppaal-SMC model checker. This would allow us to, e.g., assign different data processing rates for the map and reduce tasks and then compare the execution time estimations of two considered architectures.

5 Related Work and Conclusions

The problem of formalisation of the MapReduce framework has been studied in [12]. The authors present a formal model of MapReduce using the CSP method. In their work, they focus on formalising the essential components of the MapReduce framework: the master, mapper, reducer, the underlying file system, and their interactions. In contrast, our focus is on modelling the overall flow of control as well as the data interdependencies between the MapReduce computational phases. Moreover, our approach is based on the stepwise refinement technique that allowed us to gradually unfold the complexity of the MapReduce framework.

Formalisation of MapReduce in Haskel is presented in [9]. Similarly to our approach, it focuses on the program skeleton that underlies MapReduce computations and considers the opportunities for parallelism in executing MapReduce computations. However, in addition to that, we also reason about the involved software components – the master, map and reduce workers – that are associated with the respective map and reduce tasks.

The work [7] presents two approaches based on Coq and JML to formally verify the actual running code of the selected Hadoop MapReduce application. In our work we are more interested in formalisation of MapReduce computations and gradual building of different MapReduce models that are correct-by-construction. The performance issues of MapReduce computations have been studied in the paper [4], focusing on one particular implementation of the MapReduce – Hadoop. In contrast, we have tried to formally investigate the data interdependencies between the MapReduce phases and their effect on the degree of parallelisation, independently of a concrete MapReduce implementation.

In this paper we have proposed an approach to formalising the MapReduce framework. Our main technical contribution of this paper is two-fold. On the one hand, based on our definition of interdependencies between the processed data as well as the map and reduce stages, we have derived the conditions under which blocking between the stages can be relaxed. Therefore, we have rigorously derived constraints for implementing MapReduce with a higher degree of parallelisation. On the other hand, we have demonstrated how to use the Event Refinement Structure (ERS) technique to formally derive and verify a model of a complex system with a massively parallel architecture and complex dynamic behaviour.

The stepwise refinement approach to deriving a complex system model has demonstrated good scalability and allowed us to express system properties at different levels of abstraction and with a different degree of granularity. Moreover, combining the refinement technique with tool-assisted mathematical proofs have provided us with a scalable approach to verification of a complex system model.

Acknowledgements. The authors would like to thank the reviewers for their valuable comments. Pereverzeva's work is partly supported by the STV Grant. Butler and Salehis work is partly funded by the FP7 ADVANCE Project (http://www.advance-ict.eu).

References

1. Abrial, J.R.: Modeling in Event-B. Cambridge University Press (2010)
2. Borthakur, D.: The Hadoop Distributed File System: Architecture and Design. The Apache Software Foundation (2007)
3. Butler, M.: Decomposition Structures for Event-B. In: Leuschel, M., Wehrheim, H. (eds.) IFM 2009. LNCS, vol. 5423, pp. 20–38. Springer, Heidelberg (2009)
4. Condie, T., Conway, N., Alvaro, P., Hellerstein, J.M.: MapReduce Online. In: NSDI 2010, p. 20. USENIX Association (2010)

5. Dean, J., Ghemawat, S.: MapReduce: Simplified Data Processing on Large Clusters. In: Proceedings of the 6th Conference on Symposium on Opearting Systems Design & Implementation, pp. 137–150. USENIX Association (2004)

6. Fathabadi, A.S., Butler, M., Rezazadeh, A.: A Systematic Approach to Atomicity Decomposition in Event-B. In: Eleftherakis, G., Hinchey, M., Holcombe, M. (eds.) SEFM 2012. LNCS, vol. 7504, pp. 78–93. Springer, Heidelberg (2012)

7. Ono, K., Hirai, Y., Tanabe, Y., Noda, N., Hagiya, M.: Using Coq in Specification and Program Extraction of Hadoop MapReduce Applications. In: Barthe, G., Pardo, A., Schneider, G. (eds.) SEFM 2011. LNCS, vol. 7041, pp. 350–365. Springer, Heidelberg (2011)

8. Pereverzeva, I., Butler, M., Fathabadi, A.S., Laibinis, L., Troubitsyna, E.: Formal Derivation of Distributed MapReduce. Tech. Rep. 1099, TUCS (2014)

9. Lämmel, R.: Google's MapReduce programming model 70, 1–30 (2008)

10. Rodin: Event-B Platform, http://www.event-b.org/

11. Thusoo, A., Sarma, J.S., Jain, N., Shao, Z., Chakka, P., Anthony, S., Liu, H., Wyckoff, P., Murthy, R.: Hive - A Warehousing Solution Over a Map-Reduce Framework. Proc. VLDB Endowment 2, 1626–1629 (2009)

12. Yang, F., Su, W., Zhu, H., Li, Q.: Formalizing MapReduce with CSP. In: 17th IEEE International Conference and Workshops on the Engineering of Computer-Based Systems, pp. 358–367. IEEE Computer Society (2010)

Validating the RBAC ANSI 2012 Standard Using B

Nghi Huynh[1,2], Marc Frappier[1], Amel Mammar[3],
Régine Laleau[2], and Jules Desharnais[4]

[1] Université de Sherbrooke, Québec, Canada
[2] Université Paris-Est Créteil Val de Marne, France
[3] Institut Mines-Télécom/Télécom SudParis, France
[4] Université Laval, Québec, Canada

Abstract. We validate the RBAC ANSI 2012 standard using the B method. Numerous problems are identified: logical errors, inconsistencies, ambiguities, typing errors, missing preconditions, invariant violation, inappropriate specification notation. A clean version of the standard written in the B notation is proposed. We argue that the *ad hoc* mathematical notation used in the standard is inappropriate and we propose that a more methodological and tool-supported approach must definitely be used for writing standards, in order to avoid the issues identified in the paper. Human reviewing is insufficient to produce error-free international standards.

Keywords: Role-Based Access Control, B method, invariant preservation.

1 Introduction

RBAC is one of most cited access-control models in the scientific literature (27 300 reference in Google Scholar, 1 326 reference in ACM digital library), and one of the most widely used models in industry [8]. It is an ANSI standard developed by IN-CITS (International Committee for Information Technology Standards) [1,2,11], with a first edition produced in 2004 and a recent revision published in 2012. It is recommended by numerous governmental agencies, like Canada's Health Infoway, for controlling access to sensitive information like electronic health records. In a recent project on access control and consent management, we decided to follow these recommendations and evaluate the adequacy of RBAC for managing access to EHR. We were surprised by the number of errors and inconsistencies found in the standard. Even more surprising, all errors can be found in both editions (2004 and 2012), and the 2012 edition has been reviewed/voted by more than 141 persons (as listed in the standard).

The standard is written using mathematical definitions in the style of Z, but without strictly following the Z syntax. The mathematical definitions have not been syntax-checked nor type-checked, thus several errors could have been easily avoided. Some mathematical notations are not drawn from Z and seem rather *ad hoc*, as they are not easily found in standard mathematical textbooks, leaving the

Y. Ait Ameur and K.-D. Schewe (Eds.): ABZ 2014, LNCS 8477, pp. 255–270, 2014.

reader to guess their meaning from the context. More importantly, not sticking to the Z syntax also leads to several ambiguities, since the mathematical text interpreted with the Z semantics does not always match the natural language description. In order to make sense of the mathematical definitions, the reader must assume declarations which have been omitted in the Z schemas, relying on the natural language text to make such inferences. This is contrary to good specification practice, where the mathematical text is the definitive description, since it offers more precision than natural language The standard leaves out important concepts, which certainly do not help in reaching the objective stated in the introduction of the standard:

> *Development [of] this standard was initiated [...] in recognition of a need among government and industry purchasers of information technology products for a consistent and uniform definition of role-based access control (RBAC) features. [...] This lack of a widely accepted model resulted in uncertainty and confusion about RBAC's utility and meaning. This standard seeks to resolve this situation [...].*

The idea of using mathematics to write the standard was certainly a good idea, as it significantly helped in describing abstract concepts, and allowed us to identify inconsistencies, ambiguities and missing elements. Finding errors in a natural language text is definitely more difficult, because two many interpretations are possible, and each reader picks one, according to his personal experience, knowledge and context. For comparison, we have also evaluated the XACML standard [10] where mathematics are not used at all. We found that it is far more difficult to grasp the subtle concepts of XACML and to be reasonably sure that we could comply to it. Thus, using mathematics is a great idea, but it is insufficient to achieve the highest level of confidence in the quality of a standard. In this paper, we hope to show that the use of a formal method, which has a formal syntax and a formal semantics, supported by tools like syntax checkers, type checkers, provers, model checkers and animators, can definitely help in producing a precise and unambiguous description of a standard. We have chosen to use the B method for its rich tool set. In addition, we believe that B has helped us in detecting errors that may not be easy to find with Z, mainly because B requires proving invariant preservation, whereas in Z, invariants are typically included in the state definition and in the definition of operations through the $\Delta State$ decoration, as it was implicitly done in the RBAC standard. Proving invariant preservation helps in finding missing preconditions in operations and in reviewing the behaviour of operations when proof obligations fails.

Li *et al* published a critique of the 2004 standard in [7]. They identified several technical problems and suggested improvements to the standard, which they formulated using plain mathematics [6]. The leading authors of the standard responded to this critique in [4], without really agreeing on any of the critique of [7] (even the typos and typing errors identified by Li *et al* are still present in the 2012 version of the standard). The improvements suggested by Li *et al* in [6] do not simply correct the logical flaws, but also propose a different view of RBAC, where, among other things, the notion of session is not included in the core part of RBAC, and permissions are inherited when a role hierarchy is used. Noticing issues with the format of the specification of [6], Power *et al* [9] provided a formal Z specification of its state pace, leaving out the specification of administrative functions described in [2]. They also suggested normalization functions for permission assignment, and formalized the three interpretations of role hierarchy suggested in [6].

In this paper, our objective is to show that formal methods can significantly help avoiding errors in the specification of RBAC. We take the RBAC standard as it is described in [2], and fix all errors that we have found, to the best of our understanding of the natural language description and the accompanying mathematical text found in [2]. We do not suggest any new behaviour or feature, contrary to [6,9]. Another goal is to stress that formal methods should be used in a comprehensive manner when writing a standard. This includes specifying both the state invariant and the administrative functions; specifying only the state invariant is insufficient. Proving that administrative functions preserve the invariant provides a greater level of confidence in the standard. We have identified errors that neither [6] nor [9] identified. Using a specification animator is also crucial to validate a specification. It allows to uncover inappropriate behaviour which are not detected by invariant preservation proofs.

The paper is structured as follows. Section 2 provides relevant excerpts of the RBAC standard [2] on data structures. Section 3 describes the specification of administrative functions to update the value of RBAC data structures. Errors and omissions are identified and discussed in these two sections. We briefly describe the structure of our B specification in Section 4. The complete specification is omitted and provided in [3], due to space constraints. We conclude this paper with an appraisal of our work in Section 5.

2 Data Structures of the ANSI RBAC Standard

The RBAC standard [2] is decomposed in three components.

1. Core RBAC is the main component and is required in any RBAC system.
2. Hierarchical RBAC introduces a role hierarchy which defines role inheritance.
3. Constrained RBAC introduces separation of duties (SOD) constraints.

A compliant RBAC system is made of the Core RBAC component plus any combination of the other two.

2.1 Core RBAC

The main idea of RBAC is that permissions are assigned to roles, and users are granted these permissions by being assigned to roles. The Core RBAC component includes the following sets: *USERS, ROLES, OPS, OBS* and *SESSIONS*, which respectively stand for the set of users, the set of roles, the set of operations, the set of objects on which are applied the operations, and the set of sessions, where a user can activate a role.

The following definitions are reproduced verbatim from [2]. As a convention, all verbatim excerpts from [2] are framed, while problems are underlined within the excerpts, and numbered in superscript. Problems are explained in the text following the excerpts, numbered with Pi.

Core RBAC Reference Model

- $USERS, ROLES, OPS$ and $\underline{OBS}$ [1] (users, roles, operations and objects respectively).
- $UA \subseteq USERS \times ROLES$, a many-to-many mapping user-to-role assignment relation.
- $assigned\_users : (r : ROLES) \rightarrow 2^{USERS})$ the mapping of role r onto a set of users.
  Formally: $assigned\_users(r) = \{u \in USERS | (u,r) \in UA\}$
- $PRMS = 2^{(OPS \times OBS)}$ [2], the set of permissions.
- $PA \subseteq \underline{PERMS}$ [3] $\times ROLES$ a many-to-many mapping permission-to-role assignment relation.
- $assigned\_permissions(r : ROLES) \rightarrow 2^{PRMS}$, the mapping of role r onto a set of permissions. Formally:
  $assigned\_permissions(r) = \{p \in PRMS | (p,r) \in PA\}$
- $Op(p : PRMS) \rightarrow \{op \subseteq OPS\}$ [4], the permission to operation mapping, which give the set of operations associated with permission p.
- $Ob(p : PRMS) \rightarrow \{ob \subseteq OBS\}$ [5], the permission to objects mapping, which give the set of objects associated with permission p.
- $SESSIONS =$ the set of sessions.
- $\underline{session\_users}$ [6] $(s : SESSIONS) \rightarrow USERS$, the mapping of session s onto the corresponding user.
- $session\_roles(SESSIONS) \rightarrow 2^{ROLES}$, the mapping of session s onto a set of roles.
  Formally: $session\_roles(s_i) \subseteq \{r \in ROLES | (session\_users(s_i), r) \in UA\}$
- $avail\_session\_perm$ [7] $(s) \rightarrow 2^{PRMS}$, the permissions available to a user in a session $= \bigcup\limits_{r \in session\_roles(s)} assigned\_permissions(r)$

Description of Problems

P1 Typo: all functions of Section 7 (Functional Specification Overview) of [2] use set $OBJS$ instead of OBS. OBS is declared here and used everywhere in this section, but not in the rest of the standard.

P2 Type error: functions of Section 7 (Functional Specification Overview) of [2] use this set as if it was defined as $OPS \times OBS$. Note that this set is never updated in any administrative functions of Section 7. This leads us to conclude that *PRMS* is a type, and that all operations on objects are possible, that is, the standard does not provide means for controlling which operations are valid for which objects. On the other hand, functions *Op* and *Ob*, declared afterwards, but undefined, hint at the usage of a subset of operations on objects; otherwise, they would be useless. But these functions are not used in the rest of the standard.

P3 Typo: this symbol is not declared so far. One could presume that it is a typo for *PRMS* defined above, and used everywhere in the rest of the data structure declarations, but *PRMS* is not used in Section 7; *PERMS* is used instead.

P4 Noise: this function is not used in the rest of the specification. By its description, it is a derived function, but its definition is not provided; only its type.

P5 Noise: same issues as for *Op*. This function is undefined and not used in the rest of the specification.

P6 Noise: this function is not used in the rest of the standard. The function *user_sessions*, which maps users to sessions is used instead (and undeclared anywhere).

P7 Noise: this function is not used in the rest of the standard. Administrative function **CheckAccess** provides the same information.

Appraisal of the Definitions

None of these definitions clearly emphasizes under what conditions a user can use an operation on an object. This is quite surprising, because this is the core purpose of the standard. The definition of *avail_session_perm* describes the permissions *available* in a session, but it does not explicitly state that it determines if a user can execute an operation on an object. The reader has to wait until Section 7, page 17, where function **CheckAccess**, buried among other administrative functions, nails it down in a decisive manner:

> This function returns a Boolean value meaning whether the subject of a given session is allowed or not to perform a given operation on a given object.

The standard introduces a number of symbols (sets, relations, functions), but does not state whether they are state variables, specification parameters, or sets used only for typing. For instance, no distinction is made on the nature of sets *USERS*, *ROLES*, *OPS* and *OBS*. The first two are state variables (since they are updated by some administrative functions of Section 7); the last two are never updated and can be considered as parameters of the specification used for typing only. These distinctions would be made if a formal specification like B, Z or ASM was used. The use of derived functions like *assigned_users*, *assigned_permissions* and *avail_session_perms* can create some confusion ands inconsistencies when writing administrative functions. For instance, *UA* and *assigned_users(r)* are both updated and kept consistent in administrative functions updating them. Similarly for *PA* and *assigned_users*. On the other hand, function *avail_session_perm* is never maintained in the administrative functions. In the B method, these derived functions would be not be included as state variables, since they do not contain any new information. Their inclusion would only complicate the invariant preservation proof and the specification of operations They would be included as DEFINITIONS, which are similar to LET constructs in programming languages. Li *et al* [7] also suggested not to use derived functions.

Finally, ad hoc mathematical notations are used (*e.g.*, declaration of function *Op*), while in Section 7, the Z notation is said to be used for specifying operations. For the sake of uniformity, the Z notation could have also been used to define functions.

2.2   Hierarchical RBAC

This component introduces role hierarchies which define an inheritance relation among roles.

> This relation has been described in terms of permissions: r_1 "inherits" role r_2 if[8] all privileges of r_2 are also privileges of r_1. [...]
> This standard recognizes two types of role hierarchies—general role hierarchies and limited role hierarchies. General role hierarchies provide support for an arbitrary partial order to serve as the role hierarchy, to include the concept of multiple inheritances of permissions and user membership among roles. Limited role hierarchies impose restrictions resulting in a simpler tree structure (i.e., a role may have one or more immediate ascendants, but is restricted to a single immediate descendent).

General Role Hierarchy Specification

> - $RH \subseteq ROLES \times ROLES$ is a partial order on $ROLES$ called the inheritance relation written as $\succeq$, where $r_1 \succeq r_2$ only if[9] all permissions of r_2 are also permissions of r_1, and all users of r_1 are also users of r_2, ie. , $r_1 \succeq r_2 \Rightarrow authorized\_permissions(r_2) \subseteq authorized\_permissions(r_1)$.
> - $authorized\_users(r : Roles) \rightarrow 2^{USERS}$, the mapping of role r onto a set of users in the presence of a role hierarchy. Formally:
> $authorized\_users(r) = \{u \in USERS | r' \succeq r, (u, r') \in UA\}$
> - $authorized\_permissions$[10]$(r : ROLES) \rightarrow 2^{PRMS}$, the mapping of role r onto a set of permissions in the presence of a role hierarchy. Formally:
> $authorized\_permissions(r) = \{p \in PRMS | r' \succeq r$[11]$, (p, r') \in PA\}$

[...]

Roles in a limited role hierarchy are restricted to a single immediate descendent.

[...]

Node r_1 is represented as an immediate <u>descendent</u> [12] of r_2 by $r_1 \succ\succ r_2$, if $r_1 \succeq r_2$ but no role in the role hierarchy lies between r_1 and r_2. That is, there exists no role r_3 in the role hierarchy such that $r_1 \succeq r_3 \succeq r_2$, where $r_1 \neq r_2$ and $r_2 \neq r_3$ [13].

Limited Role Hierarchy Specification

General Role Hierarchies [14] with the following limitation:

- $\forall r, r_1, r_2 \in ROLES, \underline{r \succeq r_1 \wedge r \succeq r_2}$ [15] $\Rightarrow r_1 = r_2$.

Description of Problems

P8 Confusion: this is the first sentence where the inheritance relation is described, and the standard uses a sufficient condition (*"all privileges of r_2 are also privileges of r_1"*) to describe it; the reader shall later understand that this is instead a necessary condition (a consequence of stating $r_1 \succeq r_2$).

P9 Noise: this is the formal declaration of the inheritance relation, but it is provided in a necessary condition referring to two functions not declared yet (*authorized_users* and *authorized_permission*), leading the reader to question whether he has overlooked some definitions involving $\succeq$ in the previous sections.

P10 Noise (major!): this function is never used in the rest of the standard. Moreover, it leads the reader to believe that the permissions of a role include the permissions inherited by the role, but this is not the case. The reader shall later learn, after reading the definition of **CheckAccess** page 17 and **CreateSession** and **AddActiveRole** page 21, that a user only gets the permissions of his active roles, and the inheritance hierarchy has no effect on the permissions of a role. The inheritance hierarchy only changes the users authorized to activate a role. For instance, following the definition of the two aforementioned administrative functions, if $r_1 \succeq r_2$ and $u \mapsto r_1 \in UA$, then user u is allowed to activate r_1 and r_2. By activating r_1, user u only gets the permission granted to r_1 in PA; the permissions of r_2 can be exercised only if u also activates r_2. Li *et al.* [7] claim that inheritance as presented in the standard can be interpreted in three different ways, but we do not agree with them. If the reader sticks to the mathematical definitions of the standard, then there is only one plausible interpretation. Of course, the natural language text, the errors and superfluous definitions like *authorized_permissions* create confusion, diverting the reader from the mathematical text, which should prevail.

P11 Error: it should be $r \succeq r'$, to match the necessary condition defined for $\succeq$ just above, *i.e.*,

$$r_1 \succeq r_2 \Rightarrow authorized\_permissions(r_2) \subseteq authorized\_permissions(r_1)$$

This error was also pointed out by Li *et al.* [7].

P12 Ambiguity: The sentence

Roles in a limited role hierarchy are restricted to a single immediate descendent.

and its formal representation as the following assertion

$$\forall r, r_1, r_2 \in ROLES, r \succ\succ r_1 \wedge r \succ\succ r_2 \Rightarrow r_1 = r_2$$

(where we have corrected the error P15 on $\succeq$ explained below) entail that r_2 is the descendent in $r_1 \succ\succ r_2$. This usage is also consistent with the formal definition of operation **AddInheritance** provided on page 19 of [2].

AddInheritance(r_asc, r_desc)

This commands establishes a new immediate inheritance relationship $r\_asc \succ\succ r\_desc$ between existing roles r_asc, r_desc.

However, the following sentence defines r_1 as the descendent:

Node r_1 is represented as an immediate descendent of r_2 by $r_1 \succ\succ r_2$, if $r_1 \succeq r_2$ but no role in the role hierarchy lies between r_1 and r_2.

Thus, there is confusion in the usage of the word "descendent".

P13 Error: the standard claims to define the covering relation of an ordered set, which they call immediate descendent, and which is typically used in Hasse diagrams. A third condition is missing to do so, namely $r_1 \neq r_3$. This error was also pointed out by Li *et al.* [7], but their suggested correction is incorrect: they suggest to replace $r_1 \neq r_2$ by $r_1 \neq r_3$, which is insufficient, because the intent of the authors is to define the covering relation of a partial order. All three inequalities are required.

P14 Version change: we have reproduced the 2004 version of the standard [1] here, because the 2012 version [2] uses Definition 2a instead, but there is no definition labeled with 2a in the standard.

P15 Error: the standard claims to define the notion of single immediate descendent in a partial order, *i.e.*, the partial order is a tree, as claimed in the following sentence:

> Limited role hierarchies impose restrictions resulting in a simpler tree structure (i.e., a role may have one or more immediate ascendants, but is restricted to a single immediate descendent).

To do so, the standard should use instead $r \succ\succ r_1 \wedge r \succ\succ r_2$. This error was also pointed out by Li *et al.* [7].

Appraisal of the Definitions

Given all these problems, this section of the standard is quite hard to understand. The meaning of relation $\succeq$ is unclear until the specification of the administrative functions are provided in Section 7 of the standard. This is where the reader learns the *indirect* effect of $\succeq$ on the **CheckAccess** predicate, which describes if a user can perform an operation on an object in a given state of the RBAC system. Describing the connection between $\succeq$ and the active sessions would help clarify the meaning of $\succeq$. The following assertion, which is the body of function **CheckAccess**, would show that $\succeq$ does not directly impact the access of a user has in a given state.

$$CheckAccess(s, op, ob) \Leftrightarrow s \in SESSIONS \wedge op \in OPS \wedge ob \in OBJS \wedge$$
$$\exists r \bullet r \in ROLES \wedge r \in session\_roles(s) \wedge$$
$$(op \mapsto ob) \mapsto r \in PA$$

This assertion shows that what is accessible is determined by the roles activated by a user in a session. One then has to find out how variable *session_roles* is updated, by looking at the administrative functions updating it. This is where $\succeq$ comes into play. Function **AddActiveRole**(u, s, r) says that user u can activate role r in session s if $u \in authorized\_users(r)$.

2.3 Constrained RBAC

Constrained RBAC adds Separation of Duty relations to the core RBAC model. **Static Separation of Duty** is specified by a role set rs and an integer n such as $2 \leq n \leq card(rs)$. That type of constraint specifies that a user can be assigned at most $n - 1$ roles of rs. Formally, let SSD be the set of the static separation of duty constraints :

> - $SSD \subseteq 2^{ROLES} \times \mathbb{N}$
> - $\forall (rs, n) \in SSD, \forall t \subseteq rs : |t| \geq n \Rightarrow \bigcap_{r \in t} assigned\_users(r) = \emptyset$
> - In presence of role hierarchy
> $\forall (rs, n) \in SSD, \forall t \subseteq rs : |t| \geq n \Rightarrow \bigcap_{r \in t} authorized\_users(r) = \emptyset$

Dynamic Separation of Duty is specified by a role set rs and an integer n such as $2 \leq n \leq card(rs)$. That type of constraint specifies that a user can simultaneously hold at most $n - 1$ roles of rs, *during one session*. Formally, let DSD be the set of dynamic separation of duty constraints :

- $DSD \subseteq (2^{ROLES} \times \mathbb{N})$
- $\forall rs \in 2^{ROLES}, n \in \mathbb{N}, (rs, n) \in DSD \Rightarrow n \geq 2, |rs| \geq n$ and
  $\forall s \in SESSIONS, \forall rs \in 2^{ROLES}, \forall role\_subset \in 2^{ROLES},$
  $\forall n \in \mathbb{N}, (rs, n) \in DSD, role\_subset \subseteq rs,$
  $role\_subset \subseteq session\_roles(s) \Rightarrow |role\_subset| < n.$

We didn't find any problem with this part of the specification. The constraint could be formulated in a simpler manner, which we have done in our B specification.

3   Administrative Functions

Administrative functions describe how the RBAC system state evolves. The standard claims to use the Z notation for specifying administrative functions.

The notation used in the formal specification of the RBAC functions is a subset of the Z notation. The only change is the representation of a schema as follows:

Schema-Name (Declaration) $\lhd$ Predicate; ...; Predicate $\rhd$

Most abstract data types and functions used in the formal specification are defined in Section 3, RBAC Reference Model. New abstract data types and functions are introduced as needed.

Some examples of such specification are provided below to illustrate problems with the adapted Z notation used in the standard.

AddUser
This command creates a new RBAC user. [...]

AddUser (user : NAME) [16]

$\lhd$

   $user \notin USERS$
   $Users' = Users \cup \{user\}$
   $\underline{user\_sessions}$ [17]$' = user\_sessions \cup \{user \mapsto \emptyset\}$

$\rhd$

DeleteUser

This command deletes an existing user from the RBAC database. [...]

DeleteUser(user : NAME)

◁

$user \in USERS$

$[\forall s \in SESSIONS \bullet s \in user\_sessions(user) \Rightarrow DeleteSession(s)\ ^{18}]$

$UA' = UA \setminus \{r : Roles \bullet user \mapsto r\}$

$assigned\_users' = \{r : Roles \bullet r \mapsto (assigned\_users(r) \setminus \{user\})\}$

$USERS' = USERS \setminus \{user\}$

▷

DeleteSession($user,session$)

This function deletes a given session with a given owner user. [...]

DeleteSession (user,session : NAME) [19] $=$

◁

$user \in USERS; session \in SESSIONS; session \in user\_sessions(user)$

$user\_sessions' = user\_sessions \setminus \{user \mapsto user\_sessions(user)\} \cup$
$\qquad\qquad\qquad \{user \mapsto user\_sessions(user) \setminus \{session\}\}$

$session\_roles' = session\_roles \setminus \{session \mapsto session\_roles(session)\}$

$SESSIONS' = SESSIONS \setminus \{session\}$

▷

DeleteRole

This commands deletes an existing role from the RBAC database. [...]

DeleteRole (role : NAME) [20] $=$

◁

$role \in ROLES$

$[\ \forall s \in SESSIONS \bullet role \in session\_roles(s) \Rightarrow DeleteSession(s)]$

$UA' = UA \setminus \{u : USERS \bullet u \mapsto role\}$

$assigned\_users' = assigned\_users \setminus \{role \mapsto assigned\_users(role)\}$

$PA' = PA \setminus \{op : OPS, obj : OBJ \bullet (op, obj) \mapsto role\}$

$assigned\_permissions' = assigned\_permissions \setminus$
$\qquad\qquad\qquad \{role \mapsto assigned\_permissions(role)\}$

$ROLES' = ROLES \setminus \{role\}$

▷

Description of Problems

P16 The notation used in the standard omits important elements of a Z operation schema. First, it does not identify the state space of the operation. A typical Z operation schema will include a $\Delta State$ declaration, introducing unprimed and primed variables, to denote the before and after states, and their associated invariant. The predicate part should describe the relationship between unprimed and primed variables. Primed variables which are not subject to any condition are allowed to take any value. Obviously, this convention has not been followed in the standard, because we do not expect operation **AddUser** to let all other state variables take any value after execution. Thus, we must assume that the standard uses the convention that primed variables x' which are not occurring in the operation specification are preserved with the equality $x' = x$. However, this convention has not been followed everywhere. For instance, symbol $\succ\!\!\succ$ is used in operation **AddInheritance** where $\succeq$ is updated, but $\succ\!\!\succ$ isn't. But since $\succ\!\!\succ$ is supposed to be the covering relation of $\succeq$, we can't assume the equality $\succ\!\!\succ' = \succ\!\!\succ$, because it would break the invariant linking $\succeq$ and $\succ\!\!\succ$. This may suggest that the standard assumes that derived functions need not to be explicitly updated since their definition acts like a state invariant which is assumed to be maintained by operations, as is the case in Z when $\Delta State$ is used. But the standard doesn't follow this convention either. For instance, in operations maintaining variable UA, which maps users to roles, variable *assigned_users* is also maintained, which is not needed, since *assigned_users* is derived from UA.

P17 Variable *user_sessions* has not been declared in the data structures in the previous section. Variable *session_users*, which has been declared in the data structure section, is not updated by this operation. So the assumption we made in P16 to make sense of the notation used is broken here, because it makes *session_users* inconsistent with *user_sessions*. Luckily, *session_users* does not seem to be used at all in the specification of administrative functions, so we deduce that its declaration is superfluous in the data structure section of the standard, which solves the inconsistency problem.

P18 This is one example of operation call which does not follow and the Z syntax and that is logically unsatisfiable. The reader must suppose that a more "programming language" view is used here. There are other cases in the standard (*e.g.*, **AddAscendant**, **AddDescendant**, where two calls are represented implicitly as a conjunction, but sequential composition should have been used, to make sense out of it).

P19 Signature inconsistency: **DeleteSession** is declared with parameters (**user, session: NAME**), but called as **DeleteSession(session)** in **DeleteRole** and **DeleteUser**. Since a session is related to a single user, as provided by the unused function *session_users*, there is no need for parameter **user**. Note also that updating function *session_users* is simpler than updating its functional inverse *user_sessions*, *i.e.*,

$$session\_users' = \{session\} \lhd session\_users .$$

The Z domain subtraction is not used in the standard, and that makes the specification harder to read.

P20 Operation **DeleteRole** does not update relation "$\succeq$" and separation of duty constraints SSD and DSD. The last two raise more serious issues to deal with. We see two options:

- remove the deleted role from all the constraint role sets where it appears;

- restrict the operation to a role which is not used in SSD/DSD constraints.

The first option raises the issue of updating the cardinality. Recall that an SSD/DSD constraint (RS, n) states that at most $n-1$ roles of RS can be assigned to/activated by a user. It is subject to the invariant $n \geq 2 \wedge |RS| \geq n$. If $|RS| < n$, then the constraint can never be violated and it is useless. After deleting a role, we have the following cases:

- $n > 2 \wedge |RS'| = n - 1$: n must be decremented by 1, in order for the constraint to satisfy the state invariant;

- if $n = 2 \wedge |RS'| = 1$, the constraint is deleted because it is does not satisfy the state invariant $|RS'| \geq n$ and n cannot be fixed by decrementing n, since $n \geq 2$ is required by state invariant;

- $|RS'| \geq n$: n could be decremented by 1 or left unchanged; it depends on the particular access-control requirements of the application.

In any case, the constraint could be deleted if it does not make sense in the security requirements of the application. Furthermore, removing a role in a constraint role set may introduce constraint redundancy: if two constraints have the same role set, the one with the bigger cardinality is redundant. Then, `DeleteRole` should in addition remove the redundant constraint. Given these cases, it seems safer to let the RBAC manager manually adjust SSD/DSD affected by a role deletion before deleting a role. Hence, we have added a precondition in our specification of **DeleteRole** to check that a role is not used in any SSD/DSD constraint.

We have discovered this issue by proving that operations preserve state invariants and it hasn't been raised in [6,9].

4   The B Specification of the RBAC Standard

Due to space limitation, our B specification is omitted and fully provided in [3]. Each RBAC component has its own machine, and the Core RBAC machine is included in the other two components. The only modification to the standard was to use an acyclic directed graph RH such that $\succeq = RH^*$, as suggested in [6]. This greatly simplifies the maintenance of the role hierarchy, while preserving the intent of the RBAC standard.

The modeling phase in B allowed us to discover most of the noises and typo problems. Then the validation phase hilighted major problems by animation and proof. ProB[1] has been used to animate the model and discover invariant violation. Once the model corrected, all proofs have been discharged using Atelier B[2]. We also tried to prove an invariant which was not in the standard, the acyclicity of the role hierarchy, expressed as $RH^+ \cap \mathrm{id}(Roles) = \emptyset$. Since Atelier B has no rule about *closure*, it was impossible to prove without adding new rules in the prover. Instead, we use abstract relation algebra [12] and Kleene algebra [5], of which binary relations are models, to formally prove preservation of acyclicity when adding a new pair in an acyclic relation. The proof is provided below. For the sake of concision, we adopt some of the conventions of abstract relation algebra. For instance, we write PQ instead of $P;Q$ for relational composition. Let $L = Roles \times Roles$ denote the universal relation, $\overline{P} = L - P$, $I = \mathrm{id}(Roles)$ denote the identity relation, $A \subseteq Roles \times Roles$ denote the role hierarchy, $B = \{x \mapsto y\}$ where $x \neq y$ and $\{x, y\} \subseteq Roles$, be a new pair to add to A.

Theorem 1. *Assuming*

$$A^+ \cap I = \emptyset \;(1) \qquad B \cap I = \emptyset \quad (2) \qquad B^{-1} \cap A^+ = \emptyset \;(3)$$
$$BB = \emptyset \quad (4) \qquad BLLB^{-1} \subseteq I \;(5)$$

then $(A \cup B)^+ \cap I = \emptyset$

To this end, we use the following laws of [5,12], which include the laws of Boolean algebra, since a relation algebra is a Boolean algebra.

$$PQ \subseteq R \;\Leftrightarrow\; P^{-1}\overline{R} \subseteq \overline{Q} \;\Leftrightarrow\; \overline{R}Q^{-1} \subseteq \overline{P} \tag{6}$$
$$P \cup RQ \subseteq R \;\Rightarrow\; PQ^* \subseteq R \tag{7}$$

We also need the following two lemmas, which follow from (1) to (5).

$$A^*BA^* \cap I = \emptyset \tag{8}$$
$$(A \cup B)^+ \subseteq A^+ \cup A^*BA^* \tag{9}$$

PROOF of (8)

$$
\begin{aligned}
&A^*BA^* \cap I = \emptyset \\
\Leftrightarrow\; &A^*BA^* \subseteq \overline{I} \\
\Leftrightarrow\; &B^{-1}A^{*-1}I \subseteq \overline{A^*} \\
\Leftrightarrow\; &A^*A^* \subseteq \overline{B^{-1}} \\
\Leftrightarrow\; &A^* \cap B^{-1} = \emptyset \\
\Leftrightarrow\; &(I \cup A^+) \cap B^{-1} = \emptyset \\
\Leftarrow\; &I \cap B = \emptyset \wedge A^+ \cap B^{-1} = \emptyset
\end{aligned}
$$

$$P \cap \overline{Q} = \emptyset \Leftrightarrow P \subseteq Q, \; \overline{\overline{P}} = P$$
$$(6)$$
$$PI = P, (6)$$
$$P^*P^* = P^*, \; P \cap \overline{Q} = \emptyset \Leftrightarrow P \subseteq Q, \; \overline{\overline{P}} = P$$
$$P^* = I \cup P^+$$
$$(2), (3)$$

PROOF of (9)

---

[1] http://www.stups.uni-duesseldorf.de/ProB
[2] http://www.atelierb.eu/

$$(A \cup B)^+ \subseteq A^+ \cup A^*BA^*$$
$$\Leftrightarrow (A \cup B)(A \cup B)^* \subseteq A^+ \cup A^*BA^*$$
$$\Leftarrow A \cup B \cup (A^+ \cup A^*BA^*)(A \cup B) \subseteq A^+ \cup A^*BA^* \qquad \text{by (7)}$$
$$\Leftrightarrow A^*BA^*B \subseteq A^+ \cup A^*BA^* \qquad \text{Distributivity, } A \subseteq A^+, \ B \subseteq A^*BA^*,$$
$$A^+A \ \subseteq \ A^+, \ A^+B \ \subseteq \ A^*BA^*,$$
$$A^*BA^*A \subseteq A^*BA^*$$
$$\Leftarrow BA^*B \subseteq \emptyset$$
$$\Leftrightarrow BB \subseteq \emptyset \wedge BA^+B \subseteq \emptyset \qquad BA^*B = BA^+B \cup BB$$
$$\Leftarrow B\overline{B^{-1}}B \subseteq \emptyset \qquad (4), \ A^+ \subseteq \overline{B^{-1}} \text{ by (3)}$$
$$\Leftrightarrow LB^{-1}\overline{B} \subseteq \overline{B} \qquad (6)$$
$$\Leftrightarrow BLB \subseteq B \qquad (6)$$
$$\Leftrightarrow BLLB^{-1}B \subseteq B \qquad LL = L, \ LB^{-1}B = LB \text{ by (6)}$$
$$\Leftarrow BLLB^{-1} \subseteq I \qquad (5)$$

PROOF of Theorem 1

$$(A \cup B)^+ \cap I = \emptyset$$
$$\Leftarrow (A^+ \cup A^*BA^*) \cap I = \emptyset \qquad (8)$$
$$\Leftarrow A^+ \cap I = \emptyset \wedge A^*BA^* \cap I = \emptyset \qquad (1),(9)$$

5 Conclusion

RBAC is a widely adopted access-control model and is also widely used in commercial products, such as database management systems or enterprise management systems. The RBAC model has been published as the NIST RBAC model[11] and adopted as an ANSI/INCITS standard in 2004, which has been revised in 2012. In this paper, we have pointed out a number of technical errors identified using formal methods, by modelling in a B machine the RBAC specifications, then animating it and proving it. The main types of errors found in the RBAC standard are the following: i) typing errors, ii) unused definitions which add noise and detract attention from the specification of the essential concepts of RBAC, iii) inconsistencies in operation signatures and specification conventions, iv) inappropriate operation specification style, which leads to ambiguities and unsatisfiable specifications, v) invariants broken by operations. All these types of errors can be easily avoided by using a formal specification method. Using mathematics without a methodological framework and a supporting tool set is bound to open the door to errors. The B method seems to be particularly appropriate for specifying standards of dynamic systems like RBAC. The fact that B makes a clear distinction between the specification of operations and the state properties that these operations must satisfy (i.e., invariant preservation) proved to be very useful in validating the RBAC standard. The example of role deletion (problem P20) is a nice illustration of this. This case study also shows that human-based reviews are insufficient to detect errors in a standard. Mechanical verification is essential; animation, model checking and theorem proving are complementary in finding errors in specification.

References

1. ANSI. Role Based Access Control, INCITS 359-2004 (2004)
2. ANSI. Role Based Access Control, INCITS 359-2012 (2012)
3. Huynh, N., et al.: B Specification of the RBAC 2012 Standard (2014), http://info.usherbrooke.ca/mfrappier/RBAC-in-B
4. Ferraiolo, D., Kuhn, R., Sandhu, R.: RBAC Standard Rationale: Comments on "A Critique of the ANSI Standard on Role-Based Access Control". IEEE Security Privacy 5(6), 51–53 (2007)
5. Kozen, D.: A completeness theorem for Kleene algebras and the algebra of regular events. Information and Computation 110, 366–390 (1994)
6. Li, N., Byun, J.W., Bertino, E.: A critique of the ANSI Standard on Role-Based Access Control. Technical Report TR 2005-29, Purdue University (2005)
7. Li, N., Byun, J.W., Bertino, E.: A Critique of the ANSI Standard on Role-Based Access Control. IEEE Security Privacy 5(6), 41–49 (2007)
8. O' Connor, A.C., Loomis, R.J.: Economic Analysis of Role-Based Access Control. RTI International (2010)
9. Power, D., Slaymaker, M., Simpson, A.: On Formalizing and Normalizing Role-Based Access Control Systems. The Computer Journal 52(3), 305–325 (2009)
10. Rissanen, E.: eXtensible Access Control Markup Language (XACML) Version 3.0. OASIS (2010)
11. Sandhu, R., Ferraiolo, D., Kuhn, R.: The NIST model for role-based access control: Towards a unified standard. In: 5th ACM Workshop on Role-based Access Control, RBAC 2000, pp. 47–63. ACM (2000)
12. Schmidt, G., Ströhlein, T.: Relations and Graphs: Discrete Mathematics for Computer Scientists. EATCS Monographs on Theoretical Computer Science. Springer (1993)

Invariant Guided System Decomposition

Richard Banach

School of Computer Science, University of Manchester,
Oxford Road, Manchester, M13 9PL, UK
banach@cs.man.ac.uk

Abstract. We re-examine the problem of decomposing systems in Event-B. We develop a pattern for cross-cutting events and invariants that enables the core dependencies in multi-machine systems to be tracked. We give the essential verification conditions.

Keywords: System Decomposition, Cross-Cutting Invariants, Event-B.

1 Introduction

In top down model based development methodologies, especially the B-Method, the issue of composition and decomposition of (sub)systems has received a lot of interest. See e.g. [1,3,2,5]. For us, the main issue may be illustrated in a simple example.

Suppose there is a machine M with variables x, y. Suppose M needs to be partitioned into two machines, $M1$ and $M2$. Suppose that x needs to go into $M1$ and y needs to go into $M2$. Suppose that there is an invariant of M involving both variables, $InvM(x, y)$. If the partitioning is to go ahead, what are we to do about $InvM(x, y)$?

Sometimes it is suggested that an invariant like $InvM(x, y)$ might be replaced by $InvM1(x) \equiv (\exists y \bullet InvM(x, y))$ in $M1$, say. However although $InvM(x, y) \Rightarrow InvM1(x)$, the converse does not hold. Therefore, recognising that $InvM$ and $InvM1$ are inequivalent, if $InvM(x, y)$ is a critical safety invariant, then the suggested partitioning strategy would render the system incapable of discharging its most important duty. The usual approach if $InvM$ is important enough, is simply to not partition. However, such an approach does not scale.

The remainder of the paper is as follows. Section 2 introduces our approach to decomposition in generic terms. Section 3 covers verification issues, while Section 4 covers machine decomposition. Section 5 looks at refinement. Section 6 concludes.

2 Variable Sharing via INTERFACEs

We note that in typical embedded systems, connections are invariably unidirectional, often mirroring physical connections such as wires. We exploit this unidirectionality to design a methodology for handling a useful class of invariants that cut across subsystem boundaries. We first introduce a concept of INTERFACE, rooted in the work of Hallerstade and Hoang [4], which we extend, just enough to achieve what we desire.

An interface is a syntactic construct that declares some variables, and (going beyond [4]), some invariants that interrelate them, and their initialisations. Any machine that

Y. Ait Ameur and K.-D. Schewe (Eds.): ABZ 2014, LNCS 8477, pp. 271–276, 2014.
© Springer-Verlag Berlin Heidelberg 2014

needs to access any of these variables must refer to the interface. The interface mechanism is the only permitted way for more than one machine to have access to the same set of variables. Our use of interfaces is based on the following principles.

Consider a set of variables $\mathcal{V}$, a set of invariants $\mathcal{I}$ that mention some of those variables (and no others), and a set of events $\mathcal{E}$ that read and update some of those variables (and no others). Suppose the set of variables can be partitioned into subsets $A, B, C \ldots$, such that for every invariant $Inv \in \mathcal{I}$:

[•1] **either** all variables mentioned in Inv belong to some subset, eg. A;
[•2] **or** the invariant Inv is of the form $U(u) \Rightarrow V(v)$, where there are distinct subsets of the partition A and B say, such that u and v refer to variables in A and B respectively.

We call these type [1] and type [2] invariants respectively (t1i and t2i). For a t2i, the A and B subsets are the local and remote subsets (containing the local variables u and remote variables v). We observe that unless a system already consists simply of two unconnected, completely independent subsystems, in which all properties split into a conjunctionof properties of the two subsystems, there will be, in general, an infinity of properties that couple the two subsystems nontrivially. Referring to the discussion of the Introduction, the problem of what to do about cross-cutting invariants is unavoidable. Our thesis is that, in the kind of embedded systems we spoke of, the unidirectionality of the connections between subsystems implies that t2is are adequate to capture a sufficiently rich class of inter-subsystem properties for practical use.

Henceforth we restrict to collections of variables/invariants/events conforming to these restrictions, calling them pre-systems. Note that *any* collection of variables and invariants is as a pre-system with a sufficiently coarse variable partition, e.g. a singleton partition. We can organise a pre-system into machines and interfaces as follows.

Every subset of variables of the partition can consist of variables that, exclusively:

[•3] **either** are declared as the variables of a single machine;
[•4] **or** are declared as the variables of a single interface.

Each interface:

[•5] must contain all the type [1] invariants that mention any of its variables;
[•6] may contain type [2] invariants for which the interface's variables are in the local subset; in each such case the interface must contain a READS *ReadItf* declaration for the interface *ReadItf* that contains the remote variables.
[•7] may contain REFERS *RefItf* declarations, whenever any of its variables are the remote variables of a type [2] invariant declared in an interface *RefItf*.

Each machine:

[•8] may declare the variables belonging to a subset of the partition as local (i.e. unshared) variables;
[•9] may contain one or more CONNECTS *Itf* declarations giving access to the variables of the interface;
[•10] may contain one or more READS *Itf* declarations giving read-only access to the variables of the interface;
[•11] must contain all the type [1] invariants that mention any of its local variables;

Each event:

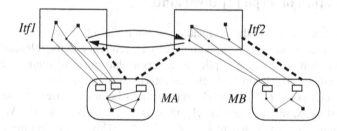

Fig. 1. An illustration of the constraints [•1]–[•15]

[•12] may read and update variables that are declared locally in the machine containing the event, or that are introduced via CONNECTS *Itf* declarations in the machine containing the event;

[•13] may read (in its guards or in the expressions that define update values) variables that are introduced: either via READS *ReadItf* declarations in the machine containing the event, or via READS *ReadItf* or REFERS *RefItf* declarations contained in an interface *Itf* that the machine containing the event CONNECTS (to).

[•14] must preserve all invariants that are declared in the machine that contains it, or that are declared in any CONNECTS *Itf* declarations of the machine, or that are contained in any READS *ReadItf* or REFERS *RefItf* declarations contained in an interface *Itf* that the machine containing the event CONNECTS (to).

Each invariant:

[•15] must be contained in the interface or machine which declares all its variables (if it is a type [1] invariant), or must be contained in the interface which declares its local variables (if it is a type [2] invariant).

By a system, we mean a collection of machines satisfying [•1]–[•15] above. We note that the keywords we introduced, CONNECTS, READS, REFERS, have no semantic connotations other than the ones we mentioned. We can see fairly readily that in a system, verifying that all the invariants are preserved by all event executions (provided the initial state satisfies them all), can be readily accomplished using verification conditions that depend on information that is easily located from the syntactic context of the event, namely, from the interfaces explicitly mentioned in the machine that defines the event. We examine verification conditions in more detail in the next section.

In Fig. 1 we show an illustration of the constraints [•1]–[•15]. Dots represent variables, while small squares represent type [1] invariants. Small rectangles represent events. Events and invariants are connected to the variables they involve by thin lines. Interfaces are large rectangles containing the variables and invariants they encapsulate — there are two in Fig. 1, *Itf*1 and *Itf*2. Machines are large rounded rectangles, containing their events and local variables — again there are two, *MA* and *MB*. The CONNECTS relationship is depicted by thick dashed lines. Finally, type [2] invariants are represented by arrows from the local to the remote interface.

3 Verification of Type [2] Invariants

In this section we focus on the verification of nontrivial t2i invariants, assuming that t1i invariants can be handled unproblematically by reference to the relevant interfaces during verification. (The same applies to an event that must maintain a t2i if it can access and update variables in *both* relevant interfaces (simultaneously).)

Consider a t2i $(*) \equiv U(u) \Rightarrow V(v)$, where u and v belong to different interfaces. We prime after-state expressions generically, thus: $(*') \equiv U'(u') \Rightarrow V'(v')$. We write the events of interest as $EvXYZ$ where $X, Y, Z \in \{U, V\}$. This means that the guard g_{UVV} of $EvUVV$ mentions the variables u, v and the before-after relation BA_{UVV} of $EvUVV$ updates variable v. The shorter notation $EvUV$ means that the guard mentions only u and the update is to variable v alone.

We assume that for events $EvUU$ and $EvVV$, verification would be restricted to variables u and v of invariant $(*)$ respectively, while for $EvUVU$ and $EvUVV$, both parts of $(*)$ could participate in verification, since both sets of variables are read via the relevant interfaces. Read access to additional variables is obviously harmless and is not considered further.

Theorem 1. *Assuming that initial states are invariant, and that all events preserve all type [1] invariants declared locally and in CONNECTS Itf declarations on reachable states, the following proof obligations (POs) are sufficient to preserve reachable invariance for type [2] invariants.*

$$EvUVU : g_{UVU}(u, v) \wedge \neg V(v) \Rightarrow g_{UVU}(u, v) \wedge \neg U(u) \Rightarrow BA_{UVU}(u, u') \Rightarrow \neg U(u')$$

$$\text{(obvious analogue for EvVU)} \tag{1}$$

$$EvUU : g_{UU}(u) \wedge \neg U(u) \Rightarrow BA_{UVU}(u, u') \Rightarrow \neg U(u') \tag{2}$$

$$EvUVV : g_{UVV}(u, v) \wedge U(u) \Rightarrow g_{UVV}(u, v) \wedge V(v) \Rightarrow BA_{UVV}(v, v') \Rightarrow V(v')$$

$$\text{(obvious analogue for EvUV)} \tag{3}$$

$$EvVV : g_{VV}(v) \wedge V(v) \Rightarrow BA_{VV}(v, v') \Rightarrow V(v') \tag{4}$$

The above gives a selection of POs which can be used for verifying the preservation of cross-cutting invariants of the t2i kind that we have considered, based on the occurrences of the relevant variables in the events that access those variables.

4 Machine Decomposition

The account so far permits us to assemble a large system by composing a number of machines together via a collection of interfaces that obey [•1]–[•15]. Equally interesting though for the B-Method in general, is the problem of the *decomposition* of a machine into a collection of smaller (sub)machines $M_1 \ldots M_k$, the development of which can subsequently be pursued (at least relatively) independently. We examine this issue now.

We approach the decomposition problem by positing that decomposition should be a syntactic manipulation whose correctness ought to be demonstrable generically. In this light, the principle constraining decompositions of a machine can be described as follows:

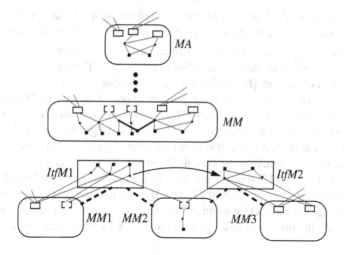

Fig. 2. An illustration of the decomposition mechanism. *MA*, refined to a larger machine *MM*, is decomposed into smaller machines and interlinking interfaces.

[•16] Regarding the variables and invariants (and events) declared in a machine *M* as a pre-system (but not including the variables or invariants of any interface accessed by *M*), any decomposition of *M* into submachines and interfaces is considered valid provided: firstly, it conforms to restrictions [•1]–[•15]; secondly, any submachine M_j that includes an event of *M* that uses a variable accessed (directly or indirectly) via an interface of *M*, must access the same interface appropriately.

It is clear that adhering to [•16] refines the partition of variables when *M* is part of a larger system already adhering to [•1]–[•15], without spoiling [•1]–[•15] overall.

Fig. 2 shows the decomposition mechanism at work. Machine *MA* from Fig. 1 is first refined to a larger machine *MM*, containing more local variables and invariants, as well as some new events shown using broken small rectangles. One new invariant is connected to its variables using slightly thicker lines. Machine *MM* is now decomposed into a collection of smaller machines and interfaces, *MM1*, *MM2*, *MM3*, and *ItfM1*, *ItfM2*. The connections from *MA* events to previously existing interfaces are retained, while the decomposition of the new ingredients conforms to constraints [•1]–[•15]. The invariant connected using slightly thicker lines becomes a type [2] invariant with *ItfM1* and *ItfM2* as its local and remote interfaces respectively (on the presumption that it was of the correct syntactic shape at the outset).

5 Refinement

We turn to the crucial issue of refinement. As for decomposition, there is a key guiding principle behind the way that refinement is handled in our scheme.

[•17] The variables of an interface *Itf* must be refined to the variables of its refining interface *ItfR* via a retrieve relation that mentions only the variables of *Itf* and *ItfR*.

[•18] The variables of a machine M must be refined to the variables of its refining machine MR via a retrieve relation that mentions only the variables of M and MR.

The independence of refinement of machines and interfaces prevents the inadvertent falsifying of refinement relations in situations such as the following.

Suppose each of M_1 and M_2 CONNECTS Itf; these constructs being refined to M_1R, M_2R and $ItfR$ respectively. Suppose the joint invariant of the M_2 to M_2R refinement involves the variables of Itf and $ItfR$ too. Then when concrete machine M_1R executes an event, faithful to some abstract event of M_1, there is no guarantee that the new state in M_1 and M_1R and Itf and $ItfR$ still satisfies the joint invariants of M_2 and M_2R via the coupled joint invariants linking the state in M_2 and M_2R to the state in Itf and $ItfR$.

Adhering to **[•17]**–**[•18]** though, it is easy to see that the problem described cannot arise. The decoupling of variables of M_2 and M_2R on the one hand, from those of Itf and $ItfR$ on the other, means that when the variables of Itf and $ItfR$ change at the behest of M_1 and M_1R, the invariants linking the M_2 and M_2R variables remain true.

6 Conclusions

In this paper we have proposed, rather tersely, an Event-B decomposition scheme inspired by the INTERFACE idea of [4]. This was broadly in the shared variables tradition, but was driven primarily by the structure of a system's invariants. Although ostensibly a shared variable approach, there are strong influences from the shared events approach too, since a key feature of both ours and the shared events approach is the desire to communicate values between machines. In this brief treatment, we just gave a minimal description of the technical details of our approach, of which a kind of pattern for cross-cutting events and invariants was the key element, and we outlined the requisite verification machinery. In a more extended treatment, we will be able to describe the mechanisms more fully, we will be able to formulate the statements as theorems, and, crucially, we will be able to illustrate the technique using examples and case studies.

References

1. Abrial, J.R.: Event-B: Structure and Laws. In: Rodin Project Deliverable D7: Event-B Language, http://rodin.cs.ncl.ac.uk/deliverables/D7.pdf
2. Butler, M.: Decomposition Structures for Event-B. In: Leuschel, M., Wehrheim, H. (eds.) IFM 2009. LNCS, vol. 5423, pp. 20–38. Springer, Heidelberg (2009)
3. Hoang, T.S., Abrial, J.-R.: Event-B Decomposition for Parallel Programs. In: Frappier, M., Glässer, U., Khurshid, S., Laleau, R., Reeves, S. (eds.) ABZ 2010. LNCS, vol. 5977, pp. 319–333. Springer, Heidelberg (2010)
4. Hallerstede, S., Hoang, T.S.: Refinement by Interface Instantiation. In: Derrick, J., Fitzgerald, J., Gnesi, S., Khurshid, S., Leuschel, M., Reeves, S., Riccobene, E. (eds.) ABZ 2012. LNCS, vol. 7316, pp. 223–237. Springer, Heidelberg (2012)
5. Silva, R., Pascal, C., Hoang, T., Butler, M.: Decomposition Tool for Event-B. Software Practice and Experience 41, 199–208 (2011)

Understanding and Planning Event-B Refinement through Primitive Rationales

Tsutomu Kobayashi[1], Fuyuki Ishikawa[2], and Shinichi Honiden[1,2]

[1] The University of Tokyo, Japan
[2] National Institute of Informatics, Japan
{t-kobayashi,f-ishikawa,honiden}@nii.ac.jp

Abstract. Event-B provides a promising feature of refinement to gradually construct a comprehensive specification of a complex system including various aspects. It has unique difficulties to design complexity mitigation, while obeying Event-B consistency rules, among the potentially large possibilities of refinement plans. However, despite of the difficulties, existing studies on specific examples or high-level and intuitive guidelines are missing clear rationales, as well as principles, guidelines or methods supported by the rationales. In response to this problem, this paper presents a method for refinement planning from an informal/semi-formal specification. By defining primitive rationales, the method can eliminate undesirable plans such as the ones failing to mitigate the complexity. In a case study on a popular example from a book, we derived an enough small number of valid plans only by using the general and essential rationales while explaining the one presented in the book.

1 Introduction

Event-B [1] supports a flexible refinement mechanism that allows a comprehensive formal specification to be constructed and verified, by gradually introducing various elements of the target system to an abstract model. This kind of refinement mechanism is essential, as nowadays specifications are abstract but enough complex, handling various elements in system's surrounding environment.

As the refinement mechanism of Event-B deals with the essence of complexity mitigation, it poses unique difficulties. The decision space of possible refinement plans is potentially very large: what aspects are handled in each space, in which order the whole aspects are introduced, what concrete elements (e.g., variables) are introduced for each aspect, etc. This situation is different from the refinement mechanism for deriving a program code, which consists of simple steps (automated with BART tool [6] in classical B). In addition, there are consistency rules on refinement mechanism as proof obligations. A naive plan, just following the intuition "from abstract to concrete", may fail and cause rollback.

There are thus strong demands for principles, guidelines and methods to help human developers to plan refinements for an efficient use of Event-B, especially for widely uses in the industry. Even with active efforts, it is left almost untouched how to construct a refinement plan.

Y. Ait Ameur and K.-D. Schewe (Eds.): ABZ 2014, LNCS 8477, pp. 277–283, 2014.
© Springer-Verlag Berlin Heidelberg 2014

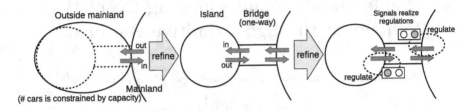

Fig. 1. Example: Cars on Island and Bridge

Available teaching materials for Event-B provide examples of refinement, e.g., the book by the method founder [1]. These examples are probably the best and seem widely accepted. However, the rationales and reasons why the refinement plans are considered "good" have not been explicitly explained, leaving questions such as "why not others?" and "how can we proceed for our problems?".

There are some studies that provide systematic ways to construct Event-B models. However, they do not provide any systematic or generic methods or only provide domain-specific knowledge. Moreover, they do not discuss the essence of why the proposed guidelines can derive "good" refinement plans.

All these backgrounds strongly motivate us to explicitly discuss rationales for refinement plans and provide a systematic method supported by the rationales.

It may be helpful to consider an analogy with object-oriented design. There are many intuitive and high-level guidelines for "good" design with separation of concerns, reusability, etc. As the "good/bad" designs, rationales can often explain like "here is a dependency between these classes, but this class is likely to change and affect the others, while this good design eliminates this direct dependency". Experienced developers may omit such explicit details, and the actual situation forms a mixed problem of various issues. However, the rationales in an extracted and idealized problem are essential enabler of education for beginners and systematic engineering more independent of individual experiences, as well as supporting methods and tools.

Similarly, this paper explicitly defines and discusses an extracted problem of refinement planning. The problem is how to construct a refinement plan, given an informal/semi-formal specification (e.g., natural language with UML) as input to rigorous modeling and verification in Event-B.

2   Refinement Planning

In Event-B, a comprehensive formal specification is constructed and verified gradually, by iteratively introducing various elements of a target system to a simple, abstract model. Fig. 1 illustrates a simplified version of the first system explained in the book by the method founder [1]. The system defines control of cars that enter and leave a bridge that connects an island and a mainland. Traffic lights realize the control of the constraints about the capacity on the whole bridge and island, as well as the one-way bridge. The initial model (left in the figure) considers the capacity constraint by modeling cars leaving and entering the mainland, with the combined model of outside (bridge and island). The second model considers the one-way constraint by separating the bridge and the island. The third model considers the constraints about how traffic lights change to satisfy the previous constraints.

A model at each step is verified in terms of consistency inside the model as well as consistency with the previous models. We consider verification of constraints such as described above the key of modeling in Event-B, which is distributed into multiple steps.

This paper calls the application constraints (requirements, assumptions or environmental knowledge) as *artifacts*. Each *refinement step* is considered to *introduce* artifact(s) while inheriting artifacts already introduced in the previous steps. The sequence of such refinement steps composes a *refinement plan*. An example of a refinement plan is $[\{a, b\}, \{c\}]$, which introduces artifacts a and b in the first setup, and c in the second step.

Each artifact is composed of primitive elements for concepts and state transitions, such as "capacity" and "traffic light at the mainland side turns red". We call such elements as *phenomena*. Formal specification of artifacts requires that of phenomena as variables, constants, and so on. Note that this paper does not put specific assumptions on the input of artifacts and phenomena, e.g., semi-formal notations such as Problem Frames or UML.

This paper explicitly discusses how to make decisions on refinement plans. Potentially, three artifacts a, b, and c lead to the solution space of possible refinement plans, such as $[\{a\}, \{b\}, \{c\}]$, $[\{b\}, \{a\}, \{c\}]$, $[\{a, b\}, \{c\}]$, and so on. Rather than intuitive and high-level claims to select one plan, this paper tries to apply a set of primitive rationales for eliminating invalid plans from the solution space.

3   Primitive Rationales for Elimination of Plans

In order to prevent obviously failing to pass type check or discharge proof obligations, developers should take care of dependencies among phenomena. For example, a variable of a traffic light's color requires a carrier set of "colors of traffic lights". Another aspect is *equality of preserved variables* imposed as proof obligations. Since it is impossible to broaden the scope in which a state variable changes through a refinement, for example, a variable of "the number of cars outside the mainland" must be specified together with all of its possible transitions "incremented (a car enters)" and "decremented (a car exits)".

Given these dependencies, an artifact a depends on phenomena dep(a), i.e., introduced in the same step, such that
dep$(a) = \bigcup_{p \in \text{inc}(a)} \{$phenomenon $q | p$ depends on $q\}$ where inc(a) denotes phenomena that directly appear in a. A set of artifacts A depends on phenomena DEP$(A) = \bigcup_{a \in A}$ dep(a).

In a refinement plan $PL = [A_1, A_2, \cdots, A_n, \cdots]$, the phenomena introduced in the step n are: intro$(PL, n) = $ DEP$(A_n) - \bigcup_{i=1 \cdots n-1}$ DEP(A_i). This is a set of phenomena required for the artifacts A_n and have not been introduced yet.

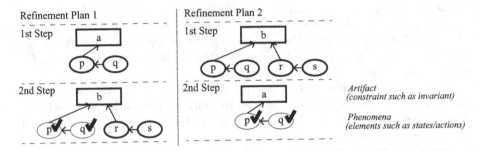

Fig. 2. Modeling of Refinement Plans

Suppose a simple system includes two constraints a and b, illustrated in Fig. 2. Dependencies are represented as arrows in the figure. The left side shows one refinement plan $[\{a\}, \{b\}]$ (i.e., a is introduced in the first step and b in the second step). The first step introduces elements p and q according to the dependencies $(\text{intro}([\{a\}, \{b\}], 1) = \text{DEP}(\{a\}) = \{p, q\})$. The second step introduces (only) r and s (enclosed by bold lines) $(\text{intro}([\{a\}, \{b\}], 2) = \text{DEP}(\{b\}) - \text{DEP}(\{a\}) = \{r, s\})$. The another plan $[\{b\}, \{a\}]$ includes a step that do not introduce any phenomena. Since $\text{intro}([\{b\}, \{a\}], 2) = \text{DEP}(\{a\}) - \text{DEP}(\{b\}) = \emptyset$, the second step only requires to verify a without extending the previous model (the verification can be actually done in the first step simultaneously). This plan can be said *invalid*, failing to mitigate the complexity through several steps.

It is notable that although this elimination may seem artificial, the authors found it often explains intuitively natural ordering or merging: "abstract artifacts (dependent on fewer phenomena) first" and "conceptually-close artifacts (dependent on the same phenomena) together".

One more point to consider is how to prevent empty steps. Suppose a situation $\text{DEP}(B) \subset \bigcup_{i=1\cdots n} \text{DEP}(A_i)$. Then any plans including a step only for B later than n ($[A_1, \cdots, A_n, \cdots, B, \cdots]$) make the step empty. These may be trivially prevented by introducing B together with other artifacts that require some phenomena ($[A_1, \cdots, A_n, \cdots, (B \cup C), \cdots]$). However, in this way the step becomes strange to talk about B that has nothing to do with the phenomena introduced in that step. Thus the solution should be $[A_1, \cdots, (A_n \cup B), \cdots]$, where $\text{DEP}(B) \subset \bigcup_{i=1\cdots n} \text{DEP}(A_i)$ and $\text{DEP}(B) \not\subset \bigcup_{i=1\cdots n-1} \text{DEP}(A_i)$. Intuitively, there is no step in which an artifact is introduced but no phenomena are introduced for the artifacts. In other words, artifacts are introduced as soon as all the phenomena required for them are introduced.

Further rationales can be considered to eliminate invalid plans in terms of common refinement intentions. These rationales are represented as *ordering* or *merging* rules of phenomena or artifacts. An ordering rule "x no later than y" eliminates plans that introduce y earlier than x. A merging rule "x and y together" eliminates plans that introduce them in different steps.

```
1: function NEXTPLANS(plan)                1: function TOGETHER(plan, artifact)
2:     next_plans ← ∅                      2:     together ← ∅
3:     for all artifact a not introduced do 3:     new_plan ← plan + [{artifact}]
4:         next_step ← {a}∪ TOGETHER(plan, a) 4:     for all artifact a not introduced
5:         next_plan ← plan + [next_step]   5:         s.t. a ≠ artifact do
6:         if next_plan does not violate any spec- 6:         new_plan' ← plan + [{artifact, a}]
        ified ordering rules then           7:         if PHENOMENA(new_plan)
7:             Add next_plan to next_plans  8:             = PHENOMENA(new_plan') then
8:         end if                           9:             Add a to together
9:     end for                             10:         end if
10:     return next_plans                  11:     end for
11: end function                           12:     return together
                                           13: end function
```

Fig. 3. Core of Our Proposed Method

4   Integration into Proposed Method

As the preliminary phase, the proposed method expects extraction and analysis of artifacts, phenomena, and dependencies among them. Although a requirements document usually just describes concrete artifacts, Event-B models often have abstract version of artifacts described in the requirements document and *gluing invariants* that state relationships among variables or constants of different steps. Since abstracted artifacts and gluing artifacts are not usually specified in a requirements document, it is also necessary to explicitly prepare them.

Refinement plans can be seen as a tree with a root corresponding to a plan of zero-steps and leaves corresponding to plans that include all artifacts. Our method is based on depth-first search of the tree, in which at each node at least one artifact is introduced that has not been introduced. Fig. 3 describes the core of our method. The function NEXTPLANS generates next steps of a plan. This corresponds to generating child nodes of a node in a search tree. Suppose the search process is looking at a specific node in the tree that has a partial plan $[A_1, \cdots, A_n]$. To construct the next step, each of the artifacts that have not been introduced are checked as a candidate one by one (Line 3 in NEXTPLANS). At this point, the artifacts are identified and introduced together such that they can be introduced without introducing any additional phenomena (Lines 4–11 in TOGETHER). The function PHENOMENA calculates all the phenomena introduced together when an artifact is introduced, reflecting the given merging rules. In this way, the search proceeds by efficiently generating the plans in complying with the core rationale and rules.

Although the definition of refinement planning started with artifacts, the result plans also define phenomena introduced in each step. Sometimes it is easier to understand by looking at phenomena, e.g., this step introduces traffic lights and constraints supported by them. Thus the result view can be in terms of artifacts, phenomena, or both. Fig. 4 shows the result view of the case study to be presented. Each node shows the whole set of phenomena introduced so far, and specific sets of phenomena introduced in each step are attached with the edges. As shown in the right bottom of the figure, the view is simplified by aggregating a set of choices about ordering and merging.

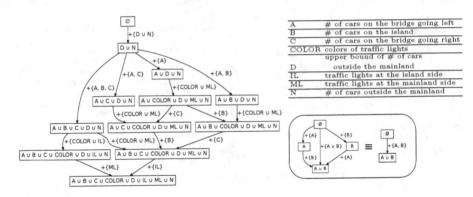

Fig. 4. Result View (introduced phenomena so far)

5   Case Study and Discussion

We applied our method to several examples including the island one (Shown in Fig.1). For the island example, we prepared 41 artifacts, 22 phenomena, and 11 dependencies carefully from the specification. To narrow down plans, we also considered several common refinement intentions such as subsystem decomposition. By running the search algorithm, 68 plan candidates were derived and presented in the simplified phenomena view in the left side of Fig. 4. This number is not so large given the fact it includes some arbitrary ordering. The candidate included the ideal plan explained in the book. Fig.4 is also simplified by labels to denote a meaningful set of phenomena, as the right side of Fig. 4 shows. (e.g., B denotes a set of phenomena related to the number of cars on the island). For example, the leftmost path of the view indicates a plan PL such that $\text{intro}(PL, 1) = D \cup N$, $\text{intro}(PL, 2) \cup \cdots \cup \text{intro}(PL, i) = A \cup B \cup C$, $\text{intro}(PL, i+1) = COLOR \cup IL$, and $\text{intro}(PL, i + 2) = ML$ (where $2 \le i \le 4$).

Thus we succeeded to derive reasonable refinement plans with acceptable amount of efforts. Although the method often requires developers to give smart suggestions such as introducing the conceptually same phenomena together, the method supports iterative procedure for human developers to modify suggestions after looking at derived plans or even after trying modeling and proof.

There are some studies have leveraged methods (Problem Frames [4], UML diagrams [7], BPMN models [2], and KAOS goal models [5]) for the preceding input to Event-B. Although they provide good candidates for representation of refinement plans, they do not discuss how to make decisions on the plans. Our work will be complemented by such high-level intuitions as well as usable notations for the preceding phase.

The authors of this paper had investigated the order in which artifacts are added to the model in order to find refinement plans that distribute phenomena most widely [3]. The study presented in this paper is an extension of the method.

Our work can also be seen as a theoretical foundation to:
- Explain and validate high-level guidelines by intuitions or experiences.
- Evaluate refinement plans with metrics, such as the maximum number of phenomena introduced in one step (i.e., peak complexity) in the plan.
- Mine characteristics of refinement plans.
- Integrate expression-level patterns (e.g., [8]) to generate model templates.

Future work includes support for the input by considering specific preliminary analysis methods based on existing ones such as Problem Frames or object-oriented domain modeling. This will allow for naturally embedding the analysis to create the input into the efforts that are originally necessary. A practical tool with a good interface will be also attractive to support workflows and interactions for our method. Further experiments with developers are also necessary.

References

1. Abrial, J.-R.: Modeling in Event-B: System and Software Engineering. Cambridge University Press (2010)
2. Bryans, J.W., Wei, W.: Formal Analysis of BPMN Models Using Event-B. In: Kowalewski, S., Roveri, M. (eds.) FMICS 2010. LNCS, vol. 6371, pp. 33–49. Springer, Heidelberg (2010)
3. Kobayashi, T., Honiden, S.: Towards Refinement Strategy Planning for Event-B. In: Proceedings of Workshop on Experience and Advances in Developing Dependable Systems in Event-B at The 14th International Conference on Formal Engineering Methods, pp. 69–78. Springer (November 2012)
4. Loesch, F., Gmehlich, R., Grau, K., Jones, C., Mazzara, M.: Report on Pilot Deployment in Automotive Sector. A Deliverable of Deploy Project (2010)
5. Matoussi, A., Gervais, F., Laleau, R.: A Goal-Based Approach to Guide the Design of an Abstract Event-B Specification. In: 16th IEEE International Conference on Engineering of Complex Computer Systems (ICECCS), pp. 139–148. IEEE (2011)
6. Requet, A.: BART: A tool for automatic refinement. In: Börger, E., Butler, M., Bowen, J.P., Boca, P. (eds.) ABZ 2008. LNCS, vol. 5238, pp. 345–345. Springer, Heidelberg (2008)
7. Said, M.Y., Butler, M., Snook, C.: Language and Tool Support for Class and State Machine Refinement in UML-B. In: Cavalcanti, A., Dams, D.R. (eds.) FM 2009. LNCS, vol. 5850, pp. 579–595. Springer, Heidelberg (2009)
8. Traichaiyaporn, K., Aoki, T.: Refinement Tree and Its Patterns: a Graphical Approach for Event-B Modeling. In: Artho, C., Ölveczky, P.C. (eds.) FTSCS 2013. CCIS, vol. 419, pp. 246–261. Springer, Heidelberg (2013)

Templates for Event-B Code Generation

Andrew Edmunds

University of Southampton, UK

Abstract. The Event-B method, and its tools, provide a way to formally model systems; Tasking Event-B is an extension facilitating code generation. We have recently begun to explore how we can configure the code generator, for deployment on different target systems. In this paper, we describe how templates can be used to avoid hard-coding 'boilerplate' code, and how to merge this with code generated from the formal model. We have developed a lightweight approach, where tags (i.e. tagged mark-up) can be placed in source templates. The template-processors we introduce may be of use to other plug-in developers wishing to merge a 'source' text file with some generated output.

1 Introduction

Rodin [2] is a platform for the rigorous specification of critical systems with Event-B [1]. Tasking Event-B [3,4,5,6] is an extension to Event-B that facilitates generation of source code. We can generate Java, Ada, C for OpenMP [9], and C for the Functional Mock-up Interface (FMI) standard [8]. The work reported in this paper has been undertaken during the ADVANCE project [7], which is primarily concerned with co-simulation of Cyber-Physical Systems. This paper introduces an approach that uses templates, with code injection, to facilitate the re-use of boilerplate code.

Often, when a software system is being implemented, much of the code is related to a particular target implementation; and is independent of the state, and behaviour, of the part of the system being formally modelled. Example include the code for system life-cycle management, system health monitoring, or task scheduling. We introduce a simple Eclipse extension, to facilitate the use of templates, with tagged mark-up. We can then merge the code, generated from the formal model, and the templates. This facilitates re-use of existing code, and most importantly, avoids the need to hard-code such details in the translator. The template creator can add tags to the boiler-plate code. These define locations where other templates are expanded; or define code injection points, and meta-data generators. The tags are associated with pre-defined code fragment-generators. The approach is suitable for use with any text-based source and target. To validate the approach, it was used in a C code generator, which was used to generate C code for our work with FMI in Advance. We provide a brief overview of FMI, and code generation with Tasking Event-B, in Sect. 2. We introduce templates, and show an example of their use, in Sect. 3, and conclude in Sect. 4.

Y. Ait Ameur and K.-D. Schewe (Eds.): ABZ 2014, LNCS 8477, pp. 284–289, 2014.

2   Background

We illustrate the approach, using the example of our Event-B-to-FMI translator. So we provide some background on the Functional Mock-Up Interface (FMI)
standard [8]. It is a tool-independent standard, developed to facilitate the exchange, and re-use, of modelling components in the automotive industry. It is
a C-based standard, defining an interface for re-usable components, known as
Functional Mock-up Units (FMUs). FMUs play the role of slave simulators in
simulations that are coordinated by a simulation master. The master simulator
is not defined in the FMI standard, but its job is to coordinate the simulation
e.g. by stopping and starting slaves. It also manages the communication; where
all the slaves' input and output values are communicated via the master, never
directly between slaves. To target the FMI co-simulation framework, we generate code for an FMU from the Event-B model. An FMU is a compressed file
containing an XML description of the model being simulated, and the shared
libraries required to run the simulation. In our work the shared libraries, and
model description, are generated from the Event-B model. To conform to the
FMI standard, FMU implementers must provide API functions for simulation
life-cycle management, such as instantiating a slave, initialising a slave's variables, and terminating the slave. Many of these functions are not dependent on
the particular model being simulated; the code is the same for all models. We
wish to avoid hard-coding the translation where possible; so, templates provide
a place to define the boilerplate code, and code injection can be used for the
model specific parts.

Tasking Event-B [5] is an extension to the Event-B language; an
implementation-level, specification language. When annotations are added to
a machine, it provides additional information to assist with code generation.
When generating code, it is usually necessary to work with a subset of implementable Event-B constructs. Machines can be implemented as task/thread-like
constructs; shared, monitor-like constructs; or provide simulations of the environment. The machine *Types* are *Autotask*, *Shared* and *Environ* respectively.
In embedded systems, *autotask* Machines typically model *controller* tasks (of
the implementation). An *autotask* machine has a task body which contains flow
control (algorithmic) constructs. The syntax of the *Task body* follows,

> Task Body ::= TaskBody ; TaskBody
> ‖ **IF** Event [**ELSEIF** Event]* **ELSE** Event **END**
> ‖ **DO** Event **END** ‖ Event ‖ EventSynch ‖ output

These elements have program-related Event-B semantics. The *Sequence* (;) construct is used for imposing an order on events, and maps to a sequence operator
in programming languages. **IF** provides a choice, with optional sub-branches, between a number of events (it can only be used with events with disjoint guards,
and where completeness must be shown). It maps to branching program statements, where guards are mapped to conditions and actions map to assignments.
DO specifies event repetition while its guard remains true. It maps to a looping
statement, with the loop condition derived from the event guard. *Event* is a

single event, where just its action is mapped to a program statement (assignment), and guards are not permitted. *EventSynch* describes synchronization between an event in an *autotask* machine and an event in a *shared* machine. Synchronization must be implemented as an atomic subroutine call. The *EventSync* construct facilitates subroutine parameter declarations, and substitution in calls, by pairing ordered Event-B parameter declarations.

Fig. 1 shows how an abstract model may be refined, decomposed, and then refined again to the implementation-level (i.e. above the horizontal line annotated with *Event-B*). The code generation phase (below this line) is a two-step

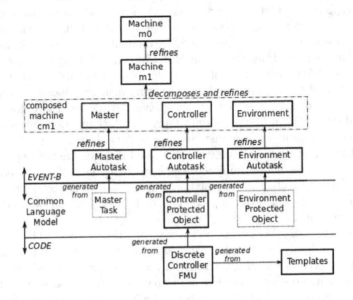

Fig. 1. The Code Generation Process

process; although only a single step is visible to the user. The first step is to translate the Event-B machine to a language-neutral model, the Common Language Model (CLM). During the second step, when the source code is being generated, the templates contribute to the generated code.

3   Using Templates

An architectural overview of our template-driven approach can be seen in the diagram of Fig. 2. We see the artefacts involved in template processing; namely text-based templates, code-fragment generators, text output, meta-data output and a template-processor that does the work. The templates may contain plain-text (which is copied verbatim to the target during processing) and tags. The tags may refer to other templates, or code-fragment generators. The code-fragment generators are hard-coded generators that relate to certain aspects of the final output; for instance, a fragment generator inserts the variable initialisations as

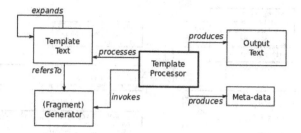

Fig. 2. The Template Processor and Artefacts

```
//## <addToHeader>
fmiStatus fmiInitializeSlave(fmiComponent c,
  fmiReal relativeTolerance, fmiReal tStart,
  fmiBoolean stopTimeDefined, fmiReal tStop){
    fmi_Component* mc = c;
    //## <initialisationsList>
    //## <stateMachineProgramCounterIni>
    return fmiOK;
}
```

Fig. 3. An Example Template

specified in a template. We can see an example of this in Fig. 3. The template shows part of an implementation of the FMI API's *fmiInitializeSlave* function; the code in the template is common to all of the FMUs that we will generate for a particular target configuration. The tags accommodate variability between models; e.g. FMUs keep track of state-variables, which may be different for each model. These state-variables correspond exactly to the variables of the system that have been modelled in Event-B. In the function shown, the first parameter is the *fmiComponent*, the 'instance' of the FMU that is to be initialised. The other parameters relate to the simulation life-cycle. In the template, we insert a place-holder (which we call a *tag*), where we want variable initialisation to occur. The tags in our example begin with the character string, //##. The line continues with an *identifier*, <*identifier*>. A tag is usually (but not always) used as an insertion point; its *identifier* can relate to another template (to be expanded and processed in-line); or the name of a fragment-generator. The fragment-generator is a Java class that can be used to generate code; or meta-data that is stored for later use, in the code generation process (see Fig. 2). In the example we have three tags. The first tag *addToHeader* identifies a generator that creates meta-data, which are used at a later stage, for generation of a header file.

It is possible to categorize the users of Rodin into several types of users. One such type are the 'ordinary' modellers, using Event-B in smaller organisations. But for large scale use, one may have meta-modellers (to develop product lines for instance), and another level of user may instantiate models (of the product line). There may also be platform developers, that provide platform tools for use by meta-modellers, modellers and product-line implementers. The extension

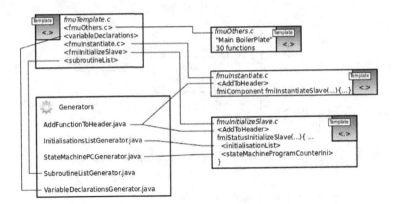

Fig. 4. The Templates and Generators in the FMI Code Generator

```
public class InitialisationsListGenerator implements IGenerator {
  public List<String> generate(IGeneratorData data){
    List<String> outCode = new ArrayList<String>();
    Protected prot = null;
    IL1TranslationManager tm = null;
    //(1) Un-pack the GeneratorData
    List<Object> dataList = data.getDataList();
    for (Object obj : dataList) {
      if (obj instanceof Protected) {prot = (Protected) obj; }
      else if(obj instanceof IL1TranslationManager){
        tm = (IL1TranslationManager) obj;}}
    ...
    //(2) Get the Declarations
    EList<Declaration> declList = prot.getDecls();
    //(3) Process each Variable Declaration/Initialisation
    for (Declaration decl : declList) {
      ...
      String initialisation = FMUTranslator.updateFieldVariableName(...);
      outCode.add(initialisation);
    }
    // (4) return the new fragment
    return outCode;}}
```

Fig. 5. An Example Fragment-Generator

points allow the platform developer to provide template utilities for the other users. They can define new tags and fragment-generators. An overview of the templates and generators used in the FMI translation, can be seen in Fig. 4 (much of the detail is omitted for brevity). The *root* template is *fmuTemplate.c*, from this we can navigate to all of the other templates, and generators. The root template generates variable declarations and the subroutines, and expands the main boilerplate functions in *fmuOthers.c*. The *fmuInstantiate* and *fmuIni-tialise* templates generate the corresponding FMI API function implementations. From the diagram we can see that these rely on generators to do some of the translation. The template-processor scans each line, and copies the output; or inserts new text, or meta-data as required, until we reach a generator tag. The class's *generate* method is invoked, to begin the process of text insertion. A fragment of the *InitialisationListGenerator* class can be seen in Fig. 5. The main steps are highlighted using numbered comments in the code. In step 1, the data is un-packed; in step 2, the declarations are obtained from the *Protected* object; in step 3, the initialisation are translated, and add to an array of initialisation statements; in step 4, the initialisations are returned to the template-processor.

4   Conclusions

Using the approach that we have described in this paper, we are able to perform target configuration prior to code-generation; and re-use boilerplate code, without having to hard-code it. The template-processor reads each line of a template and copies the contents, verbatim, to a target file unless a template tag is encountered. A tag can refer to another template, which is processed by expanding it in-line, or a custom fragment generator. As part of an extensible approach, a platform developer can enrich the template language, by adding new template tags and associate them with custom fragment-generators. In this way complex code generation activities can be performed, to generate text output, or to generate meta-data in other formats. The meta-data is useful for downstream code generation. We used the template-driven approach to implement part of a new code generator, translating Event-B models to FMI-C code.

Acknowledgement: Funded by FP7 ADVANCE Project www.advance-ict.eu.

References

1. Abrial, J.R.: Modeling in Event-B: System and Software Engineering. CUP (2010)
2. Abrial, J.R., Butler, M., Hallerstede, S., Hoang, T.S., Mehta, F., Voisin, L.: Rodin: An Open Toolset for Modelling and Reasoning in Event-B. Software Tools for Technology Transfer 12(6), 447–466 (2010)
3. Edmunds, A.: Providing Concurrent Implementations for Event-B Developments. PhD thesis, University of Southampton (March 2010)
4. Edmunds, A., Butler, M.: Linking Event-B and Concurrent Object-Oriented Programs. In: Refine 2008 - International Refinement Workshop (May 2008)
5. Edmunds, A., Butler, M.: Tasking Event-B: An Extension to Event-B for Generating Concurrent Code. In: PLACES 2011(February 2011)
6. Edmunds, A., Colley, J., Butler, M.: Building on the DEPLOY Legacy: Code Generation and Simulation. In: DS-Event-B-2012: Workshop on the Experience of and Advances in Developing Dependable Systems in Event-B (2012)
7. The Advance Project Team. Advanced Design and Verification Environment for Cyber-physical System Engineering, http://www.advance-ict.eu
8. The Modelica Association Project. The Functional Mock-up Interface, https://www.fmi-standard.org/
9. The OpenMP Architecture Review Board. The OpenMP API specification for parallel programming, http://openmp.org/wp/

The BWare Project: Building a Proof Platform for the Automated Verification of B Proof Obligations*

David Delahaye[1], Catherine Dubois[2], Claude Marché[3], and David Mentré[4]
(for the BWare project consortium**)

[1] Cedric/Cnam/Inria, Paris, France
[2] Cedric/ENSIIE/Inria, Évry, France
[3] Inria Saclay - Île-de-France & LRI, CNRS, Univ. Paris-Sud, Orsay, France
[4] Mitsubishi Electric R&D Centre Europe, Rennes, France

Abstract. We introduce BWare, an industrial research project that aims to provide a mechanized framework to support the automated verification of proof obligations coming from the development of industrial applications using the B method and requiring high integrity. The adopted methodology consists in building a generic verification platform relying on different automated theorem provers, such as first order provers and SMT (Satisfiability Modulo Theories) solvers. Beyond the multi-tool aspect of our methodology, the originality of this project also resides in the requirement for the verification tools to produce proof objects, which are to be checked independently. In this paper, we present some preliminary results of BWare, as well as some current major lines of work.

Keywords: B Method, Proof Obligations, First Order Provers, SMT Solvers, Logical Frameworks, Industrial Use, Large Scale Study.

1 Presentation

The BWare project is an industrial research project, funded by the INS ("Ingénierie Numérique & Sécurité") programme of the French National Research Agency (ANR), which aims to provide a mechanized framework to support the automated verification of proof obligations coming from the development of industrial applications using the B method and requiring high integrity. The BWare consortium gathers academic entities, i.e. Cedric, LRI, and Inria, as well as industrial partners, i.e. Mitsubishi Electric R&D Centre Europe, ClearSy, and OCamlPro.

The methodology used in this project consists in building a generic platform of verification relying on different automated theorem provers, such as first order

---

* This work is supported by the BWare project (ANR-12-INSE-0010), funded for 4 years by the INS programme of the French National Research Agency (ANR) and started on September 2012. For more details, see: http://bware.lri.fr/.
** The BWare project consortium consists of the following partners: Cedric, LRI, Inria, Mitsubishi Electric R&D Centre Europe, ClearSy, and OCamlPro.

Y. Ait Ameur and K.-D. Schewe (Eds.): ABZ 2014, LNCS 8477, pp. 290–293, 2014.
© Springer-Verlag Berlin Heidelberg 2014

provers and SMT (Satisfiability Modulo Theories) solvers. This generic platform is built upon the Why3 platform [2] for deductive program verification. The considered first order provers are Zenon [4] and iProver Modulo [5], while we opted for the Alt-Ergo SMT solver [1]. The variety of these theorem provers aims to allow a wide panel of proof obligations to be automatically verified by our platform. The major part of the verification tools used in BWare were already involved in some experiments, which consisted in verifying proof obligations or proof rules coming from industrial applications.

Beyond the multi-tool aspect of our methodology, the originality of BWare also resides in the requirement for the verification tools to produce proof objects, which are to be checked independently. To verify these proof objects, we consider two proof checkers: the Coq proof assistant and the Dedukti universal proof checker [3]. These backends should allow us to increase confidence in the proofs produced by the considered automated theorem provers.

To test our platform, a large collection of proof obligations is provided by the industrial partners of the project, which develop tools implementing the B method and applications involving the use of the B method.

2   Preliminary Results

Currently, the BWare platform is already available and works as shown on Fig. 1. The proof obligations are initially produced by Atelier B. They are then translated by a specific tool into Why3 files, which are compatible with a Why3 encoding of the B set theory [8]. From these files, Why3 can produce (by means of appropriate drivers) the proof obligations for the automated theorem provers, using the TPTP format for Zenon and iProver Modulo, and a native format for Alt-Ergo. This translation together with the encoding of the B set theory aims to generate valid statements that are appropriate for the automated theorem provers, i.e. whose proofs can be found by these provers. Finally, once proofs have been found by these tools, some of these provers can generate proof objects to be verified by proof checkers. This is the case of Zenon, which can produce proof objects for Coq and Dedukti [4,7], and iProver Modulo, which can also produce proof objects for Dedukti [6].

In order to assess the BWare platform, two industrial partners of the project provided proof obligations coming from several industrial applications. In particular, Mitsubishi Electric R&D Centre Europe provided the proof obligations of a complete railway level crossing system use case, while ClearSy provided the proof obligations coming from three deployed industrial projects. This constitutes an initial bench of more than 10,500 proof obligations on which we evaluate the BWare platform. The results obtained at the beginning of the project are as follows: the main prover (mp) of Atelier B (4.0) is able to prove 84% of these proof obligations, while Alt-Ergo (0.95.1) obtains a rate of 58%, iProver Modulo (over iProver 0.7) 19%, and Zenon (0.7.2) less than 1%. As can be observed, the first order provers (iProver Modulo and especially Zenon) encounter difficulties, which can be explained by the fact that these provers do not know the B set theory.

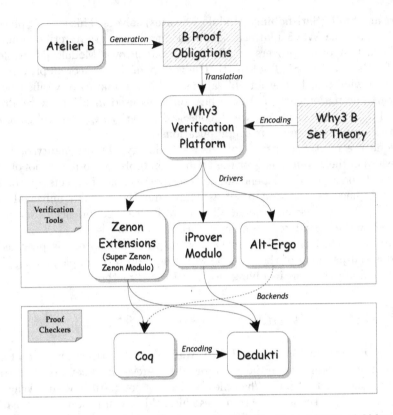

Fig. 1. The BWare Platform for the Automated Verification of B Proof Obligations

Some of the current lines of work of the project therefore focus on extending the first order provers to make them able to reason modulo the B set theory. Regarding SMT solvers, an intermediate set of results obtained with improved versions of Alt-Ergo is given at the OCamlPro blog[1]. These are very promising results: the development version of Alt-Ergo is now able to automatically discharge more than 98% of the proof obligations.

3   Current Lines of Work

The BWare project consists of several tasks, which cannot be exhaustively described in this paper due to space restrictions. We focus on two major current lines of work of the project.

The first current line of work is upstream and consists in completing the axiomatization of the B set theory in Why3 in order to be able to consider all the provided proof obligations. This is mainly carried out according to what is described in [8], i.e. by adding B constructs to the axiomatization and modifying

---

[1] Available at the following address: http://www.ocamlpro.com/blog/2013/10/22/
alt-ergo-evaluation-october-2013.html .

accordingly the translator of proof obligations from Atelier B to Why3. This line of work will allow us to consider a broad scope of proof obligations related to different application domains and test the scalability of our platform as well.

A second current line of work focuses on the first order provers to make them able to reason modulo the B set theory. To do so, we rely on deduction modulo. The theory of deduction modulo is an extension of predicate calculus, which allows us to rewrite terms as well as propositions, and which is well suited for proof search in axiomatic theories, as it turns axioms into rewrite rules. Both first order provers considered in the project have been extended to deduction modulo to obtain Zenon Modulo [7] and iProver Modulo [5]. Both tools have also been extended to produce Dedukti proofs [7,6], which is natural as Dedukti relies on deduction modulo as well. Currently, most of the efforts in this line of work consist in building a B set theory modulo, which is appropriate for automated deduction and keeps some properties such as cut-free completeness.

In the longer term and among the tasks of the project, we plan to do a more extensive benchmarking of the different provers of the project in order to determine which proof coverage ratio we can obtain from our platform (in particular, after the development of the several extensions of the provers). Ultimately, we intend to disseminate and exploit the results of our project by integrating our platform into Atelier B and therefore realizing a multi-prover output of Atelier B.

References

1. Bobot, F., Conchon, S., Contejean, V., Iguernelala, M., Lescuyer, S., Mebsout, A.: *Alt-Ergo, version 0.95.2.* CNRS, Inria, and Université Paris-Sud (2013), http://alt-ergo.lri.fr
2. Bobot, F., Filliâtre, J.-C., Marché, C., Paskevich, A.: Why3: Shepherd Your Herd of Provers. In: Leino, K.R.M., Moskal, M. (eds.) International Workshop on Intermediate Verification Languages, Boogie, pp. 53–64 (2011)
3. Boespflug, M., Carbonneaux, Q., Hermant, O.: The $\lambda\Pi$-Calculus Modulo as a Universal Proof Language. In: Pichardie, D., Weber, T. (eds.) Proof Exchange for Theorem Proving, PxTP, vol. 878, pp. 28–43. CEUR Workshop Proceedings (2012)
4. Bonichon, R., Delahaye, D., Doligez, D.: Zenon: An Extensible Automated Theorem Prover Producing Checkable Proofs. In: Dershowitz, N., Voronkov, A. (eds.) LPAR 2007. LNCS (LNAI), vol. 4790, pp. 151–165. Springer, Heidelberg (2007)
5. Burel, G.: Experimenting with Deduction Modulo. In: Bjørner, N., Sofronie-Stokkermans, V. (eds.) CADE 2011. LNCS (LNAI), vol. 6803, pp. 162–176. Springer, Heidelberg (2011)
6. Burel, G.: A Shallow Embedding of Resolution and Superposition Proofs into the $\lambda\Pi$-Calculus Modulo. In: Blanchette, J.C., Urban, J. (eds.) Proof Exchange for Theorem Proving (PxTP). EPiC, vol. 14, pp. 43–57. EasyChair (2013)
7. Delahaye, D., Doligez, D., Gilbert, F., Halmagrand, P., Hermant, O.: Zenon Modulo: When Achilles Outruns the Tortoise Using Deduction Modulo. In: McMillan, K., Middeldorp, A., Voronkov, A. (eds.) LPAR-19. LNCS, vol. 8312, pp. 274–290. Springer, Heidelberg (2013)
8. Mentré, D., Marché, C., Filliâtre, J.-C., Asuka, M.: Discharging Proof Obligations from Atelier B Using Multiple Automated Provers. In: Derrick, J., Fitzgerald, J., Gnesi, S., Khurshid, S., Leuschel, M., Reeves, S., Riccobene, E. (eds.) ABZ 2012. LNCS, vol. 7316, pp. 238–251. Springer, Heidelberg (2012)

Tuning the Alt-Ergo SMT Solver
for B Proof Obligations

Sylvain Conchon[1,2] and Mohamed Iguernelala[3,1]

[1] LRI, Université Paris-Sud, 91405 Orsay, France
[2] INRIA Saclay – Ile-de-France, Toccata, 91893 Orsay, France
[3] OCamlPro SAS, 91190 Gif-sur-Yvette, France

Abstract. In this paper, we present recent developments in the Alt-Ergo SMT-solver to efficiently discharge proof obligations (POs) generated by Atelier B. This includes a new plugin architecture to facilitate experiments with different SAT engines, new heuristics to handle quantified formulas, and important modifications in its internal data structures to boost performances of core decision procedures. Experiments realized on more than 10,000 POs generated from industrial B projects show significant improvements.

Keywords: SMT solvers, B Proof Obligations, B Method.

1 The Alt-Ergo SMT Solver

Alt-Ergo is an open-source SMT solver capable of reasoning in a combination of several built-in theories such as uninterpreted equality, integer and rational arithmetic, arrays, records, enumerated data types and AC symbols. It is the unique SMT solver that natively handles polymorphic first-order quantified formulas, which makes it particularly suitable for program verification. For instance, Alt-Ergo is used as a back-end of SPARK and Frama-C to discharge proof obligations generated from Ada and C programs, respectively.

Recently, we started using Alt-Ergo in the context of the ANR project BWare [8] which aims at integrating SMT solvers as back-ends of Atelier B. The proof obligations sent to Alt-Ergo are extracted from Atelier B as logical formulas that are combined with a (polymorphic) model of B's set theory [7]. This process relies on the Why3 platform [6] which can target a wide range of SMT solvers. However, we show (Section 2) on a large benchmark of industrial B projects that it is not immediate to obtain a substantial gain of performances by using SMT solvers. Without a specific tunning for B, Alt-Ergo together with other SMT solvers compete just equally with Atelier B's prover on those industrial benchmarks.

In this paper, we report on recent developments in Alt-Ergo that significantly improve its capacities to handle POs coming from Atelier B. Our improvements are: (1) better heuristics for instantiating polymorphic quantified formulas from B model; (2) new efficient internal data structures; (3) a plugin architecture to facilitate experiments with different SAT engines; and (4) the implementation of a new CDCL-based SAT solver.

Y. Ait Ameur and K.-D. Schewe (Eds.): ABZ 2014, LNCS 8477, pp. 294–297, 2014.

2 Benchmarks

Currently, the test-suite of BWare contains 10572 formulas[1] obtained from three industrial B projects provided by ClearSy. The first one, called DAB, is an automated teller machine (ATM). The last two, called P4 and P9, are obfuscated and unsourced programs.

The formulas generated from these projects are composed of two parts. The first one is the *context*: a large set of axioms (universally and polymorphic quantified formulas). This part contains the B's set theory model, as well as huge (in size) predicates describing the B state machines. The second part of each PO is the *goal*: a ground formula involving these predicates (see [7] for detailed explanations about the structure of these POs).

A quick look to the shape of these formulas shows that they mainly contain equalities over uninterpreted function symbols and atoms involving enumerated data types. Only a small portion of atoms contains linear integer arithmetic and polymorphic records. In comparison with our other benchmarks coming from deductive program verification platforms, the average number of axioms, as well as the size of the POs are much larger in this test suite, as shown below:

	number of POs	avg. number of axioms	avg. size (Ko)
VSTTE-Comp	125	32	8
Why3's gallery	1920	41	9
Hi-Lite	3431	125	23
DAB	860	257	236
P4	9341	259	248
P9	371	284	402

At the beginning of the project and without specific improvements, we ran some SMT solvers (z3, cvc3, and Alt-Ergo) on this test-suite. All measures were obtained on a 64-bit machine with a quad-core Intel Xeon processor at 3.2 GHz and 24 GB of memory. Solvers were given a time limit of 60 seconds and a memory limit of 2 GB for each PO. The results of our experiments are reported in the following table.

Provers Versions	Alt-Ergo 0.95.1	Alt-Ergo 0.95.2	z3 4.3.1	cvc3 2.4.1
DAB	707 82.2 %	**822** **95.6 %**	716 83.3%	684 79.5 %
P4	4709 50.4 %	**8402** **89.9 %**	7974 85.4 %	7981 85.4 %
P9	181 48.8 %	**213** **57.4 %**	162 43.7 %	108 29.1 %
Total	5597 52.9 %	**9437** **89.3 %**	8852 83.7 %	8773 83.0 %

---

[1] These benchmarks will be available for the community at the end of the project.

For every solver, we report the number (and the percentage) of solved POs for each project. Except for Alt-Ergo *0.95.1*, the other versions of SMT solvers compete equally. From what we know from our BWare partners, these results are similar to those obtained by the prover of Atelier B 4.0, which proves **84%** of this test-suite. Concerning Alt-Ergo, the low rate obtained with version *0.95.1* is due to a quantifier instantiation heuristic that disabled the use of large axioms and predicates during proof search. This was unfortunate for the BWare benchmark (especially in P4 and P9) since goals mainly involve large predicates, as explained above. The minor release *0.95.2* mainly relaxes this (misguided) heuristic.

3   Improvements

Profiling Alt-Ergo on this test-suite shows: (1) a large number of axiom instancia-tions; (2) a high activity of the congruence closure decision procedure; and (3) an important workload for the SAT engine. A more thorough investigation shows that (1) is due to some *administrative* axioms of the B model that represent properties of basic set theoretic operations (AC properties, transitive closures, etc). The structure of theses instances mixes Boolean operators and equalities over uninterpreted terms, which explains the behavior (2) and (3).

One of our main objective was to efficiently handle such axioms by limiting the number of their instances. For that, we added the possibility of disabling the generation of new instances modulo known ground equalities. We also modified the default behavior of the matching algorithm to only consider terms that are in the active branch of the SAT engine. Finally, we have reimplemented literal representation and some parts of the Formula module of Alt-Ergo to enable trivial contextual simplifications, and thus to identify equivalent formulas. In addition, we have modified the core architecture of Alt-Ergo to facilitate the use of different SAT-solvers, implemented as plugins.

The results in the following table show the impact of our optimizations. The *master branch* version of Alt-Ergo contains all the modifications described above. As we can see, this version outperforms *0.95.2*. The last column contains the result of a wrapper, called Ctrl-Alt-Ergo, that uses the time given to the solver to try different strategies and heuristics[2].

Versions	Alt-Ergo *0.95.2*	Alt-Ergo *master branch*	Ctrl-Alt-Ergo *master branch*
DAB	822	858	**860**
	95.6 %	99.8 %	**100 %**
P4	8402	8980	9236
	89.9 %	96.1 %	**98.9 %**
P9	213	234	277
	57.4 %	63.1 %	**74.7 %**
Total	9437	10072	**10373**
	89.3 %	95.3 %	**98.1 %**

---

[2] All these improvements will be available in future public releases of Alt-Ergo at
http://alt-ergo.ocamlpro.com

4   Conclusion and Future Works

As demonstrated by our experiments, the results of our first investigations and optimizations are promising. It turns out that B proof obligations have some specificities that should be taken into account to obtain a good success rate.

In the near future, we want to study how to extend the core of Alt-Ergo to handle administrative axioms (definition of union and intersection, AC properties, etc.) as conditional rewriting rules. For that, we are studying the design of a new combination algorithm which will extend our parametrized algorithms AC(X) [2] and CC(X) [3] to handle, in a uniform way, user-defined rewriting systems. This would allow us to handle a fragment of the set theory of B in a built-in way.

We also plan to investigate whether *deduction modulo* techniques, like those in Zenon Modulo [4,5] and iProver Modulo [1], would be helpful to efficiently handle quantified formulas in the SMT world. Another line of research would be the generation of verifiable traces for external proof checkers in order to augment our confidence in the SMT solver.

References

1. Burel, G.: Experimenting with Deduction Modulo. In: Bjørner, N., Sofronie-Stokkermans, V. (eds.) CADE 2011. LNCS (LNAI), vol. 6803, pp. 162–176. Springer, Heidelberg (2011)
2. Conchon, S., Contejean, E., Iguernelala, M.: Canonized Rewriting and Ground AC Completion Modulo Shostak Theories: Design and Implementation. Logical Methods in Computer Science 8(3) (2012)
3. Conchon, S., Contejean, E., Kanig, J., Lescuyer, S.: CC(X): Semantic Combination of Congruence Closure with Solvable Theories. Electronic Notes in Theoretical Computer Science 198(2), 51–69 (2008)
4. Delahaye, D., Doligez, D., Gilbert, F., Halmagrand, P., Hermant, O.: Proof Certification in Zenon Modulo: When Achilles Uses Deduction Modulo to Outrun the Tortoise with Shorter Steps. In: International Workshop on the Implementation of Logics (IWIL), Stellenbosch (South Africa). EasyChair (December 2013) (to appear)
5. Delahaye, D., Doligez, D., Gilbert, F., Halmagrand, P., Hermant, O.: Zenon Modulo: When Achilles Outruns the Tortoise Using Deduction Modulo. In: McMillan, K., Middeldorp, A., Voronkov, A. (eds.) LPAR-19. LNCS, vol. 8312, pp. 274–290. Springer, Heidelberg (2013)
6. Filliâtre, J.-C., Paskevich, A.: Why3 — Where Programs Meet Provers. In: Felleisen, M., Gardner, P. (eds.) ESOP 2013. LNCS, vol. 7792, pp. 125–128. Springer, Heidelberg (2013)
7. Mentré, D., Marché, C., Filliâtre, J.-C., Asuka, M.: Discharging Proof Obligations from Atelier B Using Multiple Automated Provers. In: Derrick, J., Fitzgerald, J., Gnesi, S., Khurshid, S., Leuschel, M., Reeves, S., Riccobene, E. (eds.) ABZ 2012. LNCS, vol. 7316, pp. 238–251. Springer, Heidelberg (2012)
8. The Bware Project (2012), http://bware.lri.fr/

Fixed-Point Arithmetic Modeled
in B Software Using Reals

Jérôme Guéry[1], Olivier Rolland[2], and Joris Rehm[1]

[1] ClearSy, 320, avenue Archimède Les Pléiades III,
Bât A, 13857 Aix-en-Provence Cedex 3, France
{jerome.guery,joris.rehm}@clearsy.com
[2] ALSTOM Transport, 48, rue Albert Dhalenne, 93482 Saint-Ouen, France
olivier.rolland@transport.alstom.com

Abstract. This paper demonstrates how to introduce a fixed point arithmetic in software developed with the classical B method. The properties of this arithmetic are specified with real numbers in the *AtelierB* formal tool and linked to an implementation written in Ada programming language. This study has been conducted to control the loss of precision and possible overflow due to the use of fixed point arithmetic in the critical software part of a communication based train control system.

Keywords: formal method, B method, fixed-point arithmetic.

1 Introduction

Our context is the industrial production of the control software involved in a new line of Communication Based Train Control (CBTC) system. As designers, we are responsible for the on-time delivery, the operational disponibility and the safety of the produced software. To ensure that, we use the *B method* [1] in order to model the software module specification and refine it formally into concrete code (and translating B0 source code into Ada source code). The B method has been successfully applied in the industry of urban transportation and is also a good and efficient way to achieve the high level of quality and confidence required by the standards and needed of a critical software.

In the railway control industry, the effective use of floating-point numbers is quite sparse. But in order to improve the precision and the safety of the control software we have decided to use a customized implementation of a fixed-point arithmetic. This allows us to represent an important range of numbers with good precision. And this allows us to express every value in the international system of units.

However the traditional B language and tools do not support real numbers or computable approximations. In a first look, this custom fixed-point arithmetic is incompatible with the use of the B method as it only includes integer numbers. But recently, types and operators for real numbers (and floating-points numbers) were added to the *Atelier B* tool-set. This allows us to specify our arithmetic

Y. Ait Ameur and K.-D. Schewe (Eds.): ABZ 2014, LNCS 8477, pp. 298–302, 2014.

operations while using an abstract set representing our custom implementation of fixed-point.

The goals we want to achieve are: to study with a high degree of confidence the possible precision loss due to the fixed-point arithmetic and the possible overflow; to formally show that the computed values satisfy properties with respect to the theoretical values and the safety issue of the CBTC system; and to apply a dimensional analysis as a strong typing discipline on the manipulated values.

2 Proposal

In railway critical systems, the program is traditionally executed on hardware architectures lacking floating-point units. A single signed implementation of fixed point arithmetic is provided to represent real data type in software. This implementation uses 64 bits: 1 bit is used to represent the sign of the number; 31 bits are used to represent the integer portion of the number; 32 bits are used to represent the fractional portion of the number.

The Ada type corresponding to this implementation is called T_Real_64. The processor used in our system does not include a floating point unit, our implementation is done with two 32 bits integer.

The implementation of the fixed point arithmetic is indeed used for every physical data. Each physical data are expressed in SI units and have its own Ada type as a partial protection against variable misuse for hand-coded source, subtyping T_Real_64. This gives us a dimensional analysis of the manipulated values.

On the specification side, experimental real numbers support was added to the *AtelierB* since version 4.1.0[1]. The *REAL* type has been added as a basic type: *AtelierB* considers *INTEGER* and *REAL* as two distinct types. As a consequence, it is not possible to mix real numbers and integers in a same equation without type casting. Arithmetic operators for *REAL* have been added as textual operators (such as *rle* for $\leq$).

We define $Real\_64\_To\_Real$ as a total injective function from the finite set T_Real_64 to $REAL\_64$, the set of reals between the minimum and maximum implementable reals:

$$MAX\_REAL\_64 = 2.0^{31} - 2^{-32}$$
$$MIN\_REAL\_64 = -2.0^{31} + 2^{-32}$$
$$REAL\_64 = MIN\_REAL\_64..MAX\_REAL\_64$$
$$Real\_64\_To\_Real \in T\_Real\_64 \rightarrowtail REAL\_64$$

This function creates a relation between the abstract set, representing implemented values, to their interpreted real number values.

---

[1] http://www.atelierb.eu/wp-content/uploads/AtelierB-4.1.0-release-notes.pdf

Each arithmetic operations (addition, subtraction, multiplication, division, square) must be modelled as an operation in a B basic machine..

All operations compute approximation of real numbers, but all properties are defined with real numbers. So two essential properties on those operations have to be specified: the protection against overflow of the result; and the properties on the precision of the result.

This overflow protection is defined as a pre-condition to the operations. The theoretical result of an operation on the operands should be defined in the set of implementable real numbers.

For example, for the addition $Res \leftarrow Add\_Real\_64(Left, Right)$, the protection against overflow can be expressed as:

$$Real\_64\_To\_Real(Left) + Real\_64\_To\_Real(Right) \in REAL\_64$$

in the precondition of the operation.

Therefore, the addition can be modelled in B with:

```
Res <-- Add_Real_64 (Left, Right) =
PRE
    Left   : T_Real_64 & Right : T_Real_64 &
    Real_64_To_Real (Left) rplus Real_64_To_Real (Right) : REAL_64
THEN
    Res := Real_64_To_Real~ (Real_64_To_Real (Left) rplus
             Real_64_To_Real (Right))
END
```

In fixed point arithmetics, some operations, like multiplication or division, differ from the theoretical value computation. The project uses multiple implementations of the same operation for different platforms. This leads us to specify the following precision for each operation:

$$Res = Real\_64\_To\_Real^{-1}(Real\_64\_To\_Real(Left) \cdot Real\_64\_To\_Real(Right))$$
$$\pm 2^{-32}$$

For example, the multiplication can be modelled in B with the same pattern than the addition but with a more complex definition of the resulting value due to the possible precision loss.

```
Res <-- Mul_Real_64 (Left, Right) =
(...)
    ANY l_Value WHERE
        l_Value : REAL_64 &
        l_Value rle Real_64_To_Real (Left) rmul
            Real_64_To_Real (Right) rplus 2.0 rpow -32 &
        l_Value rge Real_64_To_Real (Left) rmul
            Real_64_To_Real (Right) rminus 2.0 rpow -32
    THEN Res := Real_64_To_Real~(l_Value) END
END
```

2.1   Physical Types and Dimensional Analysis

Each physical type is defined as a subset of T_Real_64, limited to plausible values in the railway context of the project. As an example, $30m/s = 108km/h$ is a maximum plausible speed of a subway.

In Ada T_Speed is defined as a *new type* of T_Real_64:

```
type T_Speed is new T_Real_64 of range -30.0..30.0;
```

In a B specification we re-use the same pattern as before by defining a set for the physical unit and a injective function between this set and the acceptable range of real numbers.

The only operations needed in basic machines are operations for casting from or to real numberss.

Arithmetic operations on physical types can be entirely modelled in B, implemented in B0 and automatically translated in Ada.

3   Conclusion

With this model, we are able to develop and prove a software with a specification containing real number arithmetic and a concrete code containing a custom fixed-point computation. In addition to that, the arithmetic computations are strongly typed with a dimensional analysis expressed in the international system of unit. The types of the B model can be translated into Ada types, which allows us to mix the B coded module with hand made Ada coded module (input-output or unsafe parts, for example). This strong typing policies lets the Ada run-time perform dynamic checks during the program execution (for example the check of the numeric range of types). The type cast between numerical of physical type is explicit and allows easier safety analysis of the program behavior.

On the other side, we can study statically the precision loss of actual computation and show formally that the computed values are safe approximations of the theoretical result values. This specification expressions allows again an easier safety analysis of the CBTC system issues. We can also study statically the possible numerical overflow by taking realistic hypotheses on the input interval values and combining those hypotheses along with the computation.

We can see here that our overall goals can be achieved with this methodological pattern, but also we experiment difficulty with the formal proof as there is a lack of tool support for the real arithmetic. Due to this, this study remains an experiment but is representative of the goals we follow and the method we aim to apply. Therefore we are working on a better support of the proof obligations containing real arithmetic. The next step for that will be to design a set of *proof rules* (deduction rules or axioms that ground the formal mechanism of the *AtelierB* tool) for the real numbers. Using basic machines (B machine without implementation) to specify the Ada packages of fixed-point number is another possible issue. Indeed, with this kind of interface between Ada and B, the specification found in the machine is just, in fact, a set of hypothesis taken over Ada code and thus is only manually validated. It could be possible to build formally the core fixed-point code in B also but from an industrial point of view we need to precisely control the low-level code that runs over our embedded computer platform.

Reference

1. Abrial, J.-R.: The B-Book. Cambridge University Press (1996)

Bounded Model Checking
of Temporal Formulas with Alloy[*]

Alcino Cunha

HASLab – High Assurance Software Laboratory
INESC TEC & Universidade do Minho, Braga, Portugal
alcino@di.uminho.pt

Abstract. Alloy is a formal modeling language based on first-order relational logic, with no native support for specifying reactive systems. We propose an extension of Alloy to allow the specification of temporal formulas using LTL, and show how they can be verified by bounded model checking with the Alloy Analyzer.

1 Introduction

Alloy is a formal modeling language based on first-order relational logic [5]. Its Analyzer enables model validation and verification by translation to off-the-shelf SAT solvers. Alloy's logic is quite generic and does not commit to a particular specification style. In particular, there is no predefined way to specify and verify reactive systems, and several idioms and extensions have been proposed to address this issue. However, it is rather cumbersome and error-prone to specify and verify temporal properties with such idioms. In this paper we propose the usage of standard *Linear Temporal Logic* (LTL) to specify reactive systems in Alloy, and show how bounded model checking can be performed with its Analyzer, by resorting to the technique first proposed by Biere et al [1].

This paper is structured as follows. Section 2 shows how reactive systems and temporal properties can be specified and verified using Alloy and its Analyzer. Section 3 discusses how such properties can be specified more easily in LTL and then translated to Alloy, avoiding some of the potential problems pointed out in the previous section. Finally, we discuss some related work in Section 4.

2 Verifying Reactive Systems in Alloy

In Alloy, a *signature* represents a set of atoms. An atom is a unity with three fundamental properties: it is indivisible, immutable and uninterpreted. A signature declaration can introduce *fields*, sets of tuples of atoms capturing *relations*

---

[*] This work is financed by the ERDF – European Regional Development Fund through the COMPETE Programme (operational programme for competitiveness) and by national funds through the FCT – Fundação para a Ciência e a Tecnologia (Portuguese Foundation for Science and Technology) within project FCOMP-01-0124-FEDER-037281.

Y. Ait Ameur and K.-D. Schewe (Eds.): ABZ 2014, LNCS 8477, pp. 303–308, 2014.

```
open util/ordering[State]

sig State    {}
sig Message    { to, from : one Partition }
sig Channel    { messages : Message set -> State }
sig Partition { port : one Channel, pifp : set Partition }

fact NoSharedChannels { all c : Channel | lone port.c }

pred send    [m : Message, s,s' : State] { ... }
pred receive [m : Message, s,s' : State] { ... }
pred transfer [m : Message, s,s' : State] {
  m in m.from.port.messages.s and m.to in (m.from).pifp
  messages.s' = messages.s - m.from.port->m + m.to.port->m
}

fact Init  { no Channel.messages.first }
fact Trans { all s : State, s' : s.next |
  some m : Message | send[m,s,s'] or receive[m,s,s'] or transfer[m,s,s']
}
```

Fig. 1. A PIFP specification in Alloy

between the enclosing signature and others. Model constraints are defined by
facts. Assertions express properties that are expected to hold as consequence
of the stated facts. *Commands* are instructions to perform particular analysis.
Alloy provides two commands: run, that instructs the Analyzer to search for an
instance satisfying a given formula, and check, that attempts to contradict a
formula by searching for a counterexample.

Since fields are immutable, to capture the dynamics of a state transition system a special signature whose atoms denote the possible states must be declared.
Without loss of generality, we will denote this signature as State. Every field
specifying a mutable relation must then include State as one of the signatures
it relates. There are two typical Alloy *idioms* to declare such mutable fields:
declare them all inside State, or add State as the last column in every mutable
field declaration. The former idiom is known as *global state*, since all mutable
fields are grouped together, while the latter is known as *local state*, since mutable
fields are declared in the same signature as non-mutable ones of similar type.

Figure 1 presents an example of an Alloy model conforming to the local state
idiom. It is a simplified model of the *Partition Information Flow Policy* (PIFP)
of a *Secure Partitioning Kernel*. Essentially, the PIFP statically defines which
information flows (triggered by message passing) are authorized between partitions. Signature Message declares the fields to and fom, capturing its destination
and source Partition, respectively. Communication is done via channels, here
simplified to contain sets of messages. Obviously, the messages contained in a
channel vary over time. As such, signature Channel declares a mutable relation messages, that associates each channel with the set of its messages in each

state. Signature `Partition` declares two binary relations: `port`, denoting the channel used by the partition to communicate, and `pifp`, that captures to which partitions it is authorized to send messages.

In Alloy *everything is a relation*. For example, sets are unary relations and variables are just unary singleton relations. As such, relational operators (in particular the *dot join* composition) can be used for various purposes. For example, in the fact `NoSharedChannels` the relational expression `port.c` denotes the set of partitions connected to port `c`. Multiplicities are also used in several contexts to constrain or check the cardinality of a relation. For example, in the same fact, multiplicity `lone` ensures that each channel is the port of at most one partition.

An operation can be specified using a predicate `pred op[...,s,s':State]` specifying when does `op` hold between the given pre- and post-states: to access the value of a mutable field in the pre-state (or post-state) it suffices to compose it with `s` (respectively `s'`). Our model declares three operations: `send` and `receive` message (whose specifications are omitted due to space limitations), and `transfer` message between partitions. The `send` operation just deposits the message in the sending partition port – the `transfer` operation is executed by the kernel and is the one responsible to enforce the PIFP. Inside an operation, a formula that does not refer `s'` can be seen as a pre-condition. Otherwise it is a post-condition. For example, `m in m.from.port.messages.s` is a pre-condition to `transfer`, requiring `m` to be in the port of the source partition prior to its execution. Notice that frame conditions, specifying which mutable relations remain unchanged, should be stated explicitly in each operation.

To specify temporal properties we need to model execution traces. A typical Alloy idiom for representing finite prefixes of traces is to impose a total ordering on signature `State` (by including the parameterized module `util/ordering`) and force every pair of consecutive states to be related by one of the operations (see fact `Trans`). Inside module `util/ordering`, the total order is defined by the binary relation `next`, together with its `first` and `last` states. The initial state of our example, where all channels are empty, is constrained by fact `Init`. A desirable safety property states that the port of every partition only contains messages sent from authorized partitions:

```
assert Safety { all p : Partition, m : Message | all s : State |
   m.to = p and m in p.port.messages.s implies p in m.from.pifp
}
```

Model checking this assertion with the Alloy Analyzer yields a counter-example: every partition can send a message to itself, even if not allowed by the PIFP. To correct this problem we can, for example, add to our model the fact `all p:Partition | p in p.pifp`, stating that all partitions should be allowed to send messages to themselves. Non-invariant temporal assertions can also be expressed with this idiom, but the complexity of the formulas and expertise required by the modeler increases substantially. Consider, for example the liveness property stating that all authorized messages are eventually transferred to the destination. At first glance, it could be specified as follows (using transitive closure to access the successors of a given state):

```
assert Liveness { all p : Partition, m : Message |
  all s : State | m in p.port.messages.s and m.to in p.pifp implies
    some s' : s.*next | m in m.to.port.messages.s'
}
```

A simple (but artificial) way to ensure Liveness in this model is to disallow message sending while there are still pending messages to transfer. This could be done by adding the pre-condition no Channel.messages.s to operation send. However, even with such pre-condition, model-checking assertion Liveness with the Alloy Analyzer yields a false counter-example, where a message is sent in the last state of the trace prefix. In fact, if we consider only finite prefixes of execution traces, it is almost always possible to produce a false counter-example to a liveness property. This problem is well-known in the bounded model-checking community, and the solution, first proposed in [1], is to only consider as (true) counter-examples to such properties prefixes of traces containing a *back loop* in the last state, i.e., those that actually model infinite execution traces. It is easy to define a parameterized trace module[1], that adapts util/ordering to specify potential infinite traces instead of total orders, by allowing such back loop in the last state. Module trace also defines a predicate infinite that checks if the loop is present in a trace instance: if so, next always assigns a successor to every state, thus modeling an infinite trace. For convenience a dual predicate finite is also defined. By replacing open util/ordering[State] with open trace[State], the above liveness property can now be correctly specified (and verified) as follows:

```
assert Liveness { all p : Partition, m : Message |
  all s : State | m in p.port.messages.s and m.to in p.pifp implies
    finite or some s' : s.*next | m in m.to.port.messages.s'
}
```

3 Embedding LTL Formulas in Alloy

As seen in the previous section, although we can specify and verify (by bounded model checking) temporal properties in standard Alloy, it is a rather tricky and error-prone task, in particular since the user must be careful about where to check for finitude of trace prefixes. As such, we propose that, instead of using explicit quantifiers over the states in a trace, such properties be expressed using the standard LTL operators: X for next, G for always, F for eventually, U for until, and R for release. For example, the above temporal properties could be specified as follows:

```
assert Safety { all p : Partition, m : Message |
  G (m.to = p and m in p.port.messages implies p in m.from.pifp)
}
```

---

[1] Available at http://www.di.uminho.pt/~mac/Publications/trace.als.

$\llbracket$X $\phi\rrbracket_s$ ≡ some s.next and $\llbracket\phi\rrbracket_{s.\text{next}}$
$\llbracket$G $\phi\rrbracket_s$ ≡ infinite and all $s':s$.*next $|\ \llbracket\phi\rrbracket_{s'}$
$\llbracket$F $\phi\rrbracket_s$ ≡ some $s':s$.*next $|\ \llbracket\phi\rrbracket_{s'}$
$\llbracket\phi$ U $\psi\rrbracket_s$ ≡ some $s':s$.*next $|\ \llbracket\psi\rrbracket_{s'}$ and all $s'':s$.*next & ^next.$s'\ |\ \llbracket\phi\rrbracket_{s''}$
$\llbracket\phi$ R $\psi\rrbracket_s$ ≡ $\llbracket$G $\psi\rrbracket_s$ or some $s':s$.*next $|\ \llbracket\phi\rrbracket_{s'}$ and all $s'':s$.*next & *next.$s'\ |\ \llbracket\psi\rrbracket_{s''}$

$\llbracket$not $\phi\rrbracket_s$ ≡ not $\llbracket\phi\rrbracket_s$ $\llbracket\Phi$. $\Psi\rrbracket_s$ ≡ $\llbracket\Phi\rrbracket_s$. $\llbracket\Psi\rrbracket_s$
$\llbracket\phi$ and $\psi\rrbracket_s$ ≡ $\llbracket\phi\rrbracket_s$ and $\llbracket\psi\rrbracket_s$ $\llbracket\Phi$ & $\Psi\rrbracket_s$ ≡ $\llbracket\Phi\rrbracket_s$ & $\llbracket\Psi\rrbracket_s$
$\llbracket\phi$ or $\psi\rrbracket_s$ ≡ $\llbracket\phi\rrbracket_s$ or $\llbracket\psi\rrbracket_s$ $\llbracket\Phi$ + $\Psi\rrbracket_s$ ≡ $\llbracket\Phi\rrbracket_s$ + $\llbracket\Psi\rrbracket_s$
$\llbracket$all $x:\Phi\ |\ \phi\rrbracket_s$ ≡ all $x:\llbracket\Phi\rrbracket_s\ |\ \llbracket\phi\rrbracket_s$ $\llbracket\Phi$ -> $\Psi\rrbracket_s$ ≡ $\llbracket\Phi\rrbracket_s$ -> $\llbracket\Psi\rrbracket_s$
$\llbracket$some $x:\Phi\ |\ \phi\rrbracket_s$ ≡ some $x:\llbracket\Phi\rrbracket_s\ |\ \llbracket\phi\rrbracket_s$ $\llbracket*\Phi\rrbracket_s$ ≡ *$\llbracket\Phi\rrbracket_s$
$\llbracket\Phi$ in $\Psi\rrbracket_s$ ≡ $\llbracket\Phi\rrbracket_s$ in $\llbracket\Psi\rrbracket_s$ $\llbracket$none$\rrbracket_s$ ≡ none

$$\llbracket x\rrbracket_s \equiv \begin{cases} x.s & \text{if } x \text{ is the id of a mutable field declared with the local state idiom} \\ s.x & \text{if } x \text{ is the id of a mutable field declared with the global state idiom} \\ x & \text{otherwise (i.e., a variable or the id of an immutable field)} \end{cases}$$

Fig. 2. Embedding of temporal formulas

```
assert Liveness { all p : Partition, m : Message |
 G (m in p.port.messages and m.to in p.pifp implies
   F (m in m.to.port.messages))
}
```

Assuming traces are specified with module `trace`, the embedding of LTL into Alloy can be done via an (almost) direct encoding of the translation proposed by Biere et al. [1] (for bounded model checking of LTL with a SAT solver). Formally, a formula ϕ occurring in a fact or `run` command should be replaced by $\llbracket NNF(\phi)\rrbracket_{\text{first}}$, where $\llbracket\phi\rrbracket_s$ is the embedding function defined in Figure 2, and $NNF(\phi)$ is the well-known transformation that converts formula ϕ to *Negation Normal Form* (where all negations appear only in front of atomic formulas). When finding a model for G ϕ, only prefixes capturing infinite traces should be considered, thus assuring that ϕ is not violated further down the trace. As clarified in [1], conversion to NNF is necessary since in the bounded semantics of LTL the duality of G and F no longer hold. The embedding of logic and relational operators is trivial, and thus only a representative subset of Alloy's logic is presented. A formula ϕ occurring in an assertion or `check` command should be replaced by not $\llbracket NNF(\text{not } \phi)\rrbracket_{\text{first}}$. Since assertions in check commands are negated in order to find counter-examples, the outermost negation ensures they still remain in NNF.

Note that, to improve efficiency (and likewise to `util/ordering`), when the `trace` module is imported the scope of the parameter signature is interpreted as an exact scope. This means that trace prefixes are bounded to be of size equal to the scope of the `State` signature. Thus, to perform bounded model checking of an assertion, the user should manually increase the scope of `State` one unit at a time up to the desired bound.

4   Related Work

Several extensions of Alloy to deal with dynamic behavior have been proposed. DynAlloy [3] proposes an Alloy variant that allows the specification of properties over execution traces using a formalism inspired by dynamic logic. Imperative Alloy [6] proposes a more minimal extension to the language, with a simple semantics by means of an embedding to standard Alloy. Unfortunately, in both these works the verification of liveness properties may yield spurious counter-examples, similar to the one presented in Section 2.

One of the advantages of our approach is that reactive systems can be specified declaratively using Alloy's relational logic, as opposed to traditional model checkers where transitions must be specified imperatively. Chang and Jackson [2] proposed a BDD-based model checker for declarative models specified with relational logic enhanced with CTL temporal formulas. The current proposal shows how the Alloy Analyzer can directly be used to perform bounded model checking of temporal formulas without the need for a new tool.

Recently, Vakili and Day [7] showed how CTL formulas with fairness constraints can be model checked in Alloy, by using the encoding to first order logic with transitive closure first proposed by Immerman and Vardi [4]. Their technique performs full model checking on state transition systems specified declaratively, but bounded to have at most the number of states specified in the scope. This non-standard form of bounded model checking can yield non-intuitive results in many application scenarios, or even prevent verification at all if the the specification cannot be satisfied by a transition system that fits in the (necessarily small) scope of State. Moreover, instead of proposing an Alloy extension, CTL formulas are expressed using library functions that compute the set of states where the formula holds. This leads to unintuitive specifications, since the user is then forced to use relational operators to combine formulas instead of the standard logical connectives.

References

1. Biere, A., Cimatti, A., Clarke, E., Zhu, Y.: Symbolic model checking without BDDs. In: Cleaveland, W.R. (ed.) TACAS 1999. LNCS, vol. 1579, pp. 193–207. Springer, Heidelberg (1999)
2. Chang, F., Jackson, D.: Symbolic model checking of declarative relational models. In: ICSE, pp. 312–320. ACM (2006)
3. Frias, M., Galeotti, J., Pombo, C., Aguirre, N.: DynAlloy: upgrading Alloy with actions. In: ICSE, pp. 442–451. ACM (2005)
4. Immerman, N., Vardi, M.: Model checking and transitive-closure logic. In: Grumberg, O. (ed.) CAV 1997. LNCS, vol. 1254, pp. 291–302. Springer, Heidelberg (1997)
5. Jackson, D.: Software Abstractions - Logic, Language, and Analysis. MIT Press (2012) (revised edition)
6. Near, J.P., Jackson, D.: An imperative extension to Alloy. In: Frappier, M., Glässer, U., Khurshid, S., Laleau, R., Reeves, S. (eds.) ABZ 2010. LNCS, vol. 5977, pp. 118–131. Springer, Heidelberg (2010)
7. Vakili, A., Day, N.A.: Temporal logic model checking in Alloy. In: Derrick, J., Fitzgerald, J., Gnesi, S., Khurshid, S., Leuschel, M., Reeves, S., Riccobene, E. (eds.) ABZ 2012. LNCS, vol. 7316, pp. 150–163. Springer, Heidelberg (2012)

Formal Verification of OS Security Model with Alloy and Event-B

Petr N. Devyanin[1], Alexey V. Khoroshilov[2], Victor V. Kuliamin[2],
Alexander K. Petrenko[2], and Ilya V. Shchepetkov[2]

[1] Educational and Methodical Community of Information Security, Moscow, Russia
peter_devyanin@hotmail.com
[2] Institute for System Programming, Russian Academy of Sciences, Moscow, Russia
{khoroshilov,kuliamin,petrenko,shchepetkov}@ispras.ru

Abstract. The paper presents a work-in-progress on formal verification
of operating system security model, which integrates control of confidentiality and integrity levels with role-based access control. The main
goal is to formalize completely the security model and to prove its consistency and conformance to basic correctness requirements concerning
keeping levels of integrity and confidentiality. Additional goal is to perform data flow analysis of the model to check whether it can preserve
security in the face of certain attacks. Alloy and Event-B were used for
formalization and verification of the model. Alloy was applied to provide
quick constraint-based checking and uncover various issues concerning
inconsistency or incompleteness of the model. Event-B was applied for
full-scale deductive verification. Both tools worked well on first steps of
model development, while after certain complexity was reached Alloy
began to demonstrate some scalability issues.

1 Introduction

Complexity of practical security models used in the industrial and government
systems makes their analysis a hard work. A well-known, but not well adopted
yet, approach to decrease the effort needed for such an analysis is usage of
formal modeling with a variety of features supported by modern tools like model
checking, constraint checking, deductive verification, etc.

In this paper we present a work-in-progress on formal analysis of mandatory
entity-role model security of access control and information flows of Linux-based
operating system (MROSL DP) [1], which includes lattice-based mandatory access control (MAC), mandatory integrity control (MIC), and role-based access
control (RBAC) mechanisms [2] and is intended to be implemented inside Linux
as a specific Linux Security Module (LSM) [3]. The analysis performed is to
check model consistency and conformance to basic correctness requirements -
ability to prevent break of integrity and confidentiality levels of data and processes. In addition a kind of data-flow analysis is executed to prove that the
model is able to preserve security in the face of certain attacks, namely, to keep
high-integrity data and processes untouched even if an attacker gets full control
over some low-integrity processes.

Y. Ait Ameur and K.-D. Schewe (Eds.): ABZ 2014, LNCS 8477, pp. 309–313, 2014.

Alloy [4] and Event-B [5] (Rodin [6]) are used for formalization of the model and for its analysis and verification. Alloy was applied to provide constraint-based checking of operation contracts and to uncover various issues concerning inconsistency or incompleteness of the model. Event-B was applied for full-scale deductive verification of the model correctness.

Further sections of the paper describe some details of the security model analyzed, provide the statistics of tool usage, summarize the results obtained, and depict further development of the project.

2   Main Features of the Model

The model under analysis MROSL DP [1] describes security mechanisms of Linux-based operating system in terms of user accounts (representing users), entities (representing data objects under control - files, directories, sockets, shared memory pieces, etc.), sessions (representing processes working with data), and roles (representing arrays of rights on usage or modification of data objects).

Each session has the corresponding user account on behalf of which it works. A session has a set of roles assigned to it and a set of accesses to various entities (which it reads or writes), both these sets can be modified. Entities can be grouped into other entities (containers) and form a Unix-like filesystem (with containers-directories and hard links making possible for an entity to belong to several containers). Roles also form a filesystem-like structure, where roles-containers are used to group the rights of all the included roles.

The main security mechanisms presented in the model are the following.

- *RBAC*. Each operation performed by a session should be empowered by a corresponding right included in some of the current roles of the session.
- *MIC*. Each entity, session, or role has integrity level - high or low. Modification of high-integrity entities or roles by low-integrity sessions or through low-integrity roles is prohibited.
- *MAC*. Each entity, session, or role has security label. Security is described by two attributes - a level of ordered confidentiality (unclassified, confidential, etc.) and a set of unordered categories (e.g., whether the corresponding information concerns financial department or research department). Security labels are partially ordered according to order of levels and inclusion of category sets. Read access to an entity is possible only for sessions (or through roles) having greater-or-equal security labels. Write access is possible only for sessions (or through roles) having exactly the same security labels.

The model defines 34 operations in form of contracts - preconditions and postconditions. Operations include actions on creation, reading, or modification of user accounts, sessions, entities, roles, rights of roles, accesses of a session. For the purpose of data-flow analysis of attacks additional 10 operations are defined, which describes control capture of a session and information flows by memory or time between sessions and entities.

The model also includes a variety of specific details, like shared containers, container clearance required (CCR) attributes of containers (integrity or confidentiality ignorance flags), write-only entities (like /dev/null), administrative roles used to operate with other roles, and so on. Space restrictions prevent us from discussing them here, but these details are responsible for a large part of model's complexity.

3 Formalization and Verification Process and Its Results

Alloy and Event-B are used for model formalization. Both formalizations are close and consist of type definitions for basic model concepts (user accounts, sessions, entities, roles, rights, accesses), state data structure, basic invariants representing well-definedness of data objects (wd-invariants), invariants representing correctness of a state or conformance to basic requirements presented in the list in the previous section (c-invariants), and operation contracts.

For now data-flow-related constraints and operations are not yet recorded in the tool-supported formal models, so the further information is provided for incomplete models.

Some statistics on models elements and size is presented in the Table 1. Lines of code numbers shown are rounded to tens.

Table 1. Models' composition and size

	Alloy model		Event-B model	
	Number	Lines of code	Number	Lines of code
Type definitions	20	100	9	1
State variables	25	40	43	1
wd-invariants	17	300	76	200
c-invariants	31	230	31	120
Operation contracts	32	2200	32	1300
Total		2900		1650

Let $I(s)$ denote that all invariants hold in the state s, for each operation one have precondition $pre(s)$ and postcondition $post(s, s')$, the latter depending on pre-call state s and post-call state s'. We try to ensure that the statement $I(s) \wedge pre(s) \wedge post(s, s') \Rightarrow I(s')$ holds for all operations specified. Alloy is used for quick finding omissions of necessary constraints in operation contracts or invariants by means of constraint checking, which can find small counterexamples for wrong statements. Several omissions and inconsistencies were found in the original model with the help of Alloy. Rodin with corresponding plugins was used to prove formally all the same statements in Event-B model. Rodin generates from 30 to 90 assertions for an operation, from which up to 25% require human intervention to prove. In total proof of about 15% of generated assertions need human aid.

When the number of invariants in the model exceeded 20, Alloy stopped to generate counterexamples for wrong statements. Sometimes we met this problem before, but then it can be resolved by increasing the number of top-level objects. Since more than 20 invariants are used, this doesn't help or just causes memory overflow. We haven't managed to use ProB [7] tool for constraint checking, mostly due to model complexity.

The following issues of Event-B make the work harder: lack of possibility to use auxiliary predicates in contracts, and lack of possibility to introduce some abbreviations, which can help to ease the notation rigidity, for example, to use more suitable notation for second and third elements of triples. This can be implemented with lambda-expressions, but they hinder automated proofs. Additional problem for possible work splitting is provided by the structure of proof log stored. Since it is represented as XML, its size in our example is about 200 MB, and usually a little change in the proof of some assertion strangely result in many changes in the XML-file stored. So it is hard to split the proof elaboration between several persons, since merging significantly changed big files is hard.

4   Conclusion and Future Work

The presented in the paper work on formalization and verification of the security model of Linux-like OS is not finished yet, but the results obtained already can be used to make some conclusions. First, the project as usual shows that formalization of requirements helps to uncover some bugs that can become more serious on the implementation phase. Second, the project demonstrates that sometimes tools supporting formal analysis of models are not scalable enough for industrial use, but other suitable tools can be usually found.

On the next steps of the project we are going to finalize model formalization and verification with the help of Event-B. Then, the security mechanisms modeled are to be implemented as an LSM and the implementation code is to be verified on conformance to the model with the help of Why [8] platform.

References

1. Devyanin, P.N.: The models of security of computer systems: access control and information flows. Hot line - Telecom, Moscow (2013) (in Russian)
2. Bishop, M.: Computer security: art and science. Pearson Education Inc., Boston (2002)
3. Wright, C., Cowan, C., Smalley, S., Morris, J., Kroah-Hartman, G.: Linux Security Modules: General Security Support for the Linux Kernel. In: Proc. of the 11th USENIX Security Symposium, pp. 17–31 (2002)
4. Jackson, D.: Software Abstractions: Logic, Language, and Analysis. MIT Press (2006)
5. Abrial, J.-R.: Modeling in Event-B: System and Software Engineering. Cambridge University Press (2010)

6. Abrial, J.-R., Butler, M., Hallerstede, S., Hoang, T.S., Mehta, F., Voisin, L.: Rodin: An Open Toolset for Modelling and Reasoning in Event-B. International Journal on Software Tools for Technology Transfer 12(6), 447–466 (2010)
7. Leuschel, M., Butler, M.: ProB: A Model Checker for B. In: Araki, K., Gnesi, S., Mandrioli, D. (eds.) FME 2003. LNCS, vol. 2805, pp. 855–874. Springer, Heidelberg (2003)
8. Bobot, F., Filliâtre, J.-C., Marché, C., Paskevich, A.: Why3: Shepherd Your Herd of Provers. In: Proc. of Boogie 2011: 1st Intl Workshop on Intermediate Verification Languages, pp. 53–64 (2011)

Detecting Network Policy Conflicts Using Alloy

Ferney A. Maldonado-Lopez[1,2], Jaime Chavarriaga[1], and Yezid Donoso[1]

[1] Universidad de los Andes, Bogotá, Colombia
[2] Universitat de Girona, Girona, Spain
{fa.maldonado1897,ja.chavarriaga908,ydonoso}@uniandes.edu.co

1 Introduction

In Computer Networks, several studies show that 50 to 80% of infrastructure downtime is caused by misconfiguration [1]. Current approaches are aimed to check the configuration of each device and detect conflicts, inconsistencies and bugs, other approaches focus on the specification of the intended behaviour of a network and the automatic configuration of each one of its elements [2].

Path-Based Policy Language (PPL) [3] is a language for *Policy-based network management*. Using PPL, network administrators specify network policies, i.e. a set of rules about how the network must deal with specific types of traffic. Moreover, they can use a PPL compiler to transform these rules into operational parameters for each entity in the corresponding network.

Existing PPL compilers [3,4] perform validations on policy rules to detect conflicts such as contradicting rules for the same network segment. However, this type of compiler only detects conflicts using a subset of the language and cannot detect conflicts such as contradicting rules for security rules of users and groups. This paper presents our ongoing implementation of a PPL compiler based on Alloy with the objective of overcoming these limitations.

2 The Path-Based Policy Language

Path-Based Policy Language (PPL) [3] allows network administrators to specify 1) the network topology, nodes and links; 2) types of paths and network traffic, and 3) policy rules.

Network topology: PPL provides a set of "define" statements that can be used to define network nodes and links. For instance, Figure 1 is the PPL specification of the example network with five nodes shown in Figure 2.

```
define node Alpha , Bravo , Charlie , Delta , Echo ;
define link Alpha_Bravo <Alpha , Bravo >;
define link Alpha_Delta <Alpha , Delta >;
define link Bravo_Charlie <Bravo , Charlie >;
define link Charlie_Delta <Charlie , Delta >;
define link Bravo_Echo <Bravo , Echo >;
define link Echo_Delta <Echo , Delta >;
```

Fig. 1. Topology definition

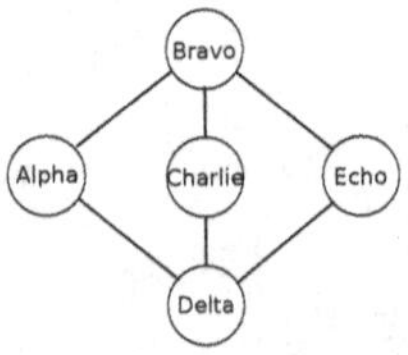

Fig. 2. Example Topology

Y. Ait Ameur and K.-D. Schewe (Eds.): ABZ 2014, LNCS 8477, pp. 314–317, 2014.
© Springer-Verlag Berlin Heidelberg 2014

Path Specifications: In addition to nodes and links, PPL supports the definition of paths, i.e. sequences of nodes and links, in order to define rules over the traffic that goes through them.

Administrators can set a path specifying all the nodes and links in the path, or by using *wildcards*. For instance, a path between *Alpha* and *Echo* may be defined specifying all the nodes using the expression <Alpha,Bravo,Echo>. In addition, a set of paths can be defined using placeholders that can be replaced by any set of nodes. Following the example, the complete set of paths between *Alpha* and *Echo* can be defined as <Alpha,*, Echo>, i.e. denoting all the paths that start in one node and ends in the other.

Policy Specification: PPL supports the definition of constraints over the traffic going through nodes, links or paths in a network. A PPL rule comprises: `policyID` a unique identifier, `userID` the ID of the policy creator, `paths` the paths affected by the policy, `target` the type of traffic and `conditions` that determine when the policy must be applied, and `action_items` describing the intended behaviour of the network.

For instance, a network administrator can define the policy *"the video traffic cannot be transmitted through Charlie"*, using the following PPL statement:

```
policy1 smith {<*,Charlie,*>} {traffic_class=video} {*} {deny}
```

Policy Conflicts: Policy conflicts occur when *two or more* policy rules cannot be satisfied at the same time. For instance, there is a conflict if another administrator define a policy *"the video traffic can be transmitted from Alpha through Charlie to Echo"* using the following statement:

```
policy2 neo {<Alpha,*,Charlie,*,Echo>} {traffic_class=video} {*} {permit}
```

Considering the above mentioned policies, there is a *contradictory* set of policies, consequently, a policy conflict.

3 Detecting Network Policy Conflicts with Alloy

In this section we show how to translate PPL statements into an Alloy model and use the Alloy Analyzer to determine conflicts.

Network Topology: Network nodes and links can be translated into Alloy using signatures and relations. Consider the above example:

```
abstract sig node{}              one sig Alpha, Bravo, Charlie, Delta, Echo
                                 extends node{}
                                 one sig topologyOne extends topology {}
abstract sig topology{           {
  nodes : some node,               nodes = Alpha+Bravo+Charlie+Delta+Echo
  links : node -> node             links = Alpha->Bravo + Alpha->Delta + ...
}                                }
```

316   F.A. Maldonado-Lopez, J. Chavarriaga, and Y. Donoso

Path Abstraction: Network paths can be translated into Alloy signatures and predicates. Informally, each path is a tuple with a set of nodes and links. Two of these nodes are the source and the target. In addition, each path must be acyclic, no loops are allowed, and the target must be accessible from the source, and all nodes in the path from the source must be included as nodes. The following is the specifications of the example paths:

```
abstract sig path {
    nodes : some node,
    links : node -> node,
    source : one nodes,
    target : one nodes
} {
    no n : nodes | n in n.^(links)
    target in source.^(links)
    source.^(links) in nodes
}
```

```
// Path <Alpha, *,Charlie, *, Echo>

pred isAlphaCharlieEcho ( p : path ) {
    isValid[ p, topologyOne ]
    p.source = Alpha
    p.target = Echo

    Charlie in p.nodes
    Charlie in p.source.^(p.links)
}
```

Policy Abstraction: Finally, policy rules can be translated into Alloy facts. Informally, above mentioned rules state that 1) there is not a network flow where the traffic type is video and *Charlie* is included in the path, and 2) exists at least one flow where the traffic type is video, and *Alpha*, *Charlie* and *Echo* are in the set of nodes. The following is the specification of these rules.

```
// smith <*,C,*> {video} {deny}
fact policy1 {
    no f: network_flow{
    some p: f.paths {
        is_C_path [ p ]
    }
    f.flow_type = traffic_type_video
    }
}
```

```
// neo <A,*,C,*,E> {video} {permit}
fact policy2 {
    some f: network_flow {
    some p : f.paths {
        isAlphaCharlieEcho [ p ]
    }
    f.flow_type = traffic_type_video
    }
}
```

Policy Conflicts: Once the Alloy model have been created, policy conflicts can be detected whenever the model is unsatisfiable. In addition, we are able to determine which policies produce the conflict using the Unsat-Core available in Alloy to detect inconsistencies in a specification.

4   Advantages of Translating to Alloy

The existing compilers use algorithms that first expand all the path definitions in sets of paths, then detect the segments where the paths overlap, and finally determine if there are conflicting conditions in the common segments [4]. In contrast, our approach translates PPL into Alloy models. This translation offers some advantages over the previous work:

- *Extended support for wildcards in paths.* The existing compilers use algorithms that support expansion of paths specifications such as {A,*,E}, but not complex path declarations like {*,B,C,*,D,*} [4]. In contrast, these paths can be easily translated into Alloy predicates.
- *Extended support for user and group policies.* Existing compilers find a conflict if some rules permit and other deny the traffic from the same user, but

not if a rule permits the traffic from a user and another rule denies the traffic of the user's group [4]. We overcome this including relations among groups and users as in other approaches for role-based access control[5].

– *Extended support for action conflicts.* The existing PPL compilers have limited support to detect when two rules try to set different values to attributes such as the *priority* or the bandwidth assigned to an specific type of traffic [4]. We improve this including additional constraints checking that two rules do not include conflicting assignations on attributes.

5   Future Work

We have implemented a PPL compiler supported on Alloy to detect conflicts in policy rules. Our compiler exploits relational logic to explore the paths in the network and reason about constraints defined on them in a simpler and a more complete way than the existing compilers.

Currently, we are focused on evaluating alternatives to optimise the conflict detection performance. Mainly, we are considering "model slicing techniques where the compiler analyses subsets of the network instead all the elements. In addition, we are working on a implementation that uses KodKod, the internal library used by Alloy that provides support for partial instances to optimise processing.

Besides PPL, there are other network policy languages focused on concerns such as fault tolerance and security such as FML and Merlin [6,7]. Future work is planned to support these languages and detect conflicts among policies aimed to deal with different concerns.

References

1. Kant, K., Deccio, C.: Security and Robustness in the Internet Infrastructure. In: Handbook on Securing Cyber-Physical Critical Infrastructure. Morgan Kaufmann (2012)
2. Stevens, M., Weiss, W., Mahon, H., Moore, R., Strassner, J., Waters, G., Westerinen, A., Wheeler, J.: IETF policy framework. Technical report, Internet Engineering Task Force, IETF (1999)
3. Stone, G., Lundy, B., Xie, G.: Network policy languages: A survey and a new approach. IEEE Network 15(1), 10–21 (2001)
4. Guven, A.: Speeding up a Path-Based Policy Language compiler. Master's thesis, Naval Postgraduate School, Monterrey, California (2003)
5. Power, D., Slaymaker, M., Simpson, A.: Automatic Conformance Checking of Role-Based Access Control Policies via Alloy. In: Erlingsson, Ú., Wieringa, R., Zannone, N. (eds.) ESSoS 2011. LNCS, vol. 6542, pp. 15–28. Springer, Heidelberg (2011)
6. Hinrichs, T.L., Gude, N.S., Casado, M., Mitchell, J.C., Shenker, S.: Practical declarative network management. In: Proceedings of the 1st ACM Workshop on Research on Enterprise Networking, WREN 2009, pp. 1–10. ACM (2009)
7. Soul, R., Basu, S., Kleinberg, R., Sirer, E.G., Foster, N.: Managing the network with Merlin. In: ACM SIGCOMM Workshop on Hot Topics in Networks, HotNets 2013 (2013)

Staged Evaluation of Partial Instances in a Relational Model Finder

Vajih Montaghami and Derek Rayside

University of Waterloo, Canada
{vmontagh,drayside}@uwaterloo.ca

Abstract. To evaluate a property of the form 'for all x there exists some y' with a relational model finder requires a generator axiom to force all instances of y to exist in the universe of discourse. Without the generator axiom the model finder will produce a spurious counter-example by simply not including an important instance of y. Generator axioms are generally considered to be expensive to evaluate, significantly limiting the scope of the analysis. We demonstrate that evaluating the generator axiom in a separate stage from the property results in substantial improvements in analysis speed and scalability.

1 Introduction

Every integer has a successor: **all** x : **Int** | **some** y : **Int** | y = plus[x,1]. This proposition is true (with overflow arithmetic) regardless of the bitwidth, because we all of the integers in the bitwidth will exist in the universe of discourse. For example, if the bitwidth is set to two the universe will contain the integers $\{-2, -1, 0, 1\}$. What if the Alloy Analyzer were allowed to arbitrarily drop integers from the universe? For example, suppose we set the bitwidth to two but got the universe $\{-2, 0, 1\}$. In this truncated universe the successor proposition does not hold.

A naïve user of Alloy might write a similar proposition for their own signatures, for example to say that every state in a transition system must have a successor: **all** x : **State** | **some** y : **State** | next[x,y]. The Alloy Analyzer will then construct a truncated universe in which some State does not have a successor, and the proposition will be reported as false. In order to make this state successor proposition true the user will need to write a *generator axiom* [11] to force all states to exist. They will probably also write an equality predicate to ensure that all of the generated states have different field values. Finally, the user will need to know how many atoms need to be generated so that they can set the scope of the analysis accordingly.

We propose a new keyword for Alloy, **uniq**, to be applied to user-defined signatures, from which we automatically synthesize the generator axiom and equality predicate. With the user's intent clearly expressed in this manner we can then automatically compute the appropriate scope.

Evaluating generator axioms can be expensive. We also show that evaluating the generator axiom and the $\forall\exists$ query in separate stages can improve speed and

Y. Ait Ameur and K.-D. Schewe (Eds.): ABZ 2014, LNCS 8477, pp. 318–323, 2014.

scalability of the analysis. First we evaluate the generator axiom, then save the result as a partial instance to be used in the evaluation of the ∀∃ query.

There is growing interest in using Alloy for analyzing ∀∃ queries. (*e.g.*, [19,3,8,5]). We show that staging makes the analyses of FORML [5] feasible.

2   Atoms and Objects

Atoms in Alloy are uninterpreted. When the user writes a generator axiom with an associated equality predicate they have an interpretation in mind: that the identity of an atom is defined by the values of its fields. The user signifies this intent with our **uniq** keyword. We shall refer to atoms that the user considers in this way as *objects*, the signatures that declare them as *classes*, and the relations in which they are the first column as *fields*. These are just conceptual terms for the purposes of this paper and are not present in the concrete Alloy syntax.

We require that there be a finite number of objects so that they can be enumerated and saved in a partial instance for subsequent analysis. Consequently, the range of each field must be finite. If the range of a field is a sig then the finite bound comes from the command's scope. If the range of a field is another class, then the objects of that class must be computed first. There is, therefore, a dependency hierarchy of classes that must be acyclic.

The specification might contain a variety of constraints that define which objects are *legal* and which are not. When generating all legal unique objects of a given class we consider only the constraints contained in the appended facts. If other constraints are subsequently found to be in conflict with these ones then the specification will be reported as inconsistent. Appended facts in Alloy are syntactically designed to refer to a single object at a time. We further restrict appended facts on classes to have no quantifiers.

3   Two Techniques for Staged Generation of Objects

We evaluate two approaches for staging evaluation of the generator axioms: using the solution enumeration feature of the underlying SAT solver, and incrementally growing the bound until all legal objects are found.

In both cases we first slice the specification to include just what is necessary to generate the objects: the class declarations and any sig declarations that they depend on, along with the associated appended facts. Other sigs are removed, as well as facts and predicates. The generated objects are saved in a partial instance block [14] for use in evaluating the ∀∃ query in the next stage.

The *incremental generation* approach starts by first solving for a small scope that is expected to be less than the total number of legal objects to be generated. The results are saved as a partial instance block. Then the scope is expanded and the solver invoked again. The newly found objects are added to the partial instance block. This process repeats until the solver returns unsat, indicating that there are no more legal objects to be generated.

Some SAT solvers, such as MiniSat [7], include a feature for *solution enumeration*. This feature works by appending the complement of the current solution to the formula and re-solving. This can be efficient because the solver retains its internal state from the previous solution. We append each returned solution to our generated partial instance block until the solver returns unsat, indicating that there are no more objects to be generated.

In order to generate all legal objects with this technique it is necessary to turn off Alloy/Kodkod's symmetry breaking feature.

4   Performance Evaluation of Staging

We evaluate the performance of the two staging techniques on a collection of common data structure specifications (sorted linked-list, binary search tree, and red-black tree), an extended version of the phone book example [11], and a specification automatically generated from the FORML [6,5] analysis tool.

In all cases the queries are of the form 'for all states there exists another reachable legal state'. For the phone book and data structure examples these queries check that the insert and remove operations take one legal object to another legal object.

The FORML [6,5] (Feature-Oriented Modelling Language) tool synthesizes an Alloy specification from a FORML specification. FORML specifications describe 'features' as finite state machines that operate on shared phenomena. These state machines (features) might interact in unintended ways. The purpose of the Alloy analysis is to detect these unintended interactions. The queries are of the same familiar form: there is always a legal next state.

As a baseline for comparison we also show the times taken to simultaneously solve the generator axioms and the ∀∃ query. This is what the Alloy user currently does. For these tests we suppose that the Alloy user magically knows the scope required to generate all of the legal objects. Both staging techniques compute this scope automatically. In current practice the Alloy user will make educated guesses, running the solver repeatedly until they find the scope that permits all legal objects to be generated.

Figure 1 shows that staging reduces analysis time and increases scalability to larger scopes for ∀∃ queries on our benchmark specifications. For sufficiently small scopes all three approaches work quickly. The non-staged approach almost always completes quickly or times out: there are only two problems (5 nodes linked-list with and 3 nodes red-black tree) that take some medium amount of time. The solution enumeration technique is fastest and most scalable. As compared to the non-staged approach, staging shifts time from SAT solving into universe generation and Kodkod translation: notice that there are no visible black bars (SAT solving time for query) in columns for techniques 2 and 3.

Table 1 shows the number of clauses and variables in the CNF file generated to check the ∀∃ query. The staging techniques produce CNF files with orders of magnitude fewer variables and clauses on all but the smallest problems.

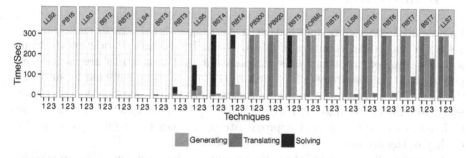

Fig. 1. Staging improves scalability and reduces runtime of ∀∃ queries. Analysis times (bars) are broken down into three components: object generation; ∀∃ query translation, which is performed by Kodkod and includes the generated objects as a partial instance; and SAT solving the ∀∃ query. The x-axis is organized by benchmark and technique. Technique 1 is not-staged; 2 incremental growth; 3 is enumeration. Lower bars are better. Bars that hit the top are time-outs. Enumeration is the only technique that solves every problem. These are the same benchmarks as in Table 1. Benchmarks available at: http://goo.gl/P7G5r6.

Table 1. Staging results in a significant reduction in the number of variables and clauses in the CNF file generated to check the ∀∃ property of interest

Specification	Size	Gen. inst.	Staged				Not-Staged	
			Enumeration		Incremental			
			Vars	Clauses	Vars	Clauses	Vars	Clauses
Sorted Linked List	2	4	19	31	18	28	1892	3964
	3	9	59	103	58	99	7089	16401
	4	25	166	323	166	323	33327	78833
	5	92	489	1159	489	1159	309974	715938
	6	458	1849	5868	TO	TO	TO	TO
	7	2987	9957	41351	TO	TO	TO	TO
Binary Search Tree	2	5	25	42	24	36	4858	9559
	3	15	138	313	138	313	28514	61258
	4	51	386	938	386	938	214564	486202
	5	188	1192	3246	TO	TO	2179446	4955942
	6	731	3345	9766	TO	TO	TO	TO
	7	2950	10883	34932	TO	TO	TO	TO
Red-black Tree	2	5	24	42	24	42	19094	46068
	3	12	80	154	80	154	143220	389576
	4	29	295	685	295	685	900413	2584570
	5	74	1669	4477	TO	TO	TO	TO
	6	201	5653	16794	TO	TO	TO	TO
	7	573	10353	30965	TO	TO	TO	TO
Phone Book	NA	15	614	912	614	912	6597	10693
	NA	300	15341	110191	TO	TO	TO	TO
	NA	920	55341	919316	TO	TO	TO	TO
FORML	NA	170	903	1582	TO	TO	TO	TO

5   Related Work

Finding all distinct consistent instances of a propositional logic formula is called #SAT, or the model counting problem. Formally, the model counting problem is a #P-complete problem; it is complete for problems which are as hard as polynomial-time hierarchy (PH) problems [17], and dramatically harder problem than an NP-complete SAT problem. Researchers tackle #SAT with both exact and approximate techniques. Biere [2, §20] reports that exact techniques scale to hundreds of variables and approximate techniques (with guarantees on the quality of the approximation) scale to a few thousand variables.

When the Alloy user needs to write generator axioms they usually also need to specify the correct scope, which can be difficult. Both of our staging approaches compute this scope for the user 'dynamically' by solving a sliced version of the specification. In Margrave [15], a security policy checker, Nelson *et alia* [16], claimed that a signature's scope can be approximated statically if predicates in the specification belong to a particular subclass of first-order logic.

The idea of generating all legal instances is also of interest in the context of test-case generation (*e.g.*, [9,1,12]). Khurshid *et alia* [12] used SAT-solver based enumeration with Alloy for this purpose.

Symmetry breaking can significantly reduce the number of generated instances when using SAT-solver based solution enumeration [13,18]. We require symmetry breaking to be turned off for now, as our experiments have shown that symmetry breaking can still lead to spurious counter-examples for ∀∃ queries.

Others have also sliced Alloy specifications (*e.g.*, [10]).

6   Conclusions and Future Work

Staging using SAT-solver based solution enumeration and partial instances enables the Alloy Analyzer to scale to larger ∀∃ queries than were previously feasible. The research community has a growing interest in analyzing these kinds of queries (*e.g.*, [19,3,8,5]). We examined on the FORML analysis tool [5]) as a case study. The FORML tool synthesizes Alloy specifications that involve ∀∃ queries for detecting unintended interactions in a FORML specification. Staging made analyzing these FORML specifications feasible.

Alloy specifications with ∀∃ queries usually require generator axioms. A common pattern for generator axioms is to treat atoms as 'objects' that are distinguished by their 'field' values. We propose a new keyword for Alloy, **uniq**, from which such generator axioms (and their associated equality predicates) can be automatically synthesized. This keyword makes it clearer and easier for the user to specify their intent and facilitates the staging techniques.

The main direction for future scalability is to incorporate symmetry breaking. Currently turning off symmetry breaking results in spurious counter-examples for ∀∃ queries. Future usability and expressiveness enhancements could be made by relaxing some of the restrictions currently in place on appended facts and signature declarations using the **uniq** keyword.

References

1. Abad, P., Aguirre, N., Bengolea, V., Ciolek, D., Frias, M.F., Galeotti, J.P., Maibaum, T., Moscato, M., Rosner, N., Vissani, I.: Improving Test Generation under Rich Contracts by Tight Bounds and Incremental SAT Solving. In: ICST 2013 (2013)
2. Biere, A.: Handbook of Satisfiability, vol. 185. Ios PressInc. (2009)
3. Cunha, A.: Bounded model checking of temporal formulas with Alloy. arXiv preprint arXiv:1207.2746 (2012)
4. Derrick, J., Fitzgerald, J., Gnesi, S., Khurshid, S., Leuschel, M., Reeves, S., Riccobene, E. (eds.): ABZ 2012. LNCS, vol. 7316. Springer, Heidelberg (2012)
5. Dietrich, D., Shaker, P., Atlee, J., Rayside, D., Gorzny, J.: Feature Interaction Analysis of the Feature-Oriented Requirements-Modelling Language Using Alloy. In: MoDeVVa Workshop at MODELS Conference (2012)
6. Dietrich, D., Shaker, P., Gorzny, J., Atlee, J., Rayside, D.: Translating the Feature-Oriented Requirements Modelling Language to Alloy. Tech. Rep. CS-2012-12, University of Waterloo, David R. Cheriton School of Computer Science (2012)
7. Eén, N., Sörensson, N.: An Extensible SAT-solver. In: Giunchiglia, E., Tacchella, A. (eds.) SAT 2003. LNCS, vol. 2919, pp. 502–518. Springer, Heidelberg (2004)
8. Fraikin, B., Frappier, M., St-Denis, R.: Modeling the supervisory control theory with Alloy. In: Derrick, et al. (eds.) [4]
9. Galeotti, J.P., Rosner, N., López Pombo, C.G., Frias, M.F.: Analysis of Invariants for Efficient Bounded Verification. In: Tonella, P., Orso, A. (eds.) Proc. 19th ISSTA, pp. 25–36. ACM (2010)
10. Ganov, S., Khurshid, S., Perry, D.E.: Annotations for Alloy: Automated Incremental Analysis Using Domain Specific Solvers. In: Aoki, T., Taguchi, K. (eds.) ICFEM 2012. LNCS, vol. 7635, pp. 414–429. Springer, Heidelberg (2012)
11. Jackson, D.: Software Abstractions: Logic, Language, and Analysis. MIT Press (2011)
12. Khurshid, S., Marinov, D.: TestEra: Specification-based testing of Java programs using SAT. Automated Software Engineering 11(4), 403–434 (2004)
13. Khurshid, S., Marinov, D., Shlyakhter, I., Jackson, D.: A Case for Efficient Solution Enumeration. In: Giunchiglia, E., Tacchella, A. (eds.) SAT 2003. LNCS, vol. 2919, pp. 272–286. Springer, Heidelberg (2004)
14. Montaghami, V., Rayside, D.: Extending Alloy with partial instances. In: Derrick, et al. (eds.) [4]
15. Nelson, T., Barratt, C., Dougherty, D., Fisler, K., Krishnamurthi, S.: The Margrave tool for firewall analysis. In: Proceedings of the 24th International Conference on Large Installation System Administration, pp. 1–8. USENIX Association (2010)
16. Nelson, T., Dougherty, D.J., Fisler, K., Krishnamurthi, S.: Toward a more complete Alloy. In: Derrick, et al. (eds.) [4]
17. Toda, S.: On the computational power of PP and (+)P. In: 30th Annual Symposium on Foundations of Computer Science, pp. 514–519 (October 1989)
18. Torlak, E., Jackson, D.: Kodkod: A relational model finder. In: Grumberg, O., Huth, M. (eds.) TACAS 2007. LNCS, vol. 4424, pp. 632–647. Springer, Heidelberg (2007)
19. Vakili, A., Day, N.A.: Temporal logic model checking in Alloy. In: Derrick, et al. (eds.) [4]

Domain-Specific Visualization of Alloy Instances

Loïc Gammaitoni and Pierre Kelsen

University of Luxembourg, Luxembourg

Motivation. Alloy is a modelling language based on first order logic and relational calculus combining the precision of formal specifications with powerful
automatic analysis features [4]. The automatic analysis is done by the Alloy
Analyzer, a tool capable of finding instances for given Alloy models in a finite
domain space using SAT solving.

The utility of this instance finding mechanism relies on the small scope hypothesis that claims that many design errors can be detected in small model
instances. Those design errors can be identified by evaluating expressions in the
instances generated, but also through the direct visualization of those instances.
In the latter case, the intuitiveness and expressiveness of the instance visualization play a key role in efficiently diagnosing the validity of an instance.

The current visualizations offered by the Alloy Analyzer produce visual diagrams that are close to the structure of instances. These low level visualizations
become difficult to read when instances become more complex.

To highlight this issue, let us consider the formalisation of a Finite State Machine (FSM). Figure 1 is the visualization of an FSM instance provided by the
Alloy Analyzer (using magic layout), while fig. 2 is an intuitive domain-specific
visualization produced by Lightning[1], a tool implementing the approach described in this paper.

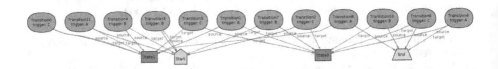

Fig. 1. Visualization of an FSM instance as given by the Alloy Analyzer

Approach Overview. In order to clearly frame the set of possible visualizations definable using our approach, we designed a Visual Language Model (VLM)
as generic as possible in Alloy, such that a large variety of domain specific visualizations can be supported. The work in [3] explains how such a visual language
can be defined. This VLM contains basic visual elements such as shapes and
text which can be composed or related to each other via connectors and can be
arranged through the definition of layouts.

The visualization of instances for a given Alloy model m is defined as a model
transformation from m to VLM. We define this transformation through a set of

Y. Ait Ameur and K.-D. Schewe (Eds.): ABZ 2014, LNCS 8477, pp. 324–327, 2014.

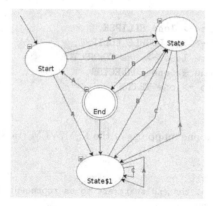

Fig. 2. Instance depicted in fig. 1 visualized using Lightning

mappings and constraints, all expressed in Alloy. Those relations and constraints enforce that any instance of the transformation contains an instance of m as well as its corresponding VLM instance. The VLM instance contained in the transformation instance can be interpreted and rendered accordingly. In fig. 3 we provide an overview of this approach. We call m2VLM the Alloy module containing the transformation specification.

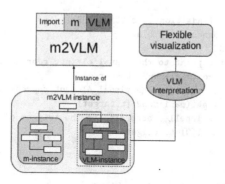

Fig. 3. Overview of our approach for visualizing an Alloy model m

This whole process is implemented by the Lightning tool[1].

Transformations. A transformation is composed of backbone mappings and constraints. Each mapping composing the backbone of the transformation is expressed via binary relations as illustrated in fig. 4.

Guard predicates, one per backbone mapping, are used to define the condition under which a given element of the input model has an image via the backbone mapping (fig. 5).

```
1   one sig Bridge{
2       map1: State one -> lone ELLIPSE,
3       map2: End one -> one DOUBLE_ELLIPSE,
4       map3: Transition lone -> one CONNECTOR,
5       map4: Start lone -> one CONNECTOR,
6       map5: State one -> one TEXT
7   }
```

Fig. 4. Backbone mappings of the FSM2VLM transformation

```
1   pred in_map1(s:State){
2 // all the state except the end state are to be represented by an ellipse (map1)
3       s in State - End
4   }
```

Fig. 5. Example of guard for mapping map1

The way the image of a given element is integrated in the result of the transformation is defined through constraints. Those constraints are rules which have the particularity to be interpretable (they can be viewed as imperative assignments or calls, rather than declarative constraints). We give an example of such rules in fig.6

```
1    pred prop_map4(t:Transition ,  c:CONNECTOR){
2 // color of the connector is black
3       c.color=BLACK
4 // connector source point to the visual element representing t.source
5       c.source=Bridge.(map1+map2)[t.source]
6 // connector target point to the visual element representing t.target
7       c.target= Bridge.(map1+map2)[t.target]
8 // the connector is labelled according to the trigger of the transition.
9       c.connectorLabel[0]=t.trigger
10   }
```

Fig. 6. Example of rules for mapping map4

From Analysis to Interpretation. We can use the Alloy Analyzer to generate the VLM instance from a given m-instance. This is done by promoting a given m-instance into an Alloy model and then solving the transformation model together with this promoted instance. Unfortunately this approach is not really practical because the analysis itself may be quite time consuming, being based on SAT solving.

In recent work [2], we discuss the use of functional Alloy modules to improve the performance of the Alloy analysis of functional transformations. In particular, we show for this case study (FSM2VLM) that the time needed to generated an instance of this transformation is much lower when using interpretation than when using analysis. As it is reasonable to assume that each instance of a given module m should have exactly one corresponding visualization, we can consider the transformation defined to achieve such a visualization as a function. Visualizations can thus be defined by a functional module.

Besides making the computation of the visualization really fast, the use of functional modules relieves us of another limitation induced of the Alloy analyzer: scope calculation. The determination of a small scope is important in order to limit the running time of the analyzer. Interpreting Alloy modules via functional modules rather than analyzing them makes scope determination unnecessary.

Conclusion. In this paper we have presented an approach to define intuitive visualizations for Alloy instances. The approach uses Alloy itself to define a visual model and a transformation from the input instance to this visual instance. We use the recently introduced notion of functional modules to make our approach practical: rather than performing SAT solving to obtain the visual instance, we interpret the input instance and the transformation specification to directly compute the visual instance.

In the future, we plan further investigation on the use of Alloy to define different aspects of modelling languages (semantics, full concrete syntax), as well as combinations of those aspects (e.g., visualization of semantics). The ultimate goal is to develop a language workbench based on Alloy.

References

1. Lightning tool website, `http://lightning.gforge.uni.lu`
2. Gammaitoni, L., Kelsen, P.: Functional Alloy Modules. Technical Report TR-LASSY-14-02, University of Luxembourg, `http://hdl.handle.net/10993/16386`
3. Orefice, S., Costagliola, G., De Lucia, A., Polese, G.: A classification framework to support the design of visual languages. Journal of Visual Languages and Computing, 573–600 (2002)
4. Jackson, D.: Software abstractions. MIT Press, Cambridge (2012)

Optimizing Alloy for Multi-objective Software Product Line Configuration

Ed Zulkoski, Chris Kleynhans, Ming-Ho Yee,
Derek Rayside, and Krzysztof Czarnecki

University of Waterloo
Waterloo, Ontario, Canada
{ezulkoski,drayside,kczarnec}@gsd.uwaterloo.ca

Abstract. Software product line (SPL) engineering involves the modeling, analysis, and configuration of variability-rich systems. We improve the performance of the multi-objective optimization of SPLs in Alloy by several orders of magnitude with two techniques.

First, we rewrite the model to remove binary relations that map to integers, which enables removing most of the integer atoms from the universe. SPL models often require using large bitwidths, hence the number of integer atoms in the universe can be orders of magnitude more than the other atoms. In our approach, the tuples for these integer-valued relations are computed outside the SAT solver before returning the solution to the user. Second, we add a checkpointing facility to Kodkod, which allows the multi-objective optimization algorithm to reuse previously computed internal SAT solver state, after backtracking.

Together these result in orders of magnitude improvement in using Alloy as a multi-objective optimization tool for software product lines.

Keywords: Product Lines, Multi-objective Optimization, Kodkod, Alloy.

1 Introduction

Alloy is used for a wide variety of purposes, from analyzing software designs to checking protocols to generating test inputs and beyond. Recently, there has been some interest in using Alloy for design exploration or product configuration [11,13]. These specifications often involve constraints on sums of integers (or other arithmetic expressions). For example, there might be a restriction on the total weight of a car, or on the disk footprint of a configured operating system kernel. Sometimes the user wishes to not only compute a viable product configuration, but an optimal one [11], often in the presence of multiple conflicting objectives.

These specifications often require solving with fairly large bitwidths, to support large metric values. In the general case, where the specification involves arbitrary constraints over relations containing integers, Alloy needs to create an atom for every integer in the bitwidth. At higher bitwidths the number of integer atoms dominate the number of other atoms, affecting solving time.

Y. Ait Ameur and K.-D. Schewe (Eds.): ABZ 2014, LNCS 8477, pp. 328–333, 2014.

```
1  one sig Car {                9
2    e : Engine,                10  abstract sig Part { x : Int }
3    f : Frame,                 11  abstract sig Engine extends Part {}
4    w : Int,                   12    one sig Petrol extends Engine {}{ x = 3 }
5  }{                           13    one sig Diesel extends Engine {}{ x = 4 }
6    w = (e.x).plus[f.x]        14  abstract sig Frame extends Part {}
7    w < 9                      15    one sig Aluminum extends Frame {}{ x = 5 }
8  }                            16    one sig Steel extends Frame {}{ x = 6 }
```

Fig. 1. An example design exploration model. The goal is to choose components (engine and frame) for a car design according to some constraints (total weight < 9).

We observe that SPL specifications are not completely arbitrary, but usually associate equality constraints with each integer-valued relation (*e.g.*, lines 12,13,15,16 of Fig. 1). We use these equality constraints to rewrite other parts of the specification that refer to these integer-valued relations. If the rewritten specification meets certain conditions (see Section 2), then most integer atoms can be removed, thus producing much smaller SAT formulas. After solving, we use the equality constraints and the solution to the modified specification to produce a model of the original specification.

The approach we use for multi-objective optimization, called the *guided improvement algorithm* [13], requires many calls to Kodkod, first by adding constraints to find optimal solutions, and then backtracking. We have enhanced Kodkod to allow the removal of constraints following a stack discipline. (Note that Kodkod 2.0 already supports incremental addition of constraints.)

Together, these two enhancements to the Alloy toolchain result in several orders of magnitude improvement for performing multi-objective optimization of SPLs. We experienced an average of over 200X speedup on our experiments.

We focus on multi-objective optimization (MOO) on software product lines (SPLs) for our experiments. The goal of SPL engineering is to facilitate the modeling and analysis of variability-rich systems [3,12]. These systems are typically represented as *feature models*: concise tree-like structures, whose *products* are valid configurations of the system [7]. Features may additionally contain *attributes*, indicating the effect of a feature on the overall *quality* of a product.

A natural analysis on attributed feature models is to identify *optimal* products with respect to the set of quality attributes. There may be many products that are considered optimal, particularly when *conflicting* objectives exist (*e.g.*, low cost *vs.* high performance). In such a case we say a product is *Pareto optimal* if increasing its value in some objective decreases its value in another. The goal of MOO is to discover all Pareto optimal solutions.

In this paper we work with a version of Alloy extended [11] with partial instances [10] and the *guided improvement algorithm* (GIA), an exact algorithm for MOO (see [13] for a full description). ClaferMOO [11] – an extension to Clafer [1] for MOO of attributed feature models – has been built using the GIA. We use a set of ClaferMOO specifications to evaluate our tool in Section 4.

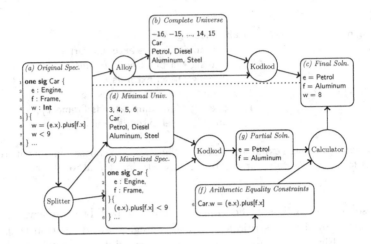

Fig. 2. Contrast of a standard Alloy run (above the dotted line) and our approach (below). The example model here is abridged from Fig. 1.

2   Eliding Integer Relations and Atoms

Fig. 2 contrasts the standard Alloy approach (above the dotted-line), and our approach to eliding integer relations and atoms through substitutions (below the dotted-line). Normally, Alloy generates the complete universe (Fig. 2b) and Kodkod specification, after which Kodkod produces our final solution (Fig. 2c).

First we divide the integer relations into dependent and independent (denoted *Splitter*). Integer-valued relations are identified as *independent* if they are bound to constants through equality constraints. *Dependent* integer relations are defined by an expression involving independent relations (*e.g.*, w in Fig. 1). Standard substitution techniques are used to remove dependent relations (Fig. 2e), however the equality constraints are retained (Fig. 2f). Integer atoms that are not explicitly named as constants in the specification may also be elided (Fig. 2d, see conditions below). A solution to the modified specification elicits values for independent relations (Fig. 2g). Dependent integer relations are computed from the retained equality constraints and the partial solution (Fig. 2c).

Conditions for when substitutions can be performed and relations can be elided:
1. The candidates for *dependent* integer-valued relations are functional (*i.e.*, *one*-multiplicity in Alloy) binary relations that map atoms to integers.
2. If the dependent relations depend on each other (and not just the independent relations), then they must do so according to some partial order.
3. The equality constraints must occur in top-level conjuncts, inside a single universal (*all*) quantifier. Alloy's *appended facts* meet this criteria.
4. The equality constraints must name just the dependent relation on one side or the other. (This constraints could be relaxed in future.)

5. If there exists multiple constraints on the same dependent variable, *e.g.*, $w = expr_1$ and $w = expr_2$, we remove both but add the constraint $expr_1 = expr_2$.

Conditions for when integer atoms can be elided from the universe:
1. All dependent integer-valued relations must be elided.
2. There can be no quantification over the integers (*e.g.*, $\{all \ x \ : Int \mid p(x)\}$).

3 Checkpointing

The GIA works through repeated calls to the solver, and then backtracking to find other Pareto optimal points. When backtracking, constraints must be removed in order to find new points. Checkpointing allows us to revert to a previously saved state of the solver, without discarding all of the work that the solver has performed. Removal of constraints can be achieved by checkpointing before every constraint addition and reverting at a later time to remove that constraint. In the case of the GIA, it is not necessary that we be able to remove arbitrary constraints from the problem. It suffices to checkpoint after finding each starting point before we begin the drive to the Pareto front. This allows us to return to a solver state that only contains the problem constraints and the exclusion constraints specified by the previously found Pareto optimal points. We can then begin our search for a new starting point by adding the exclusion constraints from the last Pareto optimal point. To test the performance benefits of adding checkpointing support to Kodkod on the guided improvement algorithm, we have added the required support to version 2.2.0 of the MiniSat solver. This implementation simply creates a copy of the entire MiniSat solver object and stores it on a checkpoint stack. While simple, it is sufficient to show the benefits of checkpointing as a concept.

4 Evaluation

Our evaluation was over a set of 93 variants of nine MOO-specifications of SPLs,[1] compiled for work in [11], and originally described in [4,14,15,16]. Each variant specification modifies the original by adding additional objectives and/or adjusting attribute values. The number of objectives ranges from one to seven.

Table 1 summarizes the speedup produced by just checkpointing, just the reductions (which include both formula changes and universe reductions), and their combination. Both techniques always result in some speedup in all experiments. Their combination results in an average speedup of over 200X, ranging from 20X to almost 1500X. Fig. 3 gives a graphical view of the data underlying the summary in Table 1. The x axis contains an entry for each of the 93 multi-objective product line specifications, ordered by their baseline solving time.

---

[1] The product line specifications can be found at: https://github.com/TeamAmalgam/
test-models/tree/f348271b005ee7d4929f73846e6ad8c4a19e0bd4/spl .

Table 1. Summary of speedups obtained on a 3.4GHz quad-core Intel i7, 16 GB RAM, 64-bit Ubuntu 12.04, Java SE 64-bit 1.7.0.12. GIA 'magnifying glass' turned off.

	Min	Max	Mean	Median	Std. Dev.
Baseline	2473 ms	3,515,676 ms	145,523 ms	15,800 ms	449758
Checkpointing	1.32 X	19.23 X	2.53 X	2.57 X	1.75
Reductions	13.25 X	1014.93 X	161.70 X	193.52 X	104.09
Combined	22.81 X	1458.82 X	221.02 X	276.18 X	134.23

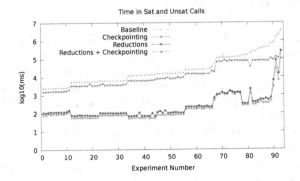

Fig. 3. Solving times of 93 multi-object product line models. Each point on the x axis represents a different model. The models are ordered by their baseline solving time.

5   Related Work

A variety of researchers have used equality constraints to rewrite formulas for improved performance. The idea is perhaps as old as the Knuth-Bendix completion algorithm [8]. In recent years the idea has been used in a number of SMT solvers [2,5,17,6] and bounded model checkers [9]. For example, STP [6] is intended to solve constraints generated from the static analysis of software that makes use of arrays. STP uses rewriting to reduce these constraints into a form suitable for a SAT solver. In addition, solvers such as Z3 [17] support checkpointing as well.

In our work the efficiency gains come more from reducing the size of the universe than from the rewriting *per se*: the rewriting is a transformation that enables the universe size reduction. Smaller universes correspond to smaller SAT formulas with faster solving times.

6   Conclusions

For Alloy specifications that characterize multi-objective product lines, rewriting based on equality constraints facilitates the elision of both integer-valued relations and integer atoms. This elision results in an average speedup of over 150X. Further, adding checkpointing to the underlying SAT solver results in a 2X speedup. The combined speedup is over 200X.

References

1. Bąk, K., Czarnecki, K., Wąsowski, A.: Feature and Meta-Models in Clafer: Mixed, Specialized, and Coupled. In: Malloy, B., Staab, S., van den Brand, M. (eds.) SLE 2010. LNCS, vol. 6563, pp. 102–122. Springer, Heidelberg (2011)
2. Brummayer, R.: Efficient SMT solving for bit vectors and the extensional theory of arrays. Ph.D. thesis, JKU Linz (2010)
3. Clements, P.C., Northrop, L.: Software Product Lines: Practices and Patterns. Addison-Wesley (2001)
4. Esfahani, N., Malek, S.: Guided Exploration of the Architectural Solution Space in the Face of Uncertainty. Tech. rep., George Mason U., Dept. of C.S. (March 2011)
5. Franzen, A.: Efficient solving of the satisfiability modulo bit-vectors problem and some extensions to SMT. Ph.D. thesis, Univ. of Trento (2010)
6. Ganesh, V., Dill, D.L.: A Decision Procedure for Bit-Vectors and Arrays. In: Damm, W., Hermanns, H. (eds.) CAV 2007. LNCS, vol. 4590, pp. 519–531. Springer, Heidelberg (2007)
7. Kang, K.C., Cohen, S.G., Hess, J.A., Novak, W.E., Peterson, A.S.: Feature-Oriented Domain Analysis (FODA) feasibility study. Tech. rep., SEI-CMU (1990)
8. Knuth, D.E., Bendix, P.B.: Simple word problems in universal algebra. In: Proc. Conf. on Computational Problems in Abstract Algebra. Pergamon Press (1970)
9. Merz, F., Falke, S., Sinz, C.: LLBMC: Bounded Model Checking of C and C++ Programs Using a Compiler IR. In: Joshi, R., Müller, P., Podelski, A. (eds.) VSTTE 2012. LNCS, vol. 7152, pp. 146–161. Springer, Heidelberg (2012)
10. Montaghami, V., Rayside, D.: Extending Alloy with partial instances. In: Derrick, J., Fitzgerald, J., Gnesi, S., Khurshid, S., Leuschel, M., Reeves, S., Riccobene, E. (eds.) ABZ 2012. LNCS, vol. 7316, pp. 122–135. Springer, Heidelberg (2012)
11. Olaechea, R., Stewart, S., Czarnecki, K., Rayside, D.: Modelling and Optimization of Quality Attributes in Variability-Rich Software. In: NFPinDSML Workshop at MODELS Conference (2012)
12. Pohl, K., Böckle, G., van der Linden, F.J.: Software Product Line Engineering: Foundations, Principles and Techniques. Springer (2005)
13. Rayside, D., Estler, H.-C., Jackson, D.: A Guided Improvement Algorithm for Exact, General Purpose, Many-Objective Combinatorial Optimization. Tech. Rep. MIT-CSAIL-TR-2009-033, MIT CSAIL (2009)
14. Siegmund, N., Kolesnikov, S., Kastner, C., Appel, S., Batory, D., Rosenmuller, M., Saake, G.: Predicting performance via automated feature-interaction detection. In: Murphy, G., Pezze, M. (eds.) Proc. 34th ICSE, Zurich, Switzerland (2012)
15. Siegmund, N., Rosenmuller, M., Kastner, C., Giarrusso, P.G., Apel, S., Kolesnikov, S.S.: Scalable prediction of non-functional properties in software product lines. In: Schaefer, I., John, I., Schmid, K. (eds.) SPLC Workshops. ACM (2011)
16. Siegmund, N., Rosenmuller, M., Kuhlemann, M., Kastner, C., Apel, S., Saake, G.: SPL Conqueror: Toward optimization of non-functional properties in software product lines. Software Quality Journal 1(3), 1–31 (2011)
17. Wintersteiger, C., Hamadi, Y., de Moura, L.: Efficiently solving quantified bit-vector formulas. Formal Methods in System Design 42(1), 3–23 (2013)

Author Index